EESES
Electric Energy Systems and Engineering Series

Editors: J. G. Kassakian · D. H. Naunin

Klemens Heumann

Basic Principles
of Power Electronics

With 242 Figures

Springer-Verlag Berlin Heidelberg New York
London Paris Tokyo

Prof. Dr.-Ing. Klemens Heumann

Institut für Allgemeine Elektrotechnik, Technische Universität Berlin
Einsteinufer 19, D-1000 Berlin 10, Fed. Rep. of Germany

Series Editors:

Prof. J.G. Kassakian

Massachusetts Institute of Technology,
77 Massachusetts Ave., Cambridge, MA 02139, USA

Prof. D.H. Naunin

Institut für Elektronik, Technische Universität Berlin,
Einsteinufer 19, D-1000 Berlin 10, Fed. Rep. of Germany

Exclusively authorized English translation of the original German book
„Grundlagen der Leistungselektronik", 3rd edition, B.G.Teubner, Stuttgart,
1985.

ISBN-13:978-3-642-82676-4   e-ISBN-13:978-3-642-82674-0
DOI: 10.1007/978-3-642-82674-0

Library of Congress Cataloging in Publication Data.
Heumann, Klemens. Basic principles of power electronics.
(Electric energy systems and engineering series)
Translation of: Grundlagen der Leistungselektronik.
Bibliography: p. Includes index.
1. Power electronics. I. Title. II. Series.
TK7881.15.H4813   1986   621.381   86-10231
ISBN-13:978-3-642-82676-4 (U.S.)

Typesetting: With a system of the Springer Produktions-Gesellschaft, Berlin.
Dataconversion: Brühlsche Universitätsdruckerei, Gießen.

2161/3020-543210

Introduction to the
Electric Energy Systems and Engineering Series

Concerns for the continued supply and efficient use of energy have recently become important forces shaping our lives. Because of the influence which energy issues have on the economy, international relations, national security, and individual well-being, it is necessary that there exists a reliable, available and accurate source of information on energy in the broadest sense. Since a major form of energy is electrical, this new book series titled *Electric Energy Systems and Engineering* has been launched to provide such an information base in this important area.

The series coverage will include the following areas and their interaction and coordination: generation, transmission, distribution, conversion, storage, utilization, economics.

Although the series is to include introductory and background volumes, special emphasis will be placed on: new technologies, new adaptations of old technologies, materials and components, measurement techniques, control – including the application of microprocessors in control systems, analysis and planning methodologies, simulation, relationship to, and interaction with, other disciplines.

The aim of this series is to provide a comprehensive source of information for the developer, planner, or user of electrical energy. It will also serve as a visible and accessible forum for the publication of selected research results and monographs of timely interest. The series is expected to contain introductory level material of a tutorial nature, as well as advanced texts and references for graduate students, engineers and scientists.

The editors hope that this series will fill a gap and find interested readers.

John G. Kassakian · Dietrich H. Naunin

Preface

Power electronics became an identifiably separate area of electrical engineering with the invention of the thyristor about 30 years ago. The growing demand for controllability and conversion of electric energy has made this area increasingly important, which in turn has resulted in new device, circuit and control developments. In particular, new components, such as the GTO and power MOSFET, continue to extend power electronic technology to new applications.

The technology embodied by the name "power electronics" is complex. It consists of both power level and signal level electronics, as well as thermal, mechanical, control, and protection systems. The power circuit, that part of the system actually processing energy, can be thought of as an amplifier around which is placed a closed loop control system.

The goal of this book is to provide an easily understood exposition of the principles of power electronics. Common features of systems and their behavior are identified in order to facilitate understanding. Thyristor converters are distinguished and treated according to their mode of commutation. Circuits for various converters and their controls are presented, along with a description of ancillary circuits such as those required for snubbing and gate drives. Thermal and electrical properties of semiconductor power devices are discussed. The line-converter and converter-load interfaces are examined, leading to some general statements being made about energy transfer. Application areas are identified and categorized with respect to power and frequency ranges. The many tables presented in the book provide an easily used reference source. Valid IEC and German DIN standards are used in examples throughout the book.

This book is designed to provide an overview of power electronics for students as well as practicing engineers. Only a basic knowledge of electrical engineering and mathematics is assumed. The list of references at the end of the book gives a survey of the field as it has developed over time. Understandably, the majority are cited from German publications.

This book was first published in German, and has been translated into Japanese, Spanish, and Hungarian. The author is pleased that an English edition has now been published.

Berlin, June 1986   Klemens Heumann

Contents

List of Principal Letter Symbols

Time variable quantities:

u,i	instantaneous values
U,I	root-mean-square values
$\hat{U},\hat{\imath}$	peak values

List of suffixes

AV, av	average (arithmetical mean)
EFF, eff	effective (root-mean-square)
M, max	maximum
N	nominal value, rated value or at rated load
b	due to converter reactors
k	commutation, short circuit
i	ideal value
L	line
t	due to converter transformer stray

Electrical and other physical quantities

letter symbol	quantity	unit
B	magnetic induction	$T = Vs/m^2$
C	capacitance	$F = As/V$
C_B	snubber capacitance	F
C_K	commutation capacitance	F
C_d	smoothing capacitance	F
D	distortion power	VA
D_r	total resistive direct voltage regulation	V
D_x	total inductive direct voltage regulation	V
d_r	total resistance direct voltage regulation (relative)	1, %
d_{rt}	resistive direct voltage regulation due to main and interphase transformer (relative)	1, %
dx	total inductive direct voltage regulation (relative)	1, %
d_{xb}	inductive direct voltage regulation due to converter reactors (relative)	1, %
d_{xt}	inductive direct voltage regulation due to converter transformer (relative)	1, %
d_{xL}	inductive direct voltage regulations due to ac system reactance (relative)	1, %
f	frequency	s^{-1}, Hz
f_1	factor at which the direct current becomes intermittent	1
f_p	pulse frequency	s^{-1}, Hz
g	number of sets of commutating groups between I_{dN} is divided	
g	content of fundamental	1, %
H	magnetic field strength	A/m
I,i	current	A

I_d	direct current (arithmetical mean)	A
I_L	current on line side	A
I_{1L}	fundamental wave of I_L	A
I_{Li}	ideal current on line side	A
I_p	branch circuit current	A
I_v	current on cell side of transformer	A
I_v	harmonic oscillation of current (order v)	A
k	relative harmonic content, distortion factor	1, %
L	inductance	H = Vs/A
L_d	smoothing inductance	H
L_k	commutation inductance	H
L_σ	stray inductance	H
M	torque	Nm
n	rotational speed	$\min^{-1}$
p	pulse number	
P,p	real power	W = VA
P_0	output power of converter	W
P_d	real power on dc side	W
P_I	input power of converter	W
P_L	real power on line side	W
P_{1L}	real power of fundamental on line side	W
P_{vt}	winding losses of converter transformer	W
Q	reactive power	VA, var
Q	short-circuit capacity of the ac system	VA
Q	electric charge	C = As
Q_L	reactive power on line side	var
Q_{1L}	reactive power on line side based on fundamental current	var
q	commutating number	
R	resistance	Ω = V/A
S	apparent power	VA
S_d	apparent power on dc side	VA
S_L	apparent power on line side	VA
S_{1L}	apparent power on line side	VA
S_{iL}	ideal apparent power on line side	VA
S_m	short-circuit capacity of the ac system	VA
s	number of series connected commutating groups	
T	period cycle; time constant	s
T_1	turn-on time	s
T_2	turn-off time	s
t	time	s
t_u	time of overlap (of commutation)	s
t_F	current conduction time	s, °, rad
t_R	blocking time	s, °, rad
t_q	circuit commutated turn-off time	s
t_c	hold-off interval	s
U,u	voltage	V
U_d	direct voltage (arithmetical mean)	V
U_{di}	ideal no-load direct voltage (at $\alpha = 0$)	V
$U_{di\alpha}$	controlled ideal no-load direct voltage	V
U_{dr}	total resistive direct voltage regulation	V
U_{drt}	resistive direct voltage regulation due to converter transformer	V
U_{dx}	total inductive direct voltage regulation	V
U_{dxb}	inductive direct voltage regulation due to converter reactors	V
U_{dxt}	inductive direct voltage regulation due to converter transformer	V
U_{dxL}	inductive direct voltage regulation due to ac system reactors	V

U_{d0}	convential no-load direct voltage	V
$U_{d0\alpha}$	controlled conventional no-load direct voltage	V
U_{d00}	real no-load direct voltage	V
$U_{d\alpha}$	U_d at delay angle	V
U_{im}	ideal crest voltage of an arm	V
U_{i0m}	ideal crest no-load voltage of an arm	V
U_k	commutation voltage	V
u_{kt}	relative short-circuit voltage of converter transformer	1, %
U_L	phase-to-phase voltage on line side	V
U_m	crest voltage between two connections of a converter set or a current circuit	V
U_{s0}	voltage to neutral, phase voltage; voltage between a cell side conductor and neutral at no load (rms value)	V
U_{v0}	phase-to-phase voltage on cell side at no load (rms value)	V
U_{xt}	inductive component of short-circuit voltage of converter transformer (rms value)	V
u_{xt}	relative value of U_{xt} at rated voltage	1, %
U_{vi}	ideal harmonic oscillation of voltage (order v)	V
u	angle of overlap	°, rad
u_0	angle of overlap at delay angle $\alpha = 0$	°, rad
W, w	energy	Ws
w_i	ideal content of voltage ripple	1, %
Z	impedance	Ω
α	control angle, delay angle	°, rad
β	angle of advance	°, rad
Δ	difference	
γ	margin angle (at inverter operation)	°, rad
δ	number of commutating groups commutating simultaneously per primary or per reactor	
η	efficiency	1, %
λ	total power factor	1
ν	order of harmonics	
ν	angular frequency of a free oscillation	s^{-1}
τ	time constant	s
φ_1	phase angle between fundamentals of ac voltage and ac current	°, rad
$\cos \varphi_1$	power factor (for fundamental waves)	1
ω	angular frequency	s^{-1}

Suffixes for power semiconductor devices

A	anode connection
K	cathode connection
G	gate connection
E	emitter connection
B	base connection
C	collector connection
D	drain connection (for MOSFETs)
S	source (for MOSFETs)
F	forward direction, on-state (for diodes)
T	forward direction, on-state (for thyristors)
R	reverse direction
D	off-state in forward direction
(B0)	forward breakover voltages
(T0)	on-state threshold voltages

H	holding state
P,p	pulse operation
(th)	thermal values
Q,q	turn-off
R (as 2nd suffix)	repetitive
S (as 2nd suffix)	surge, non-repetitive

Quantities with power semiconductor devices

u_A	voltage (in general)
i_A	current (in general)

with diodes

u_F	on-state voltage (instantaneous value)
U_F	on-state voltage (mean value)
$U_{(TO)}$	on-state threshold voltage
u_R	reverse voltage (instantaneous value)
U_R	reverse voltage (mean value)
U_{RRM}	repetitive peak reverse voltage
i_F	on-state current (instantaneous value)
I_F	on-state current (mean value)
I_N	rated current
I_{FAVM}	maximum average on-state current
i_R	reverse current (instantaneous value)
P_F, P_F	on-state conduction loss
r_F	on-state slope resistance

with thyristors

u_T	on-state voltage (instantaneous value)
U_T	on-state voltage (mean value)
$U_{(TO)}$	on-state threshold voltage
u_D	forward off-state voltage (instantaneous value)
U_D	forward off-state voltage (mean value)
$U_{(BO)}$	forward breakover voltage
$\left(\dfrac{du}{dt}\right)_{crit}$	critical rate of rise of off-state voltage
u_R	reverse voltage (instantaneous value)
U_R	reverse voltage (mean value)
U_{RRM}	repetitive peak reverse voltage
U_{RSM}	non-repetitive peak reverse voltage
U_G	gate voltage
U_{GT}	gate trigger voltage
i_T	on-state current (instantaneous value)
I_T	on-state current (mean value)
I_N	rated current
I_{TAVM}	maximum average on-state current
I_H	holding current
$\left(\dfrac{di}{dt}\right)_{crit}$	critical rate of rise of on-state current
i_D	forward off-state current (instantaneous value)
I_D	forward off-state current (mean value)

i_R	reverse current (instantaneous value)
I_R	reverse current (mean value)
I_G	gate current
I_{GT}	gate trigger current
P_T, P_T	on-state conduction loss
P_D	forward off-state power loss
P_R	reverse blocking loss
P_G	gate loss
W_T, W_T	turn-on switching energy
W_Q, W_Q	turn-off switching energy
r_T	on-state slope resistance
t_{stg}	storage time
t_{rr}	reverse recovery time
t_q	circuit commutated turn-off time
t_{gd}	gate controlled delay time
t_{gr}	gate controlled rise time
t_{gs}	gate controlled conduction spreading time
R_{th}	thermal resistance
R_{thJC}	thermal resistance, junction to case
R_{thCA}	thermal resistance, case to ambient
$Z_{(th)t}$	transient thermal impedance
ϑ	temperature (in Celcius)
$\vartheta_{(vj)}$	junction temperature
ϑ_C	case temperature
ϑ_A	ambient temperature, temperature of cooling medium

Additional terms for gate turn-off thyristors

i_{FG}	forward gate current
i_{RG}	reverse gate current
I_{TQRM}	maximum repetitive controllable on-state current
I_{TQSM}	maximum non-repetitive controllable on-state current
I_{TQT}	tail current
P_{DQ}	turn-off dissipation
t_{dq}	gate controlled storage time
t_{fq}	gate controlled fall time
t_{gq}	gate controlled turn-off time
W_{DQ}	turn-off energy

with bipolar transistors

u_{CB}	collector base-voltage (instantaneous value)
U_{CB}	collector base-voltage (mean value)
u_{CE}	collector emitter voltage (instantaneous value)
U_{CE}	collector emitter voltage (mean value)
i_e	emitter current (instantaneous value of alternating current)
i_E	emitter current (instantaneous value)
I_E	emitter current (mean value)
i_b	base current (instantaneous value of alternating current)
i_B	base current (instantaneous value)
I_B	base current (mean value)
i_c	collector current (instantaneous value of alternating current)
i_C	collector current (instantaneous value)
I_C	collector current (mean value)
t_d	delay time

t_r	rise time
t_f	fall time
t_s	storage time

with field effect transistors

U_{DS}	drain-source voltage
U_{SG}	source-gate voltage
I_D	drain current
I_{DS}	drain-source current
$R_{DS(on)}$	on state resistance

1 Introduction and Definitions

Power electronics covers the switching, control, and conversion of electrical energy using semiconductor devices and includes the associated measuring and open- and closed-loop control equipment.

The fraction of electrical energy which is switched, controlled, and converted by power electronics is constantly increasing. Power electronics thus represents an important link between power generation and the load (Fig. 1.1). It is growing in significance as the demand to control and convert electrical energy increases [1.1, 1.2, 1.3].

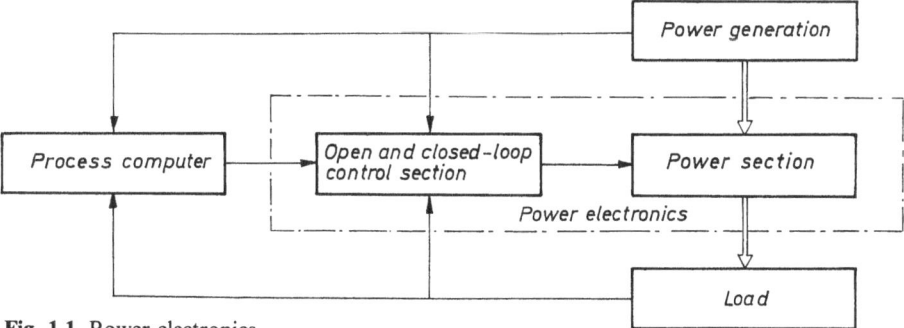

Fig. 1.1. Power electronics

It is useful to distinguish between the power section and the open- and closed-loop control section of a power electronics system. Nowadays, not only in the power section but also in the open- and closed-loop control section, components are becoming predominantly based on monocrystalline semiconductor material, i.e. rectifier diodes, thyristors, and power transistors in the power section and diodes, transistors, and integrated circuits in the open- and closed-loop control section. Using similar types of components achieves the compatibility between the subassemblies, equipment, and installation of power electronics essential for reliability.

1.1 Development History

Power electronics has developed from static converter technology which is several decades old. Even in the thirties large numbers of converter installations with mercury-arc rectifiers were in operation, chiefly as uncontrolled or controlled

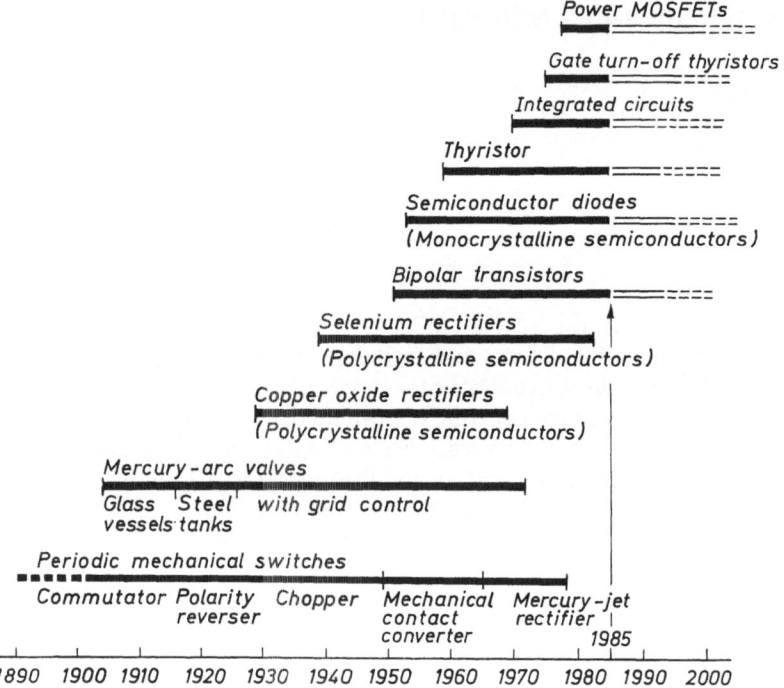

Fig. 1.2. Origin of types of rectifier valves

rectifiers with ratings in the megawatt range [1, 2]. At the beginning of this century the simplest converters i.e. uncontrolled rectifiers were developed for the purpose of battery charging from single-phase or three-phase supplies. In the course of further developments new spheres of application were found, namely, the supply of medium power dc loads (so-called light and power works) via rectifier substations and urban dc supply systems as well as the operation of dc railways and electrolytic plants. DC railway applications included urban tramways, overhead and underground railways, and suburban railways for which dc motors are employed on account of their good starting characteristics and ease of control. In a number of european countries electrification of the long-distance railways was also carried out with dc supply systems fed by mercury-arc rectifiers.

Figure 1.2 shows the origin of the different types of construction of rectifier valves.

Converter valves are functional elements which are cyclically changed between electrically conducting and non-conducting states. Genuine valves have a directional conductivity produced under certain conditions in a vacuum, in gases or in semiconductors. The types of construction of genuine converter valves are listed in Fig. 1.3. These are high-vacuum valves, gas-discharge valves, and semiconductor valves. Semiconductor valves presently dominante in power electronics.

High-vacuum valves
   with hot cathode

Gas-discharge valves
   Inert gas-filled valves
      with hot cathode and filled with inert gas
   Mercury-arc valves
      with hot cathode and mercury-vapour filling (Thyratron)
      with liquid cathode (mercury cathode)
      with continuous excitation (Excitron)
      with ignitor (Ignitron)

Semiconductor valves
   Polycrystalline semiconductors
      Copper oxide rectifiers
      Selenium rectifiers
   Monocrystalline semiconductors
      Semiconductor diodes
      Thyristors
      Transistors   **Fig. 1.3.** Types of genuine rectifier valves

In the case of non-genuine valves which have no directional conductivity a valve action is produced by cyclic actuation of mechanical contacts or similar devices. Non-genuine valves are therefore the periodic mechanical switches listed at the bottom of Fig. 1.2 which, as commutators in electrical machines, were already in use in the middle of the last century (W. Siemens discovered the Principle of Electrodynamics in 1866 and construction of the first direct current dynamo tookplace). Later came the so-called polarity reversers for calling systems in the long-distance telephone service of the Post Office and mechanical choppers to generate alternating voltage from a battery. A special place was gained for about two decades by the contact converter in the field of direct current supply for electrolytic plants. This works with periodically actuated mechanical contacts synchronously switched by an eccentric shaft in rhythm with the mains frequency. Mercury-jet rectifiers switch cyclically using a rotating mercury jet.

The first gas-filled valves with genuine valve characteristics in which the cyclic switching function is performed by electric arc discharges were developed at the beginning of the century. The first mercury-arc rectifiers were built by P. Cooper-Hewitt in 1902. At first, mercury-arc valves with liquid cathode were built as single-bulb or multiple-bulb glass vessels. Soon after P. Cooper-Hewitt and F. Conrad in the USA developed the first steel-tank rectifiers (in Europe, B. Schäfer in 1910) for which instead of glass, steel tanks which have the advantage of greater mechanical strength and better cooling were used and hence opened the way to higher ratings. Steel-tank rectifiers were later built either as a welded fabrication with bolted-on cover plate and vacuum pump for the highest currents or as hermetically sealed steel tanks without vacuum pump. They were cooled either by air or for the higher ratings with water. Mercury-arc valves with liquid cathode handle currents of some 1000 A at voltages up to several kV. For high-voltage direct-current transmission special high-reverse-voltage constructions were developed with reverse voltages of up to more than 150 kV.

A distinction may be made between mercury-arc valves with continuous excitation and those with an ignitor. The former are called excitrons and the latter ignitrons. After J. Langmuir had discovered the principle of grid control of an arc discharge in 1914, P. Toulon discovered a method of applying grid control to voltage control in 1922. This resulted in the possibility of building controllable rectifiers as well as inverters in which the energy flow is in the reverse direction. Besides the mercury-arc valves with liquid cathode valves with hot cathode were also developed. These work with either a mercury-vapour filling or a filling of inert gas (preferably argon). They are called *Thyratrons* and handle voltages up to about 15 kV at valve currents below 20 A.

With these technically mature gas-discharge valves available converter technology achieved greater technical significance from the end of the twenties onwards. The mercury-arc valves were mainly employed for the conversion of single-phase and three-phase alternating current into direct current with open or closed-loop control. Even in the thirties the problem of generating single-phase alternating current at 16 2/3 Hz to feed traction systems from the 50 Hz three-phase supply system by means of conerters was tackled and realized in an experimental installation in the Black Forest in the German Federal Republic. Considerable technical difficulties arose, however, in the generation of the necessary firing pulses using the components available at that time in the control circuits, particularly in the case of the more extensive converter connections.

The first semiconductor rectifiers were employed around 1930 for rectification purposes in the lower power range. These were copper oxide rectifiers at first and soon afterwards selenium rectifiers, the base material of which is a polycrystalline semiconductor.

Selenium rectifiers have been continuously improved and nowadays still have some applications as miniature rectifiers (e.g. high-voltage rectifiers in television sets).

The fifties saw the development of semiconductor diodes made from monocrystalline semiconductor material. These were at first germanium diodes and several years later silicon diodes which enable higher voltages to be attained. Then, in 1958, General Electric in the USA developed the first *Thyristors* which were at that time called Silicon Controlled Rectifiers (SCR). These original controllable power semiconductors initiated a development in electric power engineering comparable the discovery of transistors in communication engineering a decade previously. At the beginning of the sixties development work led to a constant improvement in the semiconductor components and the associated circuit technology resulting in rapid development and extension of the classical converter technology. Besides the circuits already fully developed technically using mercury-arc valves, novel connections, and applications were opened up.

This was promoted by two factors: first, the superior electrical characteristics of semiconductor valves e.g. lower on-state voltage no arc-back and faster switching, and second, advances in control components which made possible the realization of complex open- and closed-loop control functions. The main feature of the present development phase is the increasing use of integrated circuits into the open and closed-loop control section.

In the middle of the sixties, the term converter technology was extended to that of power electronics.

Power electronics are today in most areas consolidated technics. Since the beginning of the eighties however strong new impulses have come. More and more integrated circuits are employed in the control section of converters which causes a transition from analogue to digital circuitry. Data processing is handled by microprocessors.

In the power circuit of self-commutated converters bipolar power transistors reach up to the range of 100 kW. In the lower power range MOSFETs start to take over. GTOs improve dc power controller and inverter (smaller weight and volume, better efficiency, less audible noise) [1.4].

1.2 Basic Functions of Static Converters

Static converters (converters for short) are circuit using static valves which convert or control electrical energy. They enable the energy flow between different systems to be controlled. When ac and dc systems are coupled four basic functions are possible (Fig. 1.4):

1. Rectification, the conversion of ac into dc whereby energy flows from the ac system into the dc system.
2. Inversion, the conversion of dc into ac whereby energy flows from the dc system into the ac system.
3. DC conversion, the conversion of dc of a given voltage and polarity into that of another voltage and where applicable reversed polarity whereby energy flows from one dc system into the other.
4. AC conversion, the conversion of ac of a given voltage, frequency, and number of phases into that of another voltage, frequency, and where applicable number of phases whereby energy flows from one ac system into the other.

These four basic functions in the conversion of electrical energy are performed by corresponding types of converter (Fig. 1.5), namely the basic rectification

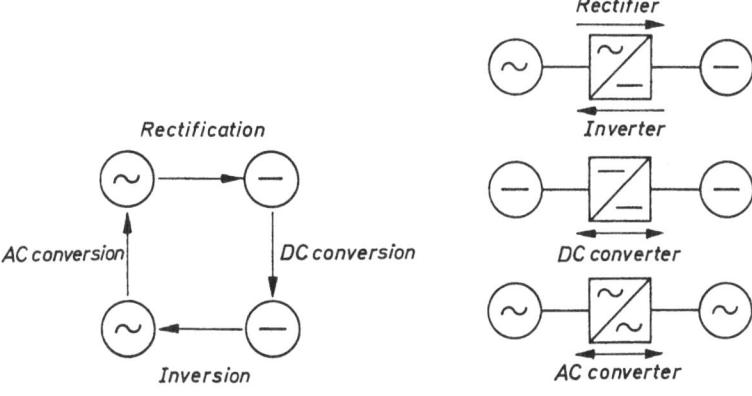

Fig. 1.4. Types of energy conversion   **Fig. 1.5.** Types of converter

function by a rectifier, the basic inversion function by an inverter, the basic ac conversion function by a dc converter, and the basic ac conversion function by an ac converter, the latter also referred to as a frequency converter.

In the case of rectifiers and inverters the energy direction is preset but with dc and ac converters the direction of energy flow can generally change.

As illustrated in Fig. 1.4 the basic functions of converters can be used when coupling ac and dc supply systems. However, they also occur in the same way when active or passive loads are fed from sources of alternating or direct current. Besides the basic functions mentioned converters are also employed for further duties e.g. to generate reactive power or switch ac and dc circuits. These duties can, however, also be treated as special cases of ac or dc conversion. The limited number of basic functions is realized by a large number of converter connections. This will be dealt with in detail later.

2 System Components

A general description of an energy conversion system using converters requires only a few elements: voltage sources, transformers, resistors, magnetic and electrical energy stores, and converter valves performing switching functions [4].

2.1 Linear Components

The assumption of idealized sources, idealized transformers, and linear passive elements produces the linear systems components of converter circuits given in Fig. 2.1.

A *general voltage source* has the characteristic $u(t)$. A sinusoidal voltage source is described by the curve of voltage against time

$$u = \hat{u} \sin \omega t \tag{2.1}$$

and a direct voltage source by the voltage

$$u = U_d. \tag{2.2}$$

An *ideal transformer* has the transformation ratio w_1/w_2 with which the value of the electrical quantities, voltage, and current are changed between the primary and secondary sides. An ideal transformer stores no electrical energy. Power

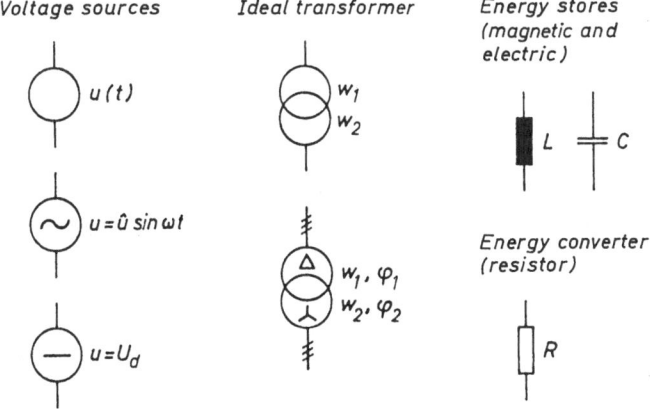

Fig. 2.1. Linear system components of converter circuits

equilibrium between the primary and secondary sides is maintained during the transformation of voltages and currents. In the case of multi-phase transformers, a phase displacement between the primary and secondary electrical quantities is also possible and depends upon the transformer connection. Energy is not stored in an ideal transformer in this case either.

A *resistor R* represents an energy converter whose relationship between current and voltage is described according to Ohm's Law by

$$i = \frac{u}{R}. \tag{2.3}$$

The electrical power converted into heat is u^2/R or Ri^2.

An *inductance L* represents a magnetic energy store. The voltage and current of an inductance are mutually linked according to

$$u = L\frac{di}{dt}. \tag{2.4}$$

The magnetic energy stored in an (constant) inductance at a current i ist $Li^2/2$.

A *capacitance C* represents an electrical energy store. The current and voltage of a capacitance are mutually linked according to

$$i = C\frac{du}{dt}. \tag{2.5}$$

The electrical energy stored in a capacitance C at a voltage u is $Cu^2/2$.

2.2 Semiconductor Switches

The functions of static converters presuppose cyclic switching processes performed with the aid of converter valves. Genuine valves use a directional electrical conductance which can be produced in a vacuum, in gases or in semiconductors. Valves labelled as non-genuine are those with which a valve action is achieved by mechanical contacts or similar devices without directional conductivity.

The converter valves of greatest importance for power electronics are semiconductor components, namely, the non-controllable semiconductor diode and the controllable thyristor. Special forms of construction of the thyristor (triode ac semiconductor switch [triac] and gate turn-off thyristor) and power transistors are gaining in significance (see Chap. 3).

In general, semiconductor switches can be differentiated according to their ability to carry current in either one or two directions and according to their ability to be turned on and off.

Figure 2.2 lists semiconductor switches which can be turned on for one and two directions of current. On the left are illustrated the appropriate symbols for a diode, thyristor, and triac with the current-voltage characteristic alongside on the right. In this context, capable of being turned on means that, ignoring the non-controllable diode with which the current begins to flow on its own when the anode voltage becomes positive, the start of current can be defined by triggering a

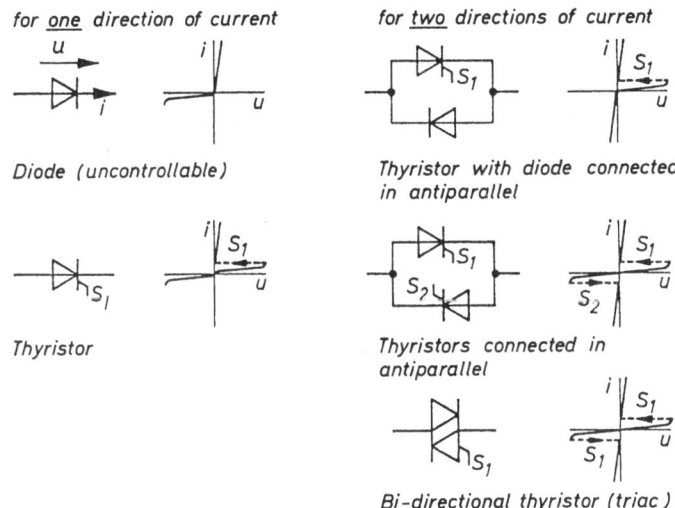

Fig. 2.2. Semiconductor switches capable of being turned on

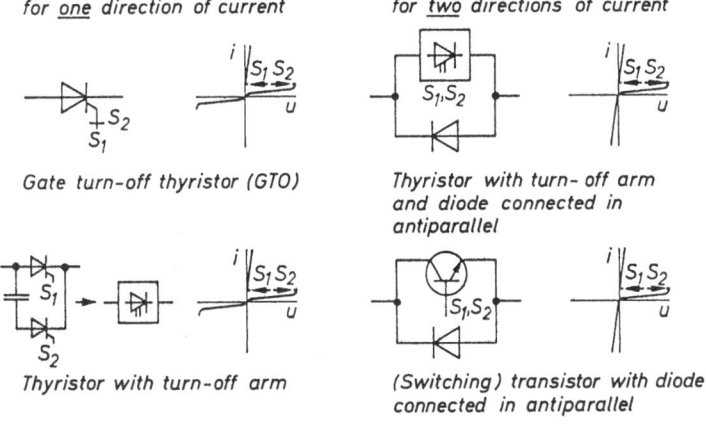

Fig. 2.3. Semiconductor switches capable of being turned on and off

gate terminal S with positive anode voltage, i.e. conduction can be initiated as a function of other quantities e.g. the time. Diodes and thyristors can carry current in only one direction because of their valve property. The diode reliably switches into the conducting state when the anode voltage becomes positive, a thyristor only when a trigger pulse is present at the same time.

A thyristor with a diode or another thyristor connected in antiparallel can carry current in two directions. In the case of the triac this property is concentrated in one semiconductor component. Its current-voltage characteristic corresponds to that of two thyristors connected in antiparallel.

Figure 2.3 shows a summary of semiconductor switches capable of being turned on and off. Again the appropriate symbols for thyristors and transistors are in each case illustrated on the left with the corresponding current-voltage characteristics on the right.

Gate turn-off means that the principal current in the semiconductor switch can be interrupted by an appropriate turn-off pulse in the gate circuit i.e. by means of a negative trigger pulse to the gate terminal. Thyristors can also be turned off via turn-off arms (see Chap. 8). The symbol for a thyristor with turn-off arm is often shown abbreviated — as drawn in Fig. 2.3 — by a boxed-in thyristor symbol with two gate terminals. GTO thyristors and thyristors with turn-off arms permit only one direction of current. Diodes connected in antiparallel produce semiconductor switches for two directions of current which can be turned on and off in one of the directions.

To investigate the behaviour of converter connections the characteristics of the semiconductor switches can be idealized at first, i.e. for the conducting state characteristic a forward voltage drop of 0 is assumed independent of the forward current, and for the blocking state a reverse current of 0 independent of the reverse voltage. Such an idealization of the semiconductor switches is of course permissible only within limits. It gives for instance no information concerning power losses, forward voltage drops, insulation ratings or dynamic switching characteristics. It is, however, permissible for the examination of the basic functions of a converter connection and produces sufficiently accurate results.

2.3 Network Simulation

Network simulation makes use of this simplified treatment of semiconductor switches whereby the two variables, thyristor voltage, and thyristor current are allocated to a thyristor and the relationship between these two electrical variables is identified by two different generally idealized branches of a characteristic curve [2.1 – 2.12].

We illustrate the procedure for simulating a network by considering the thyristor. According to Fig. 2.4a two electrical variables, voltage u_A and current i_A, can be allocated to a thyristor. The relationship between these two electrical variables is identified by two different branches of a characteristic curve (Fig. 2.4b). The holding current i_H fixes the minimum current required to maintain operation on the on-state characteristic. The voltage u_D identifies the minimum voltage at which the thyristor can still be turned on. Generally the thyristor current can be described mathematically by

$$i_A = f(u_A, z). \tag{2.6}$$

The current i_A varies not only with the voltage u_A but also with the system parameter z. Operation on the blocking-state characteristic mean $z = 0$ and operation on the conducting-state characteristic means $z = 1$. The transition from the thyristor voltage u_A to the thyristor current i_A is described by a characteristic module with two characteristic branches which is controlled by the system parameter z (Fig. 2.4c). The system parameter z is in turn supplied by a storage module which describes the "memory" of the thyristor. This store is set to be dominant when the triggering signal s is present and the turn-on condition

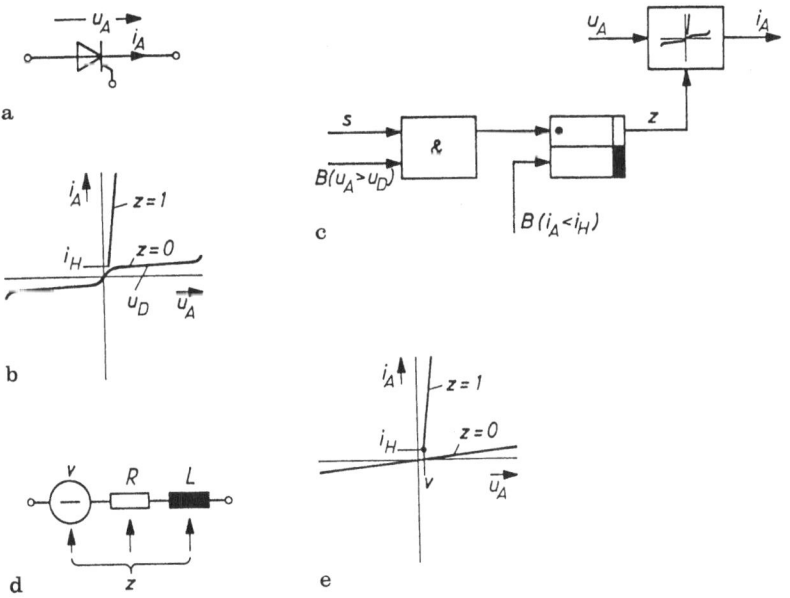

Fig. 2.4a–e. Mathematical simulation model of a thyristor

$B(u_A > u_D)$ is satisfied. It is erased when the thyristor current i_A is less than the holding current i_H. In place of the description by means of a characteristic module illustration by RL combinations and impressed voltage is also common, the values of these being present by the system parameter z (Figs. 2.4d and e).

For all semiconductor switches listed in Figs. 2.2 and 2.3 corresponding mathematical simulation models can be constructed but they can, however, be considerably more complicated, as in the case of the triac.

2.4 Non-linear Components

Besides the semiconductor switches converter circuits also include non-linear components. Non-linearity is caused particularly by saturation phenomena in the iron circuits of inductances, electrical machines or transformers. Figure 2.5 shows typical forms of hysteresis loops in which a non-linear relationship exists between the magnetic field strength H and the magnetic induction B. Hence a linear relationship between current and voltage generally exists only over restricted ranges of H or B. This can be taken into account during simulation by an angular or smoothly curved magnetizing characteristic approximately corresponding to the hysteresis loop.

Other non-linearities arising in converter circuits can be caused by the behaviour of the load, e.g. when supplying arcs, in smelting, charging batteries or operating electrical machines (see Chap. 10). Components with a non-linear relationship between current and voltage are employed in protection engineering e.g. voltage sensitive resistors, surge diverters or saturable reactors.

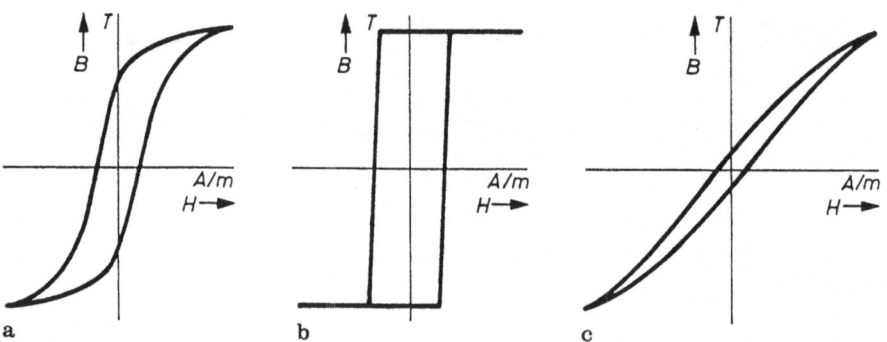

Fig. 2.5a – c. Typical hysteresis loops. **a** Normal, rounded loop; **b** rectangular loop; **c** inclined or slowly rising loop

Summarizing, it should again be said that most properties of converters and converter connections can be examined under ideal conditions using only a few system components, namely voltage sources, transformers, energy converters, and energy stores as well as idealized semiconductor switches. The general system theory of converter circuit technology can be established using these elements alone.

In the next two chapters, the electrical and thermal properties of the power semiconductors will first be illustrated in greater detail together with their snubber circuits, triggering, and cooling. Then in Chap. 5 the system components just defined will be used to examine switching processes and the internal method of operation of converters.

3 Power Semiconductor Devices

The converter valves used in solid state based on power electronic systems are designated power semiconductor devices. Here they will be abbreviated as power semiconductors. The most important power semiconductors are silicon diodes, thyristors, and power transistors [3.9, 3.15, 3.16, 3.26, 3.27].

Besides the power semiconductors there exist converter valves of other construction (see Figs. 1.2 and 1.3). These are mercury-vapour valves with hot cathodes (Thyratrons) and with liquid cathode (Excitrons with continuous excitation and Ignitrons with an ignitor). Excitrons and Ignitrons are constructed as single-anode metal-tank units. At the same time large multiple-anode rectifiers were built as welded metal-tank units with bolted-on cover plate (see Sect. 1.1). For voltages in excess of 100 kV and currents up to about 1 A high-vacuum converters with hot cathodes were employed primarily in high-frequency and X-ray applications. Except for special applications, the mercury-vapour valve has been superceeded by thyristors during the last decades.

Semiconductor rectifier valves can be divided into polycrystalline semiconductors and monocrystalline semiconductors. Polycrystalline semiconductors are copper oxide rectifiers and selenium rectifiers. Copper oxide rectifiers are now employed solely as instrumentation rectifiers.

Selenium rectifiers still have some considerable area of application. They consist of a thin layer of selenium applied to a metal plate with a soft metal alloy sprayed on as the counter-electrode. The depletion layer is formed between the selenium and the counter-electrode. Its maximum reverse voltage lies between 30 V and 50 V. High reverse voltages can be achieved by connecting many selenium rectifiers in series.

The power semiconductors mentioned above, namely silicon diodes, thyristors, and power transistors, are constructed from monocrystalline semiconductor material. The base material is almost exclusively silicon monocrystals. Previously germanium was also used in the fabrication of semiconductor devices.

The conduction mechanism of semiconductors is dealt with only briefly here since this is described in detail in other publications [5, 13, 15, 27].

pn Junction. All power semiconductors contain one or more pn junctions which have a blocking capability for one voltage polarity and can carry current with a small forward on-state voltage drop for the opposite polarity.

A silicon monocrystal has a crystal lattice of the diamond type in which each silicon atom is surrounded by four neighbouring atoms forming a tetrahedron. Each of the four outer electrons of the silicon atoms enters an electron-pair bond with an outer electron of the four neighbouring atoms.

Into such a crystal lattice impurities can be introduced in such a way that a silicon atom is replaced, for example by an element of the fifth group of the periodic system, for instance phosphorus, arsenic or antimony, having one valence electron more than silicon. The fifth valence electron is available for the conduction process owing to the loose bond at the impurity. Such impurities which make a surplus conduction electron available are called donors. An extrinsic semiconductor with surplus valence electrons is called an n-type conductor or extrinsic n-type semiconductor. If the impurities in the silicon lattice are replaced by an element of the third group of the periodic system, such as boron, aluminium, indium or gallium, a p-type conductor or extrinsic p-type semiconductor is created. Since these elements have only three valence electrons, one electron-pair bond is incomplete. A defect electron, or hole, is created which can contribute to electrical conduction because it represents a positive charge carrier. The impurities of a p-type conductor that release a defect electron i.e. accept an electron are called acceptors.

The intentional incorporation of impurities into a semiconductor is called doping. The mode of conductivity of a semiconductor depends upon the difference between the concentration of donors and acceptors; when the donors dominate an n-type conductor is created, a p-type conductor is created if there is a majority of acceptors.

Silicon and germanium can be doped to be either n-conducting or p-conducting. Selenium is known only in the p-conducting form. The most important prerequisite for the production of power semiconductors is the intentional adjustment of the impurity concentrations effecting the n-type or p-type conduction. The conductivity is approximately proportional to the impurity concentration. With p or n-conducting silicon the resistivity can be set over a range from about $10^{-3}\Omega$ cm to $10^4\Omega$ cm. This occurs in a thin monocrystalline silicon slice (several 100 µm) in alternate p and n-conducting layers in sequence. Modern manufacturing processes enable the thickness of the layers and the impurity concentrations to be chosen and varied.

Where p and n-conducting layers come into contact a pn junction is created. This represents the simplest configuration of a semiconductor rectifier. Such a pn junction blocks when the n-conducting layer has positive electrical potential compared to the p-conducting layer, because in this case the junction is depleted of moving charge carriers. If the polarity of the external voltage is reversed, electrons and holes migrate into the pn junction which thereby becomes conducting.

Semiconductor diodes have only one pn junction, thyristor, and transistors several pn junctions as well as a control electrode, or gate.

Forms of Case Construction. In all power semiconductors the silicon slice provided with variously doped layers (the actual semiconductor system) is fitted into a case for protection against mechanical damage and atmospheric effects. It also dissipates the electrical loss arising in the semiconductor system to a heat sink. For silicon diodes and thyristors in the medium and high-power ranges two standard forms of construction have evolved: the cell with flat base or threaded stud cooled from one side and the disc cell cooled from both sides.

Figure 3.1 shows a thyristor constructed as a flat-base cell. The acutal semiconductor system is pressed onto the solid copper base of the case by

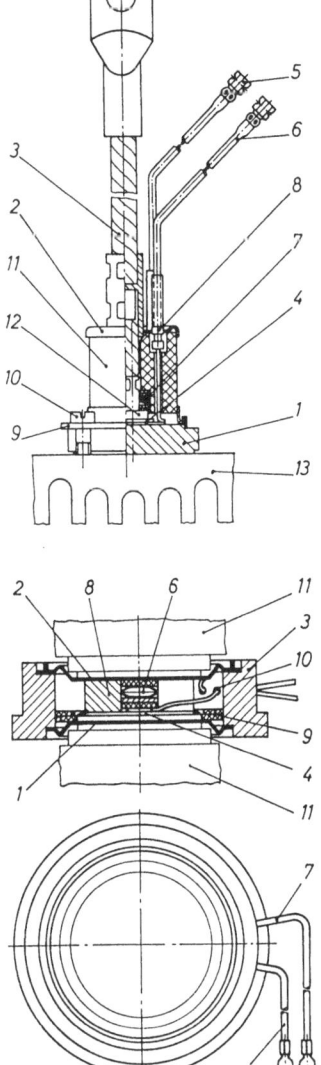

Fig. 3.1. Thyristor (flat-base cell), half-section. *1* Base (anode terminal), *2* cap, *3* cathode terminal, *4* thyristor system, *5* second cathode terminal, *6* gate terminal, *7* disc springs, *8* ceramic bushing, *9* clamping plate, *10* screw, *11* ceramic ring *12* slug, *13* heat sink

Fig. 3.2. Thyristor (disc cell). *1* Anode terminal, *2* cathode terminal, *3* ceramic ring, *4* thyristor system, *5* gate connection, *6* disc springs for gate connection, *7* second cathode terminal, *8* slug, *9* insulating washers, *10* ceramic bushing, *11* heat sink

preloaded disc springs. A ceramic ring insulates the upper cathode terminal with slug contact from the anode terminal on the base of the case. The flat-base cell is bolted to a heat sink using a clamping plate.

Figure 3.2 shows a thyristor constructed as a disc cell. Here the thyristor system lies in a disc-shaped case, the upper and lower terminals of which are insulated from one another by a ceramic ring. Contact is made by squeezing the upper and lower heat sinks together with an external clamp which in turn clamps the semiconductor system between copper slugs inside the case. The control electrode, or gate, is applied by disc springs.

With flat-base cells the heat due to the losses is dissipated to only one side but in the case of disc cells it is dissipated to the heat sinks on both sides.

The more important electrical properties of power semiconductors are discussed in the following.

3.1 Semiconductor Diodes

Figure 3.3 shows diagrammatically the construction of a semiconductor diode. This has a pn junction. When the anode is positive forward current flows from the anode to the cathode. When the cathode is positive the pn junction blocks allowing only a very small reverse current of a few mA to flow so long as the maximum permissible reverse voltage is not exceeded.

3.1.1 Characteristic Curve

Figure 3.4 illustrates the characteristic curve of a semiconductor diode. It consists of two branches: the reverse characteristic with negative anode voltage u_A and the forward characteristic with positive anode voltage. Important electrical characteristics of a semiconductor diode are the rated reverse voltage U_{RN} which is the continuously permissible peak value of reverse voltage with sinusoidal terminal voltage and the rated current I_N which is the arithmetic mean value of the continuously permissible forward current. Rated currents apply in conjunction with the associated heat sink.

With power semiconductors more information is obtained from the maximum permissible values of current and voltage than from the rated values. The rated values recommended for operation can be taken from these maximum values after the application of certain safety factors.

Maximum average on-state current I_{FAVM} is the name given to the arithmetic mean of the highest continuously permissible forward current with sinusoidal current half-cycles. The design of the protective devices is also decided by the maximum overcurrent which must result in disconnection the surge on-state current that is allowed to occur only once as a sinusoidal half-cycle at 50 Hz resp. 60 Hz from rated operation and the $i^2\,dt$ rating (the maximum allowable value of the square of the instantaneous forward on-state current integrated over the time).

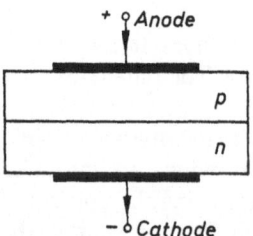

Fig. 3.3. Semiconductor diode (diagrammatic construction)

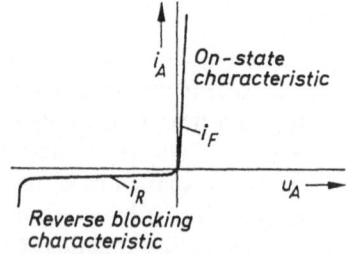

Fig. 3.4. Characteristic curve of semiconductor diode

The forward voltage u_F is the voltage appearing between the terminals of a semiconductor diode in the forward direction. For silicon diodes it is 1V to 1.5V. The maximum permissible momentary value of reverse voltage in each cycle is called the repetitive peak reverse voltage U_{RRM}. With regard to operational transient overvoltages, diodes are usually operated at a terminal voltage with a peak value that is lower than the repetitive peak reverse voltage by a safety factor of 1.5 to 2.

The forward characteristic and particularly the reverse characteristic both vary with the temperature of the semiconductor system. This is designated the junction temperature. For silicon diodes an upper junction temperature of between 150°C and 200°C is permissible. Silicon power diodes achieve reverse voltage of several kV and maximum mean forward currents of over 1000 A (see Fig. 13.24).

3.1.2 Switching Behaviour

The switching behaviour of semiconductor diodes is characterized by a forward time lag and a reverse time lag. When switching on a certain (although very short) time elapses before the forward current flows because charge carriers have first to be injected into the pn junction from the highly doped regions. This delay is called the turn-on delay. When a semiconductor diode switches off the current is not extinguished as it passes through zero, but first flows in the negative direction until the base region has been freed of charge carriers and reverse voltage can be assumed (Fig. 3.5). This is called the reverse recovery delay. After the storage time t_s has expired the peak reverse recovery current I_{RRM} decays with a high rate of change. The diode then assumes the reverse voltage. The hatched area under the current-time curve is called the stored charge Q_s (or lag charge). It gets larger with rising junction temperature, rising forward current, and increasing rate of rise of commutating current.

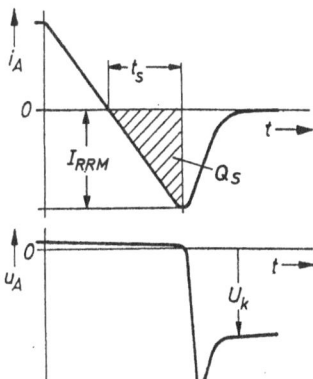

Fig. 3.5. Turn-off process of semiconductor diode

3.2 Thyristors

A thyristor is a power semiconductor with four layers of alternate conductivity p−n−p−n [3.1, 3.3, 3.7]. Figure 3.6 shows the construction diagrammatically. The anode terminal is connected to the outer p region, the cathode terminal to the outer n region. The gate terminal (on standard thyristors controlled from the cathode side) is fitted to the p region nearest to the cathode.

Thyristors were originally called silicon controlled rectifiers (SCR) but several years ago the term Thyristor was agreed internationally. (Thyristor is an artificial term like Transistor: the combination of **Thyra**tron = current gate in Greek and Trans**istor**).

3.2.1 Characteristic Curve

The characterstic curve of a thyristor shown in Fig. 3.7 has three branches: the reverse blocking characteristic, the off-state characteristic, and the on-state characteristic [3.2]. The reverse blocking characteristic corresponds to that of semiconductor diodes. Below the maximum permissible peak reverse voltage a reverse current i_R of a few mA flows and increases with rising junction temperature. So long as no gate current flows to the cathode via the gate terminal a thyristor blocks even with positive anode voltage u_A. Below the maximum permissible peak positive off-state voltage U_{AM} forward off-state current i_D of only a few mA flows.

If a thyristor with positive anode voltage is triggered by means of a gate current flowing from the gate terminal to the cathode, its locus of operation move to the on-state characteristic. This on-state characteristic corresponds to that of a semiconductor diode with the difference that as a result of three pn junctions instead of one being present a somewhat higher on-state voltage u_T of from 1.2 V to over 2 V arises. Changeover from the off-state characteristic to the on-state characteristic also occurs with no gate current when the permissible peak off-state voltage is exceeded or the rate of rise of voltage exceeds a critical value.

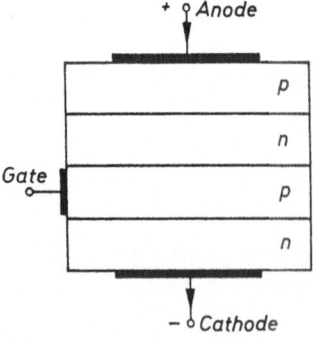

Fig. 3.6. Thyristor (four-layer triode)

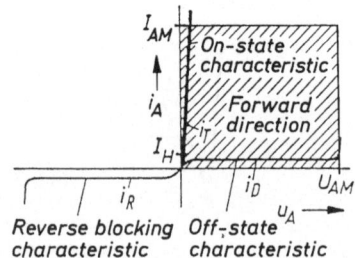

Fig. 3.7. Characteristic curve of thyristor

The forward off-state voltage at which a thyristor with zero gate current changes over from the off-state to the on-state is called the zero breakover voltage $U_{(BO)zero}$. Such triggering must not be carried out repeatedly in operation whereas occasional triggering by exceeding the zero breakover voltage in the event of a fault is permissible. In contrast to this exceeding the permissible peak reverse voltage on the reverse characteristic leads to destruction of the thyristor.

In general, once triggered, a thyristor cannot be turned off again via the gate terminal. Only when the anode current drops below the holding current I_H due to changes in the external circuit does the thyristor turn off. In this respect a thyristor behaves like a thyratron or mercury-arc valve. Recently a device of special construction called a gate turn of (or GTO) thyristor has become available in high power ratings. Such special thyristors can be turned off from the gate terminal.

Important electrical properties of a thyristor are the *repetitive peak reverse voltage* U_{RRM} and the *rated current* I_N which is the arithmetic mean value of the continuous on-state current. The maximum continuous on-state current when loaded with sinusoidal current half-cycles is called the maximum average *on-state current* I_{TAVM}. Maximum overcurrent $I_{T(OV)M}$ is the name given to the value of on-state current at which the thyristor must be turned off if damage is to be prevented. When loaded with this current, a thyristor can temporarily lose its blocking capability in the forward direction. The surge current is permissible only once as a sinusoidal half-cycle at 50 Hz resp. 60 Hz. It is quoted for previous operation at no load or rated load. The ultimate load integral $\int i^2 dt$ is used to size the protective devices.

The maximum permissible values of current and voltage are also stated for thyristors in data sheets and can be used to obtain recommended rated values for various applications. The user determines the safety factors according to the limiting data of the thyristors and the stresses occurring in his circuit.

3.2.2 Switching Behaviour

The peak reverse blocking and off-state voltages and maximum average state current represent the steady-state characteristics of a thyristor. The dynamic characteristics describe its switching behaviour [34, 3.4, 3.5, 3.8]. The most

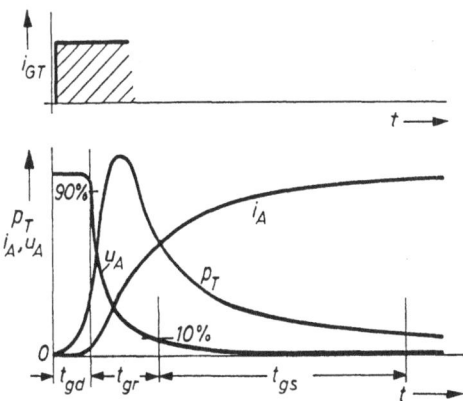

Fig. 3.8. Turn-on process of thyristor

important are the *critical rate of rise of off-state voltage* $(du/dt)_{crit}$, the *critical rate of rise of on-state current* $(di/dt)_{crit}$, and the *circuit-commutated* turn-off time t_q.

Figure 3.8 shows the turn-on process of a thyristor. Like a semiconductor diode a thyristor also needs a finite time to turn on. After application of the triggering current i_{GT} the *gate-controlled delay time* t_{gd} elapses before the thyristor voltage u_A collapses. The thyristor current i_A rises at a finite rate. Its response depends, of course, upon the impedance of the load circuit. Within the *gate-controlled rise time* t_{gr} the thyristor voltage drops from 90% to 10% of the initial value.

Then follows the *gate-controlled conduction spreading time* t_{gs} which plays an important role in the case of the large-area thyristors. It arises due to the finite time taken during the triggering process for conduction to spread from a point on the cathode near the gate terminal over the entire surface of the cathode.

The conduction spreading speed is of the order of magnitude of 0.1 mm/µs. The gate controlled delay time is between 1 µs and 2 µs, and the gate controlled rise time is of the same order of magnitude. The gate controlled conduction spreading time can exceed 100 µs depending upon the diameter of the silicon slice and the arrangement of the gate terminal.

During the turn-on process the on-state power loss

$$p_T = u_A i_A \tag{3.1}$$

converted to heat can assume considerable momentary values of several kW. As they are converted in a small volume of the silicon slice near the gate terminal there is a danger of damage if the rate of rise of current or the switching frequency are too high [3.11].

Figure 3.9 shows the turn-off process of a thyristor. As with the semiconductor diode after passing through zero the thyristor current first continues to flow unhindered in the opposite direction. Only at instant t_2 does the junction on the cathode side begin to assume reverse voltage. The thyristor voltage u_A becomes negative and is approximately the breakdown voltage of the junction on the cathode side. At instant t_4 the charge carrier concentration at the junction on the anode side is reduced to such an extent that this pn junction can also assume

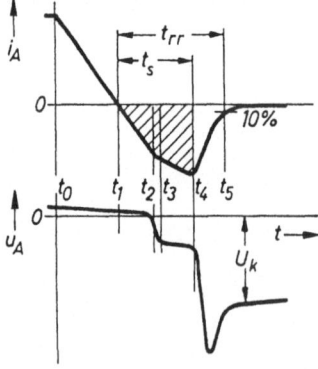

Fig. 3.9. Turn-off process of thyristor

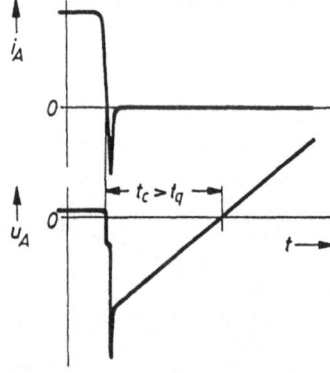

Fig. 3.10. Turn-off time of thyristor

reverse voltage. The thyristor current then drops to zero with a high initial rate of change.

The period between passage of the current through zero and its decay to 10% of its peak value is called the *reverse recovery time* t_{rr} as with the semiconductor diode. The quantity of charge carriers stored between t_1 and t_4 is called the *reverse recovery charge* Q_s. As with silicon diodes it increases with rising junction temperature, rising on-state current, and increasing rate of rise of on-state current. The rapid decay of the thyristor current from instant t_4 leads to transient overvoltages. In order to limit these to acceptable values a so-called recovery effect snubber circuit is necessary (see Sect. 4.1.1).

After the thyristor current has been turned off reverse voltage must be applied temporarily between the anode and cathode of the thyristor. The thyristor is not at first capable of blocking a forward off-state voltage. The two outer pn junctions indeed block. However, in the base regions and above all at the middle junction immediately after turning off there are still surplus charge carriers which must be removed by recombination. Only after that can be thyristor also assume forward off-state voltage without turning on.

Circuit-commutated turn-off *time* t_q is the name given to the minimum time between passage of the current through zero from the forward to the reverse direction and the earliest permissible return of an off-state voltage. Should an off-state voltage be applied before expiry of the circuit-commutated recovery time, the thyristor turns on again even with no gate current.

The period of reverse voltage after the current has passed through zero determined by a particular circuit is called the *hold-off interval* t_c. This hold-off interval is a characteristic of the circuit whereas the circuit-commutated recovery time is a characteristic of the thyristor. In every operating state the hold-off interval t_c must be longer than the circuit-commutated recovery time t_q (Fig. 3.10). So that this is also the case on temporary voltage dips or eventual overcurrents a safety factor of at least 1.3 to 1.5 is generally applied.

The circuit-commutated recovery time of a thyristor is not constant. It becomes considerably longer with rising junction temperature. Moreover, it increases slightly with an increase in the previous on-state current.

When the reverse voltage is very low during the hold-off interval as for instance in the case of thyristors with a diode connected in antiparallel, the circuit-commutated recovery time increases considerably (by a factor of up to 2 to 2.5).

When this voltage is greater than 50 V the circuit-commutated recovery time is approximately independent of the reverse voltage.

3.2.3 Thyristor Specifications

Depending upon the application, the blocking capability, on-state voltage, and circuit-commutated recovery time are important considerations when sizing thyristors.

The on-state voltage determines the maximum average on-state current for given cooling. The circuit-commutated recovery time is an important dynamic characteristic of thyristors playing a significant role particularly in applications at higher frequencies and in circuits with forced commutation (see Chap. 8).

However, only two of the stated quantities can be optimized at any time at the expense of the third. This happens because of the interaction among the specific resistance of the silicon, the carrier lifetime, and the thickness of the n-type base. A thick base and high specific resistance result in a high blocking voltage. A long carrier lifetime causes the on-state voltage drop to be small but the circuit-commutated recovery time to be larger.

Two types of thyristor have been developed: phase control, of N-type, thyristors for applications at supply frequencies of 50 or 60 Hz, and inverter grade, or F-type, thyristors exhibiting circuit-commutated recovery times between 10 μs and 60μs. These latter are required for circuits using forced commutation and at medium frequencies. N-type thyristors generally have circuit-commutated recovery times exceeding 100 μs, but they can be fabricated with voltage and current ratings exceeding those for F-type thyristors. The limiting values for N-type thyristors are 2.5 kV to 4 kV for off-state voltage and > 1000 A for maximum average on-state current. With F-type thyristors according to the circuit-commutated recovery time voltages of 1.2 kV to 2.5 kV are attained at currents of more than 500 A [3.9, 3.12].

Devices with very high current (> 2500 A) and voltage (> 2500 V) ratings rely on large diameter, uniformly doped, and defect free silicon wafer for their fabrication [3.10, 3.17].

A particularly homogeneous doping of the initial silicon material is achieved by neutron radiation in a nuclear reactor. In this way an exactly defined number of silicon atoms is converted into phosphorus atoms which are distributed extremely uniformly throughout the bulk of the crystal. Large diameter silicon crystals can thus be manufactured with highly homogenous doping. This neutron-doped silicon material (NDS) can be used to build high-current thyristors with a crystal diameter of over 100 mm.

3.2.4 Types of Thyristor

Until now the standard thyristor has been described using a characteristic curve in accordance with Fig. 3.7. To be more accurate this is a reverse blocking thyristor triode controlled on the cathode side. Besides this there is a whole series of other types of thyristors.

Generally a thyristor is defined as a bistable semiconductor component with at least three junctions that can be changed over from an off-state into an on-state or vice versa. However, the expression Thyristor may be used for the reverse blocking thyristor triode solely when misunderstandings cannot arise.

Other types of thyristors (see Fig. 3.11) of importance are thyristor diodes, (Diacs) reverse blocking triode thyristors controlled on the anode side (anode fired), reverse blocking tetrode thyristors, triode thyristor capable of being turned off (GTO-thyristor), and bidirectional triode thyristors (Triacs). Moreover, there are also reverse conducting triode thyristors controlled on the anode or cathode sides.

Diode thyristors are four-layer elements with no gate terminal. They are triggered by exceeding the breakover voltage or the critical rate of rise of voltage.

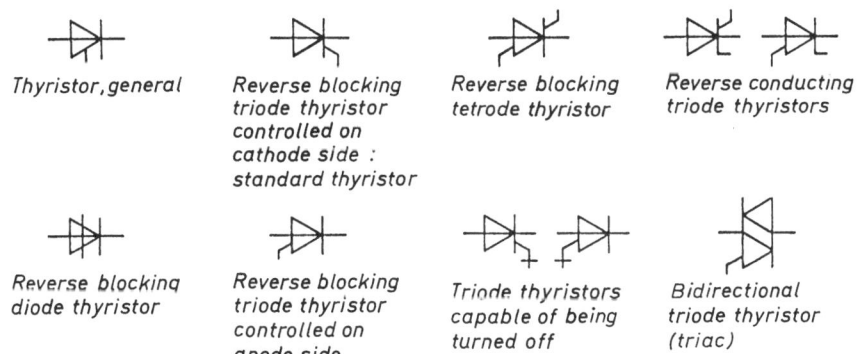

Fig. 3.11. Circuit symbols of thyristors

In the thyristor controlled on the anode side the gate terminal is connected to the n-type layer on the anode side. Thyristor tetrodes have a contact on both base regions; they can therefore be controlled on either the cathode or anode side. Next to the mentioned thyristor types the switching behaviour of the normal thyristor is improved by special gate structures. The most important are the amplifying gate and the field initiated emitter [3.6]. Such thyristors have a higher crictial rate of rise of one-state current and are suitable for circuits with higher rates of di/dt and for application in the middlefrequency range [3.13].

To improve the du/dt capability shorted emitters are used which short the cathode pn-junction on single points of the p-base zone.

In the last few years further developments in the thyristor switching times have been achieved [3.23, 3.24, 3.25]. Of special importance for choppers and inverters with forced commutation and uses in the middle frequency range are the asymmetric blocking thyristor (ASCR), the reverse conducting thyristor (RCT), the gate-assisted-turn-off thyristor (GATT), and gate turn-off thyristor (GTO). In Table 3.1 the layer construction and the technological characteristics of the above thyristors are shown [3.30].

3.2.4.1 Triac

Triacs are bidirectional thyristors which can conduct current in both directions. They contain in a silicon layer two antiparallel pnpn-zones. The polarity of the firing current is indifferent. Triacs reach peak blocking voltages between 1000 and 1500 V with currents of over 100 A. Critical is the acceptable rate of change of voltage after commutation which only allows for 10 V/μs. Triacs are used in electronic contactors and in light dimmers (see Sect. 6.1).

3.2.4.2 Asymmetrical Silicon Controlled Rectifier (ASCR)

The asymmetrical silicon controlled rectifier has a strongly reduced capability of blocking voltage in reverse direction. Through a highdoped n-zone this is limited to about 20 V which allows the n-basezone thickness to be reduced without afflicting the off-state voltage in forward direction. This improves the forward conditions and almost halves the turn-off time. An other possibility would be to

Table 3.1. Production technology for fast thyristors

Feature	Semiconductor element			
	SCR	ASCR	RCT	GTO
Circuit symbol				
Internal construction				
Technological characteristics – of the doping – of the masking	Double diffused p-zones Single sided photo process	Thin, double doped n-base Single sided photo process	Thin, double doped n-base and diffused diode ring double sided photo process	Cathode lines and short circuited anodes Precise double sided photo process
Lateral structure of the cathode	Simple, ringformed or branched cathode areas (line width 2 mm to 3 mm)	As SCR	As SCR with additioned diode ring	Numerous single cathode lines, width 0,1 mm to 0,4 mm
Contact construction	Solder or press contact, slice structure for the branched cathode	As SCR	As SCR	Overlay-technique with bond or pressure contact hight difference between gate and cathode line

double the off-state voltage in forward direction without changing any other characteristic.

3.2.4.3 Reverse Conducting Thyristor (RCT)

The reverse conducting thyristor has the same electrical characteristics as a fast thyristor with a diode in antiparallel connection [3.32]. The RCT has the same five layer structure as the ASCR and to that an outer lying diode ring. Due to the asymmetrical structure of the reverse conducting thyristor the active part can be optimized. With the same off-state voltage in forward direction the turn-off time is a factor 0.6, or the on-state voltage is a factor 0.7 smaller than with symmetrical blocking thyristors. Forward and switching losses are at constant voltage also smaller. Up to currents of 100 A the thyristor and diode parts are produced with a current ratio of 1:1.

With larger currents the required current conductance of the integrated diode is smaller than the thyristor.

To decrease the number of elements two reverse conducting thyristor-chips are often integrated on an isolated module producing saving in space and weight. Moreover this makes circuits more insightable and has definite characteristics with respect to circuite commutated turn-off time (small stray inductances).

3.2.4.4 Gate-assisted-turn-off-thyristor (GATT)

The turn-off time of a thyristor can be shortened by a negative gate-cathode voltage after the current has passed through the zero point. This process is called gate assisted turn-off. Figure 3.12 shows the typical current and voltage characteristics of a GATT [3.35]. Circuit commutated turn-off times of under 10 μs are obtainable. This allows for frequencies of 10 kHz in inverters with parallel resonant circuits (see Sect. 7.3). Typical data for a GATT are:

off-state and reverse voltage 1200 V, maximum average on-state current 400 A, $di/dt = 1000$ A/μs, circuit commutated turn-off time $t_q = 10$ μs (with a negative gate cathode voltage $U_{GK} = -15$ V).

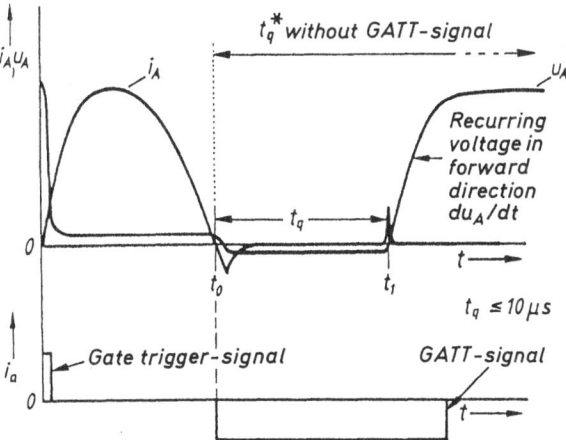

Fig. 3.12. Typical current and voltage curves of GATT

3.2.4.5 Gate Turn-off thyrisor (GTO)

Gate turn-off thyristors can be fired by a control current of one polarity and be switched off by a control current of opposite polarity [3.31].

Already at the start of the sixties GTOs (for currents of a few ampere and voltages of several hundred volts) were offered. Todays elements, however, with much higher current and voltage capability have only been available for the last few years. Off-state and reverse voltages of up to 2.5 kV and controllable on-state currents of up to 1000 A have been obtained, GTOs like conventional thyristors have a four layer structure of changing conductivity pnpn. The cathode area is split up into thin lines which are surrounded by the gate paths. Unlike conventional thyristors a smaller surface resistance of the p-basezone is required. An acceptable blocking capability of the gate-cathode depletion layer and a reduced current amplification of the pnpn-system are also required. These requirements are necessary so that the negative gate current through which the thyristor is switched off is kept as small as possible.

A typical switch off procedure of a GTO is shown in Fig. 3.13. After the injection of the negative gate current i_G the middle pn junction of the thyristor cuts off after a few μs (delay time) and the steep fall of the anode current i_A starts. The rising gradient of the returning anode voltage u_A is determined by capacitance circuitry. The permissible rate of change of voltage for GTOs lies between 100 V/μs and 1000 V/μs.

After the steep descent a decaying current tail appears for a few μs caused by the remaining charge carriers in the n-base. Its waveform is nearly independant of

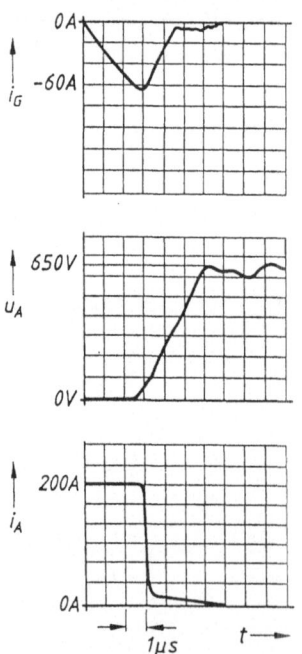

Fig. 3.13. Turn-off process of a GTO-thyristor

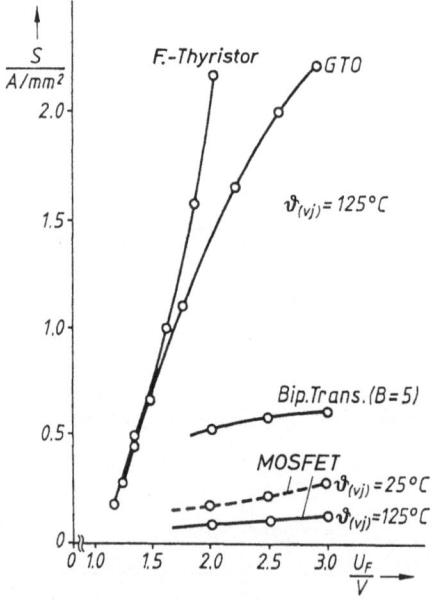

Fig. 3.14. Comparison of different controllable semiconductor elements

the gate current, but the current tail can increase switch off losses perceptible. Figure 3.14 shows the comparision of different characteristic curves of controllable semiconductors devices. Only the GTO due to its four layer structure, has such favorable values as a frequency thyristor.

An analysis of the possible applications in converter circuits leads to the result that GTOs are suitable for converters with impressed voltage and commutation on the dc side (see Sect. 11.3.2). In converters with impressed current and commutation on the ac side GTOs can only be used with extra circuitry due to the different stray inductances.

GTOs have to be suitably circuited for use in dc choppers and inverters (see Sect. 4.1.6). For the impulse control generator gate currents of both polarity are required. The required rate of rise and amplitude of current, especially for negative impulses, are far greater than that of conventional thyristors (see Sect. 4.2.3). The maximum frequency for GTO thyristors is determined by the switching times and the switch-off losses. Due to the small switch times (see Fig. 3.13) frequencies of several kHz can easily be reached. Small GTOs allows for higher switching frequencies (> 10 kHz).

3.2.4.6 Light-triggered Thyristor

Light-triggered thyristors produce the necessary charge carriers for firing by light which passes through a window and into the silicon layer. As the transferrable firing energy by light is small, gate structures with a high switch-on sensitivity are necessary. They must also, however, be insensitive to du/dt-occurences. Smaller gate surfaces and shorter light energy impulses fulfill these requirements. The light sensitive surface switches only a small part of the total volume and for this reason extra current amplifying structures must support the switch-on process.

The light signal is produced from a light emitting diode (LED). The transfer of the firing impulse by light assures a safe potential barrier between power and control parts. Direct light-triggered thyristors are thus used for high voltage converters (such as EHVDC).

Direct light triggered thyristors are available for voltages up to 4 kV and currents between 1.5 and 3 kA.

3.2.4.7 Static Induction Thyristor (SITh)

In the last few years fast switching elements with small control power needs are developed in which the forward resistance is controlled by the injection of charge carriers in a high resistance layer with the help of electrostatic induction. The principle workings are equivalent to that of the FET (see Sect. 3.3.2).

Experimental prototypes of static induction transistors (SIT) and static induction thyristors (SITh) have recently become available [3.35].

Figure 3.15 shows the structure of a SI-thyristor. In general the expression thyristor is displayed by a four layered element which remains in its switching condition without a control signal. The SI-thyristor has from the anode to cathode a diode structure i.e. in the forward direction it is selfconducting. It may, however, be switched off by a negative gate cathode voltage. The possible uses of SIThs in converters are the same as that of the GTOs, however, without control current the SITh switches on into the forward direction while the blocking state in forward

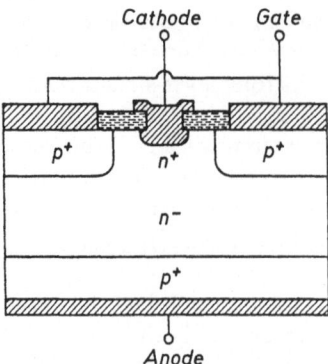

Fig. 3.15. Structures of a SI-thyristor

direction requires a constant negative control current. The switching times are even shorter than with GTOs (turn-off time about 3 µs). The acceptable di/dt is about ten times greater than with normal thyristors (> 7000 A/µs). The gate control is less complicated than that of the GTO. The on-state voltages is however some what greater.

The SI-thyristor are at the start of their development. Experimental prototypes for 600 V and 40 A, 2500 V and 300 A, 4000 V and 500 A have been built in Japan.

3.3 Power Transistors

Transistors operating as switches can also be employed as converter valves in the power section. Power transistors on a silicon basis had already been developed by the sixties. These transistors have small heat resistance between the crystal and housing, and large admissible powers losses (at 25°C ambient temperature > 1 W). The distinguishing factor between small and large power transistor is their heat resistance. The border of heat resistances is about 15 K pro Watt. The power transistor housing is normally made of metal. The cooling area must have good heat dispersion qualities.

Power transistors form two large groups, *low-frequency power transistors* (LF), and *high-frequency power transistors* (HF). The limit between the two groups lies at 30 MHz where the transistor parameters depend no longer upon the technology but rather upon the geometry. With LF power transistors the parameters depend mainly upon the chosen manufacturing techniques. With HF power transistors the geometry of the crystal has a strong influence upon the transistor parameters.

LF power transistors are employed in power electronics circuites [3.19]. Depending on application the power transistors have been developed in the last decade for higher currents or for higher voltages.

High-current transistors switch currents of up to 600 A at maximum reverse voltages of 100 V to 150 V. High reverse biasing transistors (high-voltage transistors) reach maximum inverse voltage of more than 1000 V at currents of more than 50 A (see Fig. 13.24). It is thus possible to construct converter circuits with powers ranging up to several 10 kW using power transistors.

In the last few years transistors of greater power capabilities have been developed. With doublesided cooled power transistors in disc form current-voltage combinations of 200 A, 1200 V or 400 A, 100 V have been reached.

To increase the rating the power transistors in the arm of a converter can be connected either in series or in parallel. Morcover, power modules are available in which several transistors are already connected in parallel to form an integral unit. Power modules are built with collector currents of more than 1000 A. Since 1971 there have been so called Darlington power transistors available which are constructed monolithycally from two independent transistors (generally from the same chip). Such a combination simplifies the design of the output stages in amplifier curcuits. Current amplifications of more than 1000 are attained at relatively high currents.

Types of Transistors. A transistor is a controllable electronic amplification element in which a current of charge carrier wander through semiconducting material from one electrode to the other. The current strength is controlled by a middle electrode (controllability).

From the semiconducting point-contact diode made out of germanium or silicium Bardeen and Brattain developed the first point-contact transistor (n-Transistor). Its follower was the junction transistor which consists of three different crystal zones of the same semiconducting material, however, with different doping in the orders of p–n–p or n–p–n (Figs. 3.16 and 3.17). In general the base crystal is much thinner than the outer emitter and collector crystals. The base zone has the lowest doping, the emitter the highest. Between the crystal zones lie the very thin depletion layers.

By the development of transistors several technical procedures must be carried out in order to achieve the required purity of the crystal and to prevent any surfaces inversion layers.

For the different applications different transistor have been constructed: **Planar transistors** are silicon transistors with a diffused base zone where the pn

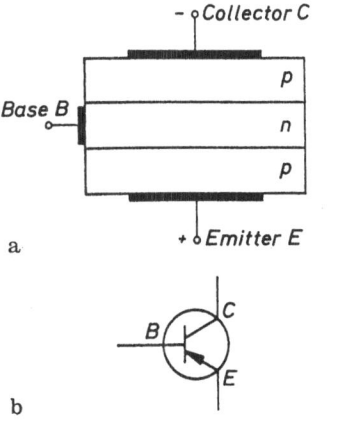

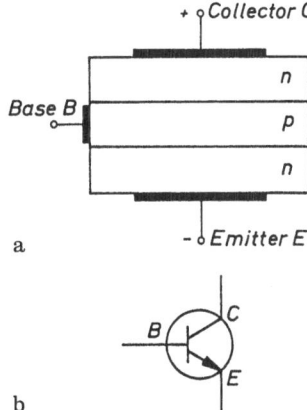

Fig. 3.16a,b. p−n−p transistor (three-layer triode). **a** Diagrammatic construction; **b** circuit symbol

Fig. 3.17a,b. n−p−n transistor (three-layer triode). **a** Diagrammatic construction; **b** circuit symbol

junction is protected by a thin SiO$_2$ coating. Most transistors have alloy junctions: **Alloy junction transistor**. In a n-type base crystal one alloys or diffuses on both sides a p-type material (acceptor) e.g. Bor, Indium or Gallium.

Other transistors include Drift-Transistors, Mesa-Transistors, and Epitaxial-Planar-Transistors. They have better frequency characteristics and are used in highfrequency circuits. For all the forenamed transistors both charge carries (majority and minority carriers) are used for carrier transport mechanism. They are all commonly referred to as bipolar transistors.

The field effect transistor is different in its principles from the junction transistor (namely the control of a diode by neighbouring charge carries injection). The semiconductor current of majority charge carriers is controlled by two opposite lying reversed biased pn junctions. This control is almost achieved with no current as the control electrode is in the reversed biased state. Field effect transistors are the most used of unipolar transistors in which (different to the bipolar transistor) only charge carriers of one polarity play a note.

Surface field effect transistors with MOS structures have been recently developed into MOS power transistors and have applications in the smallest power ranges (see Sect. 3.3.2).

3.3.1 Bipolar Power Transistors

Bipolar power transistors are alloy junction transistors with a small heat resistance and high admissible power losses.

3.3.1.1 Construction of a Transistor

A transistor is an amplifying element that unlike the thyristor can be continuously modulated by controlling the charge carriers. The charge carriers are controlled by means of a control current. As with constant modulation in the transistor itself a considerable portion of the power is converted into heat because the transistor assumes part of the voltage, switching duty is the preferred application when employed in the power section.

On switching duty a transistor is either fully blocked or fully turned on. The switching duty of a power transistor corresponds to the bistable behaviour of a thyristor whereby a transistor unlike the standard thyristor, however, can also be switching off via the gate current.

A bipolar transistor consists or three differently doped layers. The emitter E and collector C are connected to the two outer layers and the base B to the middle region. Transistors can be constructed with a layer sequence of either p − n − p or n − p − n. Figure 3.16 shows diagrammatically the construction of a p–n–p transistor and the related circuit symbol. Figure 3.17 shows diagrammatically the construction and related circuit symbol of an n–p–n transistor. Power transistors on a silicon basis are nowadays chiefly manufactured as n–p–n transistors.

3.3.1.2 Basic Connections

Transistors can be operated fundamentally in three different connections: common-emitter connection, common-base connection or common-collector connection.

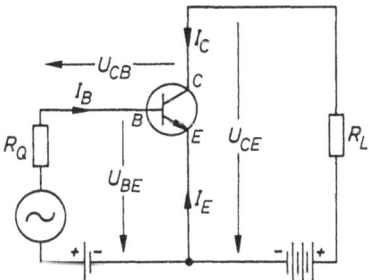

Fig. 3.18. Common-emitter circuit of n−p−n transistor

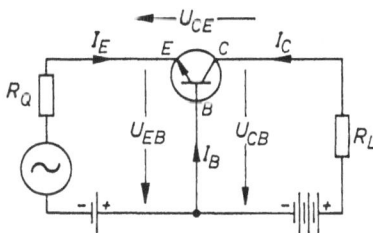

Fig. 3.19. Common-base circuit of n−p−n transistor

With the common-emitter connection the load current flows via the emitter and the base (Fig. 3.18). Current amplification factors of more than 10 can be achieved with power transistors operated in this connection. When fully turned on the load current causes an on-state voltage drop of 1 V to 1.5 V between the emitter and the collector. This corresponds to the forward voltage of semiconductor diodes or thyristors.

The common-base connection (drawn for an n−p−n transistor) is illustrated in Fig. 3.19. Again the control is via the emitter-base circuit. The load current flows via the collector, emitter, and gate voltage source. The common-base circuit is employed in high-frequency engineering.

With the common-collector connection the output circuit is formed between collector and emitter and the input circuit between base and collector. The common-collector connection is used for impedance transformation.

3.3.1.3 Characteristic Curves

Power transistors on switching duty are operated in common-emitter connections. In Fig. 3.20 the family of output characteristic curves of an n − p − n transistor in a common-emitter connection is illustrated. In the blocking zone only a small

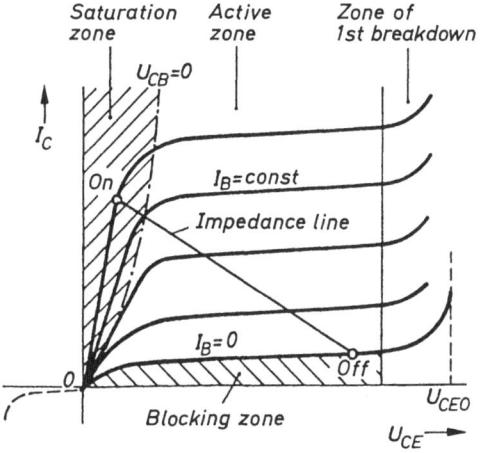

Fig. 3.20. Output characteristic curves of n−p−n transistor in common-emitter circuit

collector current I_C flows with high collector-emitter voltage U_{CE}. In the saturation zone reached by raising the base current I_B a high collector current I_C flows at low collector-emitter voltage U_{CE}.

On switching duty changing back and forth between the off and on working points occurs as quickly as possible. The distance between the two points in the family of characteristic curves depends upon the type of impedance in the load circuit. In Fig. 3.16a line of constant impedance is drawn in.

3.3.1.4 Switching Behaviour

The turn-on characteristic of a transistor on switching duty corresponds approximately to the characteristic of a thyristor (see Fig. 3.8). Figure 3.21 shows the curves of collector-emitter voltage u_{CE} and collector current i_C on turn-on using a base current i_B for the common-emitter connection whereby turn-on power losses

$$p = u_{CE} i_C \tag{3.2}$$

occur temporarily. The delay time and rise time can be defined in the same way as with the thyristor.

The turn-off process of a transistor in a common-emitter connection is illustrated in Fig. 3.22. It is performed using the base current i_B. The collector current i_C and collector voltage u_{CE} take a finite time to switch during which large turn-off power losses p occur.

The area under the power loss curve on turn-on and turn-off represents the energy losses converted in the power transistor during the switching processes. The switching losses must be taken into consideration in the loss balance equation of the transistor particularly at the higher operating frequencies.

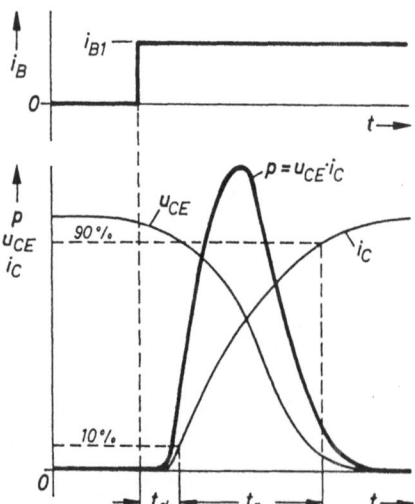

Fig. 3.21. Turn-on process of transistor

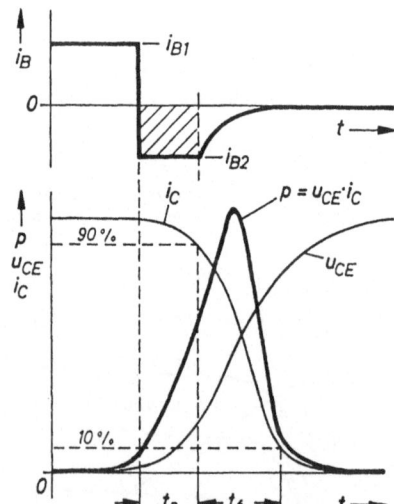

Fig. 3.22. Turn-off process of transistor

3.3.2 MOS Power Transistors

The first experiments to build an electronic device based upon the fieldeffect go back to the year 1930 (J. E. Lilienfeld in New York) and since then has led to numerous patents. After the second worldwar while examining the fieldeffect, Bardeen, Brattain, and Shockley succeeded in discovering the transistor effect in the Bell Telephone Laboratories. At the start of the sixties the MOSFET appeared (Metal-oxide-silicon-fieldeffect-transistor). Due to its low power loss the horizontally built MOSFET had for a long time only played a role in signalelectronics. Since the start of the eighties this has changed and today there are MOS-power transistors available which contain many small elements with vertical structures in a parallel layout. Modern day integration has allowed the parallel layout of some hundred transistors on one ship [3.18, 3.19, 3.20, 3.21].

An example of data achieved in 1985 were:

$$U_{DS} = 50 \text{ V}, \ I_D = 40 \text{ A}, \ R_{DS(on)} < 0{,}03 \ \Omega$$

or

$$U_{DS} = 1000 \text{ V}, \ I_D = 5 \text{ A}, \ R_{DS(on)} < 2 \ \Omega.$$

Thus new switchable semiconductor devices (on and off) with short switching times (frequency range over 50 kHz) and small control losses are available in the lower range of power electronics [3.22, 3.28, 3.29].

3.3.2.1 Construction of a MOSFET

Depending on the construction and the form of the cell (e.g. hexagonal or rightangled) different kinds of MOSFETs have been developed. The so called D-MOS Technology (Double-Diffused-MOS Technology) allow for greater current densities and higher reverse voltages than the first power MOSFETs which were built after the Surface Groove Technology.

Figure 3.23 shows the construction of a single cell built from many parallel lying elements in a SIPMOS-Transistor. The shown SIPMOS-structure displays the usual characteristics of all power MOSFETs.

The basis material is a highconducting substrate upon which an epitaxial layer is laid. The construction of the doting material and the thickness of the structure determines its voltage capability.

On the surface the source elements are arranged so that they were all connected with an Al-source-layer. Each source element consists of a n^+-ring in a p-zone, which are also connected to each other with Aluminium (Al). On the surface

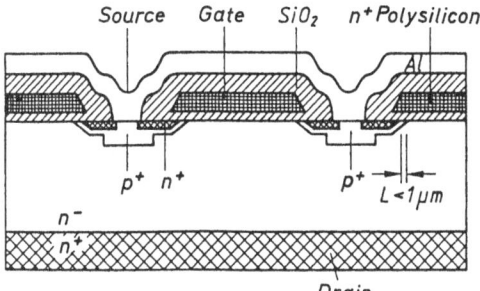

Fig. 3.23. Power field effect transistors in SIPMOS structure

between the thin gateoxide and the thicker oxide under the Al-source contact lies the embedded crossover polysilicon. Applying a positive voltage on the gate electrode electrons will be pulled to the surface by an electric field and create a conducting channel. Between the n^+-source and n^--drains and through the gate voltage the current is modulated. For this no input current is necessary (excluding capacitance discharge currents) as the gate electrode is insulated which allow control with little power. The current capability of a power MOSFET for a given voltage is greater the more elements it contains i.e. the greater its area is.

3.3.2.2 Characteristic Curves

Figure 3.24 shows for example the characteristic curve of a 36 mm² big SIPMOS transistor [3.33]. The BUZ 15 can only block 50 V, but can trigger a few hundred ampere in pulses. Its forward resistance is smaller than 30 m Ω through which it can conduct 40 A in the on-state.

The highvoltage-transistor BUZ 54 can trigger a few 10 A in impulses, but its on-state-current is only about 5 A, as its forward resistance is more than 2 Ω.

The characteristic curves show the only disadvantage of the MOS-transistors for higher power. MOSFETs have by larger voltages a greater voltagedrop and greater R_{DSon} then the samesized bipolar devices like transistors and thyristors.

3.3.2.3 Control and Switching Behaviour

As the power-MOSFET has an isolated control electrode (gate) it is able to be operated without power so long as the switching frequency and speed are low. For higher switching speeds the parasite stray capacitors must be quickly discharged which requires an input current which is by far smaller than that needed for the running of a bipolar transistor. For operation an impulse generator with a 50 Ω output resistance is sufficient. The power is, however, bigger as it could be produced by the VLSI-elements, but is smaller than that for comparable transistors. Power-MOSFETs switch quicker than bipolar powertransistors. High biasing SIPMOS-transistors have switch-off times under 50 ns (switching capability > 10 kW) [3.34].

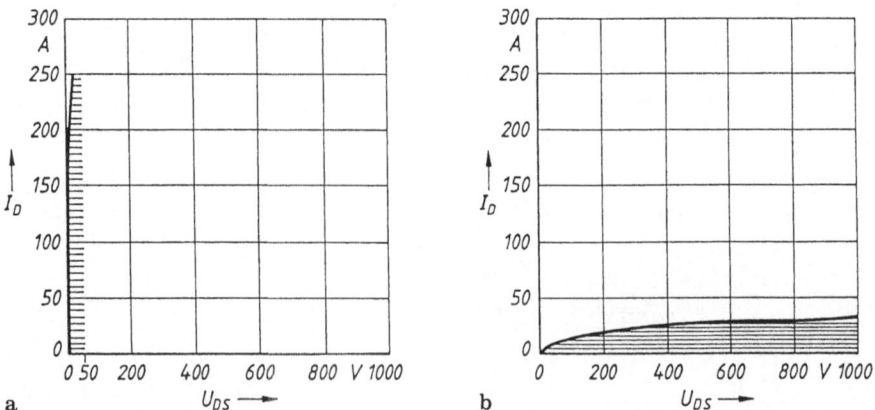

Fig. 3.24a,b. Characteristic curves if the SIPMOS-FETs (6mm × 6mm silicon area). **a** BUZ 15 (for 50 V); **b** BUZ 54 (for 1000 V)

3.3.3 Static Induction Transistor (SIT)

In 1950 Nishizawa and Watanabe obtained a Japanese Patent for a device which they called Electrostatic Induction Transistor (SIT). The basic idea is to control the forward resistance through charge injection in a high resistance wafer with help of electrostatic induction. Its principle function is equivalent to the fieldeffect transistor.

The structure and output characteristic curve of a self conducting power-SIT are shown in Fig. 3.25. The p^+-doped gate is buried in the n-channel. A negative gate-source voltage builds a positive space-charge region in the b-zone which narrows the channel. With increasing voltage the depletion region increases until the channel is pinched off.

The offered power-SITs differ neither in the forward resistance nor in the gate-source capacitance compared to comparable power-MOSFETs. Their switching times of 200 to 300 ns are somewhat higher than those of most MOSFETs.

E.G.
$U_{DS} = 800$ V; $I_{DS} = 10$
A (60 A peak);
Switch-on time: t_{on} $= 250$ ns;
Switch-off time: t_{off} $= 300$ ns;
Forward resistance: $R_{DS(on)}$ $= 0,66$ to $1,5$ Ω.

SIT with higher reverse bias voltages and smaller switching times are announced.

In converter circuits the danger of short circuits occur due to the self-conducting characteristic of these elements. For this reason the full negative pinch off voltage always has to be brought to the gate before a drain source voltage is applied.

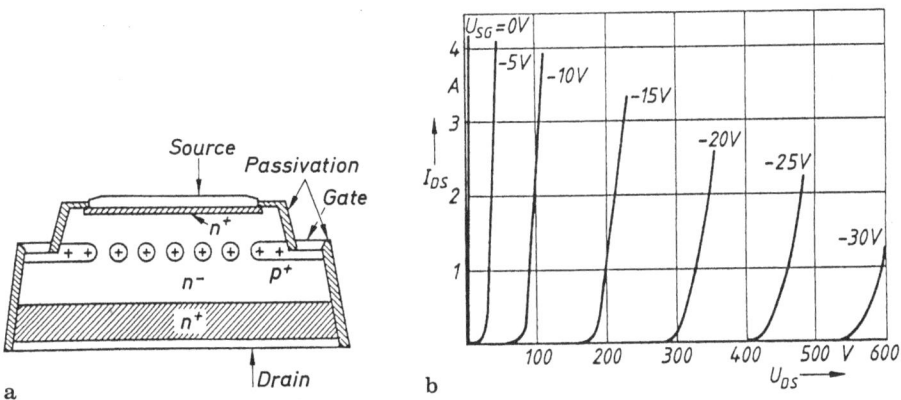

Fig. 3.25a,b. Static-induction transistor (SIT). **a** Structure; **b** output characteristic curves

4 Snubber Circuits, Triggering, Cooling, and Protection Devices

For power semiconductors to operate properly in converter connections three prerequisites must be satisfied:

1. They should be protected from undue transient voltage and current stresses.
2. Controlled power semiconductors such as thyristors or power transistors need suitable drive units to generate the necessary gate or base currents.
3. The heat due to the losses occurring in the power semiconductors must be dissipated.

The snubber circuit provides the protection against undue voltage stresses [4.15, 4.17]. Drive units generate gate or base pulses of defined rate of rise, magnitude, and duration. Cooling is ensured by mounting the power semiconductors on suitable heat sinks from which the heat due to the losses is dissipated via gases (air) or liquid (water, oil). Power semiconductors are generally unable to function proper without suitable snubber circuits, cooling, and also in the case of controlled rectifiers without suitable gate currents. Failures can almost always be traced back to faulty snubber circuits, triggering or cooling. The additional protection from undue overcurrents is provided by current limiters, fuses, high-speed switches or crowbars [4.14, 4.19, 4.20, 4.22].

4.1 Snubber Circuits

By snubber circuits is understood the application of capacitors and resistors sometimes also in combination with series-connected inductances to damp surge voltages and reduce the rate of rise of voltage or current in the power semiconductors. Because of their physical construction these have only a limited insulation strength and permit only certain maximum values of the rate of rise of voltage and of the rate of rise of current. These limiting values can be taken from the data sheets. Frequently, even temporarily exceeding the stated limiting values leads to destruction of the semiconductor elements.

The necessary RC circuits ought to be dealt with taking thyristors as an example. Similar conditions apply for semiconductor diodes and power transistors.

Figure 4.1 shows typical snubber circuits for thyristors. The basic circuit is the capacitor-resistor combination C_B and R_B in parallel with the thyristor. This suppression circuit can be augmented by an additional high-value resistor R_P.

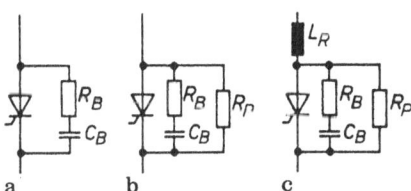

a   b   c

Fig. 4.1a−c. Thyristor snubber circuits

To limit the rate of rise of current and voltage a series inductance L_R can also be provided. This series inductance can have either a linear characteristic i.e. constant inductance independent of the current or a non-linear characteristic i.e. inductance varying with the current. In the latter case a saturable iron core is employed.

With thyristors the snubber circuit performs three duties:

1. Damping of surge voltages caused by the rapid collapse of the negative reverse recovery current. These are then called surge absorbing capacitors [4.4].
2. Limitation of the rate of rise of voltage.
3. Uniform voltage distribution in series circuits, both for the static and dynamic case i.e. on turn-on and turn-off.

In special cases limitation of the rate of rise of current is also performed by series inductances.

4.1.1 Recovery Effect Snubber Circuits

The method of operation of an RC combination as a snubber circuit in parallel with the thyristor will be explained using Fig. 4.2. The current i_A in the thyristor is interrupted (actually with a finite rate of fall) at moment t_0 at the end of the recovery time. The current commutates from the thyristor to the snubber circuit. A series resonant circuit damped by resistor R_B is formed by L_k and C_B. The current and voltage characteristics can be calculated from the differential equation

$$U_k = L_k \frac{di_C}{dt} + R_B i_C + \frac{1}{C_B} \int i_C dt. \tag{4.1}$$

The current is found from the equation

$$i_C = A e^{-\frac{t-t_0}{\tau}} \sin[v(t-t_0) - \varphi_i] \tag{4.2}$$

whereby a damped oscillation is produced with the time constant

$$\tau = \frac{2L_k}{R_B} \text{ and circuit frequency } v = \sqrt{\frac{1}{L_k C_B} - \left(\frac{R_B}{2L_k}\right)^2}.$$

The corresponding equation for the capacitor voltage u_C is

$$u_C = \frac{1}{C_B} \int i_C dt = ZA \cdot e^{-\frac{t-t_0}{\tau}} \sin[v(t-t_0) - \varphi_u] + U_k \tag{4.3}$$

with the oscillation impedance $Z = \sqrt{L_k/C_B}$.

The constants are obtained from the initial conditions.

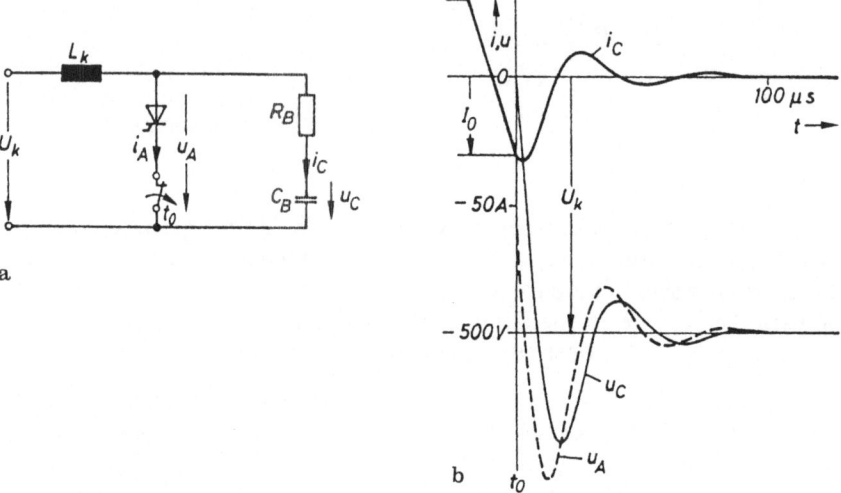

a

b

Fig. 4.2a,b. Equivalent circuit for calculation of the recovery effect snubber circuit. **a** Equivalent circuit; **b** current and voltage characteristics

In Fig. 4.2 the voltage and current curves are plotted quantitatively for the assumed values $U_k = 500$ V, $L_k = 100$ µH, $C_B = 0.5$ µF, $R_B = 10$ Ω and reverse current $I_0 = 30$ A. At moment t_0 the thyristor voltage u_A jumps to the value $R_B I_0$ and approaches in a damped oscillation the momentary value of the commutating voltage U_k. The peak value of the inverse voltage can lie considerably above the value U_k of the commutating voltage. The recovery effect snubber circuit should be designed accordingly. In fact the conditions are somewhat more favourable than illustrated in Fig. 4.2, because the thyristor current decays with a finite rate of fall at the end of the recovery time.

4.1.2 Rate of Rise of Voltage (du/dt) Limitation

In combination with series connected inductances the snubber circuit also limits the rate of rise of voltage du/dt across the thyristor at moments of switching. Figure 4.3 shows a simplified equivalent circuit for calculating the du/dt stress. It is assumed that at moment t_0 a voltage U_k is switched onto the thyristor via the commutating inductance L_k. This occurs in static converter connections due to the switching operation of other cells. The commutating inductance L_k is formed from the cable inductances and other stray inductances e.g. that of transformers. Therefore an accurate value can often only be estimated. When the equivalent switch is closed at moment t_0 a damped series resonant circuit is again formed whose response can be calculated with appropriate boundary conditions in accordance with Eqs. (4.1), (4.2), and (4.3).

In Fig. 4.3 the voltage and current curves are plotted quantitatively for the assumed values $L_k = 50$ µH, $C_B = 0.5$ µF, $R_B = 10$ Ω and the commutating voltage $U_k = 500$ V assuming that the capacitor voltage $u_C = 0$ at moment t_0. Damped oscillations result for the voltage and current.

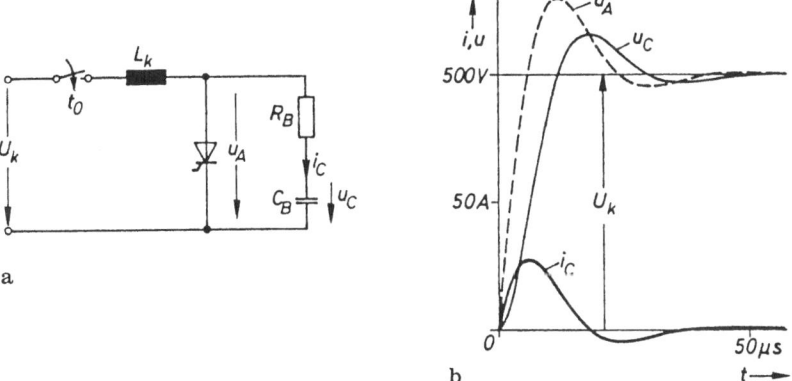

a

b

Fig. 4.3a,b. Equivalent circuit for calculation of the du/dt stresses. **a** Equivalent circuit; **b** current and voltage characteristics

The maximum rate of rise of voltage in the thyristor occurs at moment t_0. It can be calculated from Eqs. (4.2) and (4.3) as

$$\frac{du_A}{dt} = R_B \frac{di_C}{dt} + \frac{du_C}{dt}. \tag{4.4}$$

For the quoted example the value $du_A/dt_{max} = R_B U_k/L_k = 100$ V/μs is obtained. This value must be smaller than the permissible rate of rise of voltage stated in the data sheets.

4.1.3 Transformer and Load Snubber Circuits

Besides connecting the snubber circuit directly to the power semiconductors (valve suppression circuit) suppression circuits can also be applied to other components of a converter circuit. In Fig. 4.4 further suppression circuits are given. The snubber circuit connected across a transformer suppresses surge voltages which occur due either to switching the transformer or to load changes. Additionally it damps surge voltages coming in from the ac supply system.

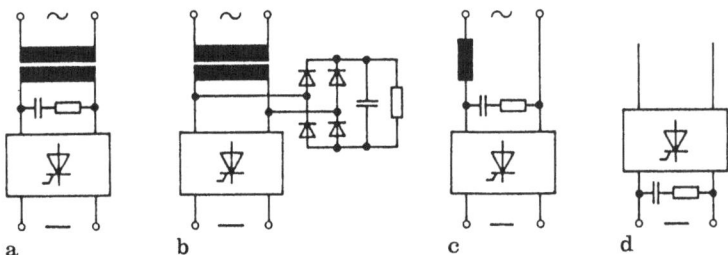

a   b   c   d

Fig. 4.4a–d. Transformer, ac supply and load snubber circuits

Instead of an RC element connected directly to the secondary side of the transformer, the transformer snubber circuit can also be connected via a rectifier bridge. In this case a unipolar dc capacitor can be used as the suppression circuit capacitor. For direct mains connection with no transformer a protective reactor is generally inserted behind which a suppression circuit can be connected. Finally, a suppression circuit can also be provided on the load side in addition to the valve suppression circuit.

4.1.4 Series Connection

When thyristors are connected in series it must be ensured that the total voltage divides itself as evenly as possible among the series-connected thyristors [4.1, 4.2]. Without a suppression circuit this is by no means guaranteed, because thyristors of the same type can have different reverse currents whereby an uneven voltage distribution corresponding to similar reverse currents would result, and also because on turning on and off, the switching points of thyristors in series can differ by microseconds as a consequence of dissimilar gate controlled delay time and uneven hole storage charge. In this case, on turning on, the voltage across the thyristor with the shortest delay time falls first causing a temporary voltage increase to occur across the other thyristors. On turning off the thyristor with the smallest reverse recovery charge temporarily assumes the full commutating voltage.

These effects must be alleviated by means of an appropriate snubber circuit.

Figure 4.5 shows the snubber circuit for static and dynamic voltage distribution. The high-value parallel resistors R_p determine the static voltage distribution. To be effective their current must be about one order of magnitude higher than the reverse current of the thyristors. The dynamic voltage distribution is undertaken by the elements of the suppression circuit. The capacitor C_B stores charge differences ΔQ resulting from dissimilar delay time on turning on and uneven hole storage charge on turning off. It is then charged up higher by a differential voltage

$$\Delta U_c = \frac{\Delta Q}{C_B}. \tag{4.5}$$

On turning off this leads to an additional voltage rise which must be kept as small as possible. According to Eq. (4.5) the suppression circuit capacitor C_B is dependent upon the charge differences ΔQ which varies with thyristor type.

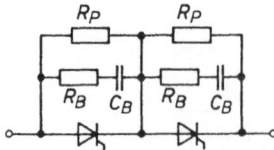

Fig. 4.5. Static and dynamic voltage distribution in a series connection of thyristors

4.1.5 Parallel Connection

When thyristors are connected in parallel as even a current distribution as possible is strived for. Here again as with the voltage distribution with series connection a distinction must be made between static and dynamic current distributions. The static current distribution is determined by the on-state characteristics of the parallel-connected thyristors as well as from the ohmic voltage drops present in the circuit. On rapid current changes e.g. during switching and commutation processes the dynamic current distribution is in addition strongly influenced by the inductances present in the parallel connection.

Figure 4.6 shows the current distribution when thyristors with different on-state characteristics are operated in parallel. In steady-state operation the voltage U_T across the parallel thyristors is equal. According to their different on-state characteristics, therefore, an uneven current distribution can result. Supplementary series resistors would equalize the current distribution but are, however, uneconomic on account of the additional losses. Thyristors (also semiconductor diodes) are therefore classified for parallel operation according to their on-state voltage class and only elements of the same on-state voltage class are connected in parallel. When fuses are employed in each thyristor arm, these can equalize the current distribution as a result of their voltage drop.

Series inductances are employed to improve the dynamic current distribution. Figure 4.7 shows possible configurations. In high-current installations with many parallel thyristor or diode arms the inductance of the individual parallel arms can be equalized by the configuration of the busbars (incoming and outgoing bars on

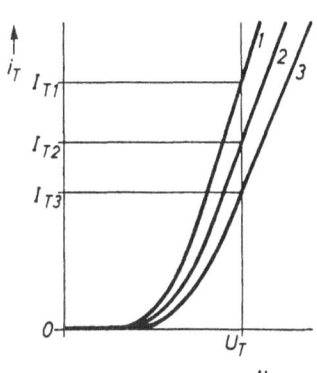

Fig. 4.6. Parallel operation of thyristors with differing on-state characteristics

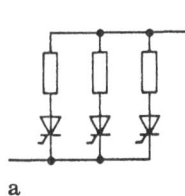 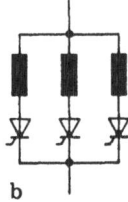 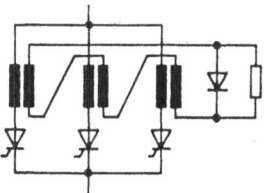

a   b   c

Fig. 4.7a–c. Improvement of the current distribution with parallel-connected thyristors. **a** Busbar configuration; **b** series inductances; **c** coupled series inductances

different sides). Additional series inductances can also be used. If these are interconnected by means of secondary windings, the current change in one thyristor arm will support corresponding current changes in the other parallel-connected arms. These coupled current-distribution reactors are, however, expensive and are therefore only employed in special cases.

4.1.6 Snubber Circuits for GTO-Thyristor

GTOs are fast switching thyristors, in which the current can be interrupted by a negative gate current impulse. Unlike other thyristors it requires no separate commutating circuit (commutating capacitor) for current switch-off. GTOs must however be circuited so that no overvoltages or critical du/dt occur when switching off with large rates of change of current.

Normally GTOs are fitted with the RCD circuit as shown in Fig. 4.8. This circuit is simple and non-complicated (few elements). The capacitor C_B is determined by the switch-off current and critical rate of change of the anode voltage.

The current interruption in the GTO applies

$$\frac{du_{CB}}{dt} = \frac{1}{C_B} \cdot i_A \approx \frac{du_A}{dt},$$ (4.6)

e.g. switch off current of 100 A and a capacitor of $C_B = 1$ μF

$$\frac{du_A}{dt} \approx \frac{100 \text{ A}}{1 \text{ μF}} = 100 \frac{\text{V}}{\text{μs}}.$$

After switching on the capacitor C_B discharges over the resistor R_B. An energy loss arises $1/2 C_B \cdot U^2$. With increasing switching frequencies this can lead to perceptible powerloss which is noticeable in the size of the resistor R_B. This reduces the efficiency of converters with GTOs slightly.

In principle it is possible to design lossfree or lowloss snubber circuits for GTOs which leads to higher switching frequencies and improves the efficiency of GTO-converters. Figure 4.9 shows a lowloss snubber circuit for GTOs in a dc

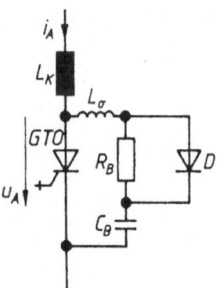

Fig. 4.8. RCD-circuit of a GTO

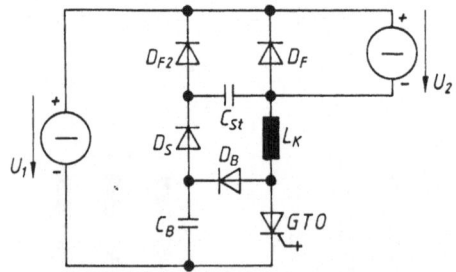

Fig. 4.9. Low loss snubber circuit of a GTO in a dc converter. Storage capacitor C_{St}, additional blocking diode D_S, additional free wheeling diode D_{F2}

chopper. Here is also a unipolar capacitor C_{st} for intermediate storing of the switch off energy. The saved energy is given up to the mains or load side when switching on again. The principle of lowloss circuits with intermediate storing capacitors can be extended to one or more phase inverters [4.23].

4.2 Triggering

Controlled power semiconductors need a gate current by means of which the cell is switched into the conducting state (from the off-state characteristic to the on-state characteristic) when off-state voltage is applied. The control characteristics should be taken into consideration for this [4.21].

For thyristors the gate current i_G is the current flowing through the gate circuit and is considered to be positive when it enters the gate electrode. The gate voltage u_G is the voltage between the gate electrode and the cathode and is considered to be positive when the gate electrode is positive with respect to the cathode. The gate electrode is often also called the gate for short.

Differentiation is made between gate current and gate voltage and gate trigger current i_{GT} and gate trigger voltage u_{GT}. The gate trigger current is the value of the gate current causing the thyristor to switch from the blocking to the conducting state in the forward direction (triggering). The gate trigger current varies both with the anode-cathode voltage and with the junction temperature. The gate trigger voltage is the voltage arising when the gate trigger current flows in the gate circuit.

4.2.1 Triggering Area

Figure 4.10 shows the triggering area of the thyristor enclosed by an upper and lower limiting curve for a forward off-state voltage greater than 6 V and a resistive main circuit. The gate trigger current and gate trigger voltage depend to a great degree upon the junction temperature $\vartheta_{(vj)}$. At low junction temperatures there is a large gate trigger current demand; at high junction temperatures a considerably lower one. The triggering zone is often specified for a single junction temperature

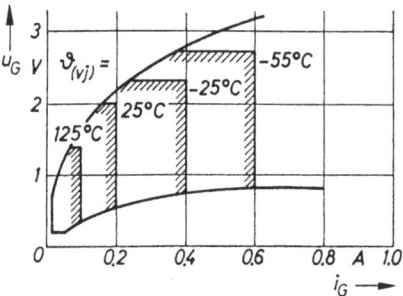

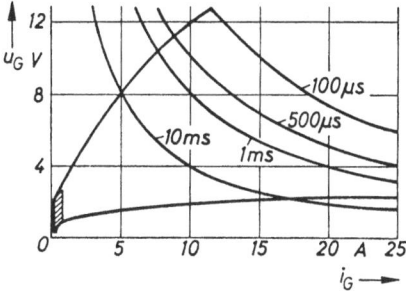

Fig. 4.10. Triggering zone at various junction temperatures

Fig. 4.11. Maximum permissible peak gate power (parameter: pulse duration)

of 25°C. The hatched lines mark the limits between the zone of possible triggering and the zone of certain triggering for the various junction temperatures.

The product of gate voltage u_G and gate current i_G is the gate power losses which must not exceed an upper limiting value. Naturally, the maximum permissible peak gate power varies with the duration of the trigger pulse. In Fig. 4.11 the maximum permissible peak gate power is plotted for a particular type of thyristor. Hyperbolas are produced with the trigger pulse duration as parameter. Exceeding the permissible peak gate power can lead to destruction of the gate circuit and thus of the thyristor.

4.2.2 Trigger Pulse

In Fig. 4.12 typical curve shapes of trigger pulses are illustrated. Figure 4.12a shows the minimum and maximum values for the gate current i_G. In order to ensure perfect triggering the trigger pulse must lie in the zone of certain triggering i.e. above the hatched lines of Fig. 4.10. Moreover, the trigger pulse must have a duration longer than the gate controlled delay time. Finally, the maximum permissible gate losses must not be exceeded.

The shape and duration of the trigger pulse can be described by the method illustrated in Fig. 4.12b. To keep down the turn-on losses at high rates of rise of current, often a high peak is superimposed upon the trigger pulse at the beginning. Such a steep pulse indicated in Fig. 4.10c (short rise time compared to pulse duration, not drawn to scale) makes for rapid turning on of the thyristor. It

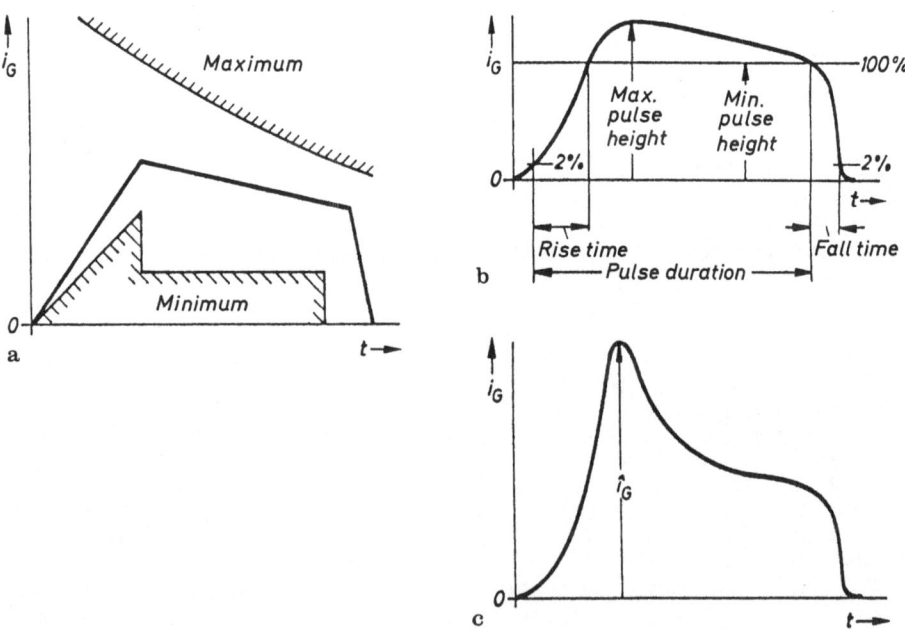

Fig. 4.12a – c. Waveshapes of trigger pulses. **a** Limiting values for trigger pulse; **b** typical trigger pulse; **c** steep pulse

reduces the differences in the gate controlled delay times which is important for the voltage distribution when turning on thyristors connected in series.

4.2.3 Trigger Pulse Generator

The gate current is generated by a trigger pulse generator. This is normally a trigger pulse amplifier whose output is fed into the gate terminal of the thyristor either directly via a trigger pulse transmitter. To trigger the thyristor the following characteristics of the trigger pulse generator are important: short-circuit current and no-load voltage, rate of rise of the trigger pulse unit short-circuit current to 90% of its final value, and duration of the trigger pulse.

4.2.3.1 Trigger Pulse Generator for Thyristor

Trigger pulse generators can be constructed in different ways. Figure 4.13 shows a typical sample circuit.

The two-stage output amplifier made up of the two transistors 3 and 4 generates a pulse current of defined magnitude and duration when the pulse shaper with the two transistors 1 and 2 is driven by the pulse generator with an imput voltage u_E. The current pulse i_{C4} is fed to the gate electrode of the thyristor via the pulse transmitter. A zener diode limits the reverse magnetization of the pulse transmitter. The complete circuit diagram of the trigger pulse generator is reproduced in Fig. 4.14.

4.2.3.2 Trigger Pulse Generator for GTO

The gate drive circuit has to produce the control impulses with defined rise time and amplitude for GTOs and to make a potential barrier between the control and power part.

In switching on and off the additional requirement arises that currents of both polarities have to be produced. The required rate of rise and amplitude of current especially for negative impulses are much greater than that for conventional thyristors. Under some conditions an extra constant control current may be necessary [4.23, 4.24].

A possibility to produce the required control impulses is that to use power transistors and potential barriered voltage sources. Additional means (e.g. optoelectronic coupling elements) for a potential barrier of the control signal are necessary. The disadvantage is the complication of such circuits for the potential barriered voltage sources and the fact that the amplitude of the negative control pulses i_G is bordered by the maximum current of the power transistors.

Another possibility to realize the impulse control generator is with pulse transformers. The direct feeding in of the control current with a pulse transformers which is normal for thyristors is also possible with GTO-thyristors. Additional problems arise as the control pulses must be of both polarity. The pulse transformers must be reverse magnetized without producing pulses of unwished polarity.

A circuit which fulfills these requirements is shown in Fig. 4.15a. The pulse transformers Tr_1 has four MOS-transistors T_1 to T_4 on the primary side. This has the advantage of the current transformation, however, a continuous firing current

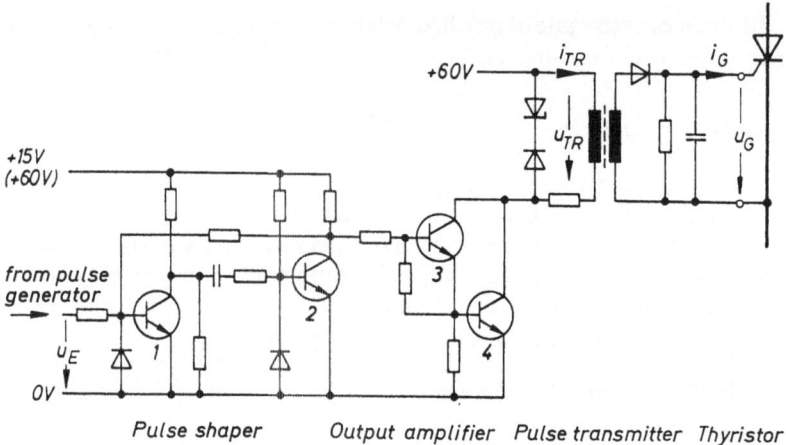

Fig. 4.13. Trigger pulse generator (sample circuit)

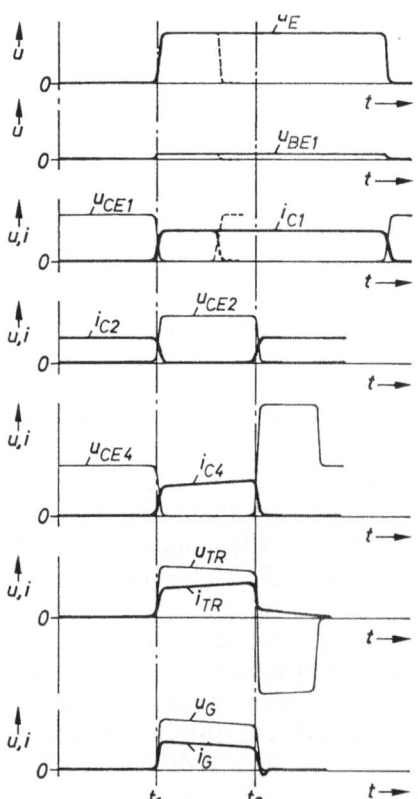

Fig. 4.14. Pulse diagram of trigger pulse generator in Fig. 4.13

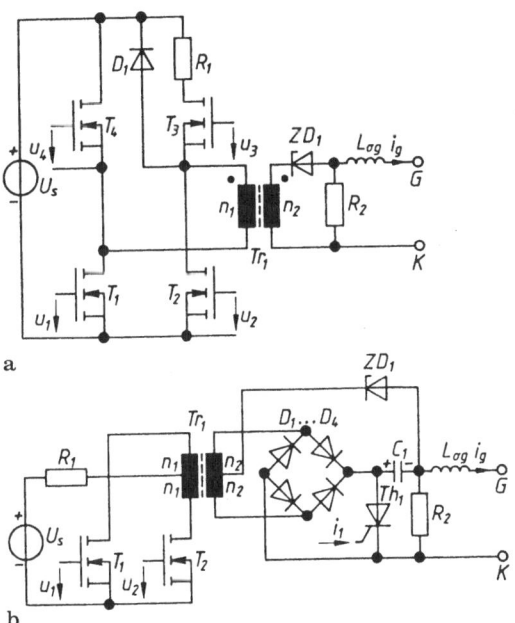

a

b

Fig. 4.15. Gate drive curcuit for a GTO. **a** With direct impulse transformation; **b** with energy and information transformation by high frequency impulse chains

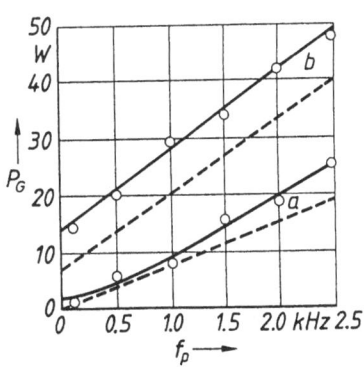

Fig. 4.16. Measured power of the gate drive circuit for GTO in Fig. 4.15. *a* Gate drive circuit Fig. 4.15a, *b* gate drive circuit Fig. 4.15b

i_{GC} cannot be supplied and the rate of change of control current di_G/dt is bordered by stray inductances of the impulse transformer. The required large rate at change of the negative gate current at up to 10 A/μs is hard to achieve with the windings technique (very low stray inductances).

Both the disadvantages can be avoided when the pulse transformer is only used to transmit high frequency impulse chains (Fig. 4.15b). The firing current is produced by the transistors T_1 and T_2 switching alternatively with high frequencies ($f \approx 100$ kHz). The charge current of C_1 makes the firing impulse whereas the switch-off impulse is achieved by the discharge of this capacitor over a small thyristor Th_1. The firing current i_1 for this thyristor can be produced on the secondary side of a simple electric circuit which detects the occurance or absense of the frequency chain on the secondary windings n_2. This curcuit' is left out for simplicity. Over a diode ZD_1 or a resistance the continuous firing current i_{GC} is fed into the gate during the period of the high frequency impulse chain. This and similar circuits are suitable for GTO-thyristors of high power.

Figure 4.16 shows the measured power of two gate drive circuits. Dotted is the lower border of the power which can be determined by simple theoretical explanations. The shown impulse control generator has been dimensioned for a GTO-thyristor with about 400 A switch-off current [4.23].

4.2.4 Trigger Equipment

In converter connections the trigger pulses must be fed to the controlled power semiconductors, thyristors or power transistors periodically at the clock frequency [4.3, 4.8, 4.9, 4.11]. Generally an adjustable control angle α is used to control the flow of energy (see Chaps. 6, 7, and 8).

Figure 4.17 shows an example for the construction of a trigger system made up of various functional units. In the pulse generator the pulse determining the moment of triggering is released periodically. The pulse shaper determines the pulse shape and pulse length. The output amplifier increases the pulse power and generally also the rate of rise of the pulse. The pulse transmitter provides isolation between the control unit and the individual converter cells. Moreover, it serves as

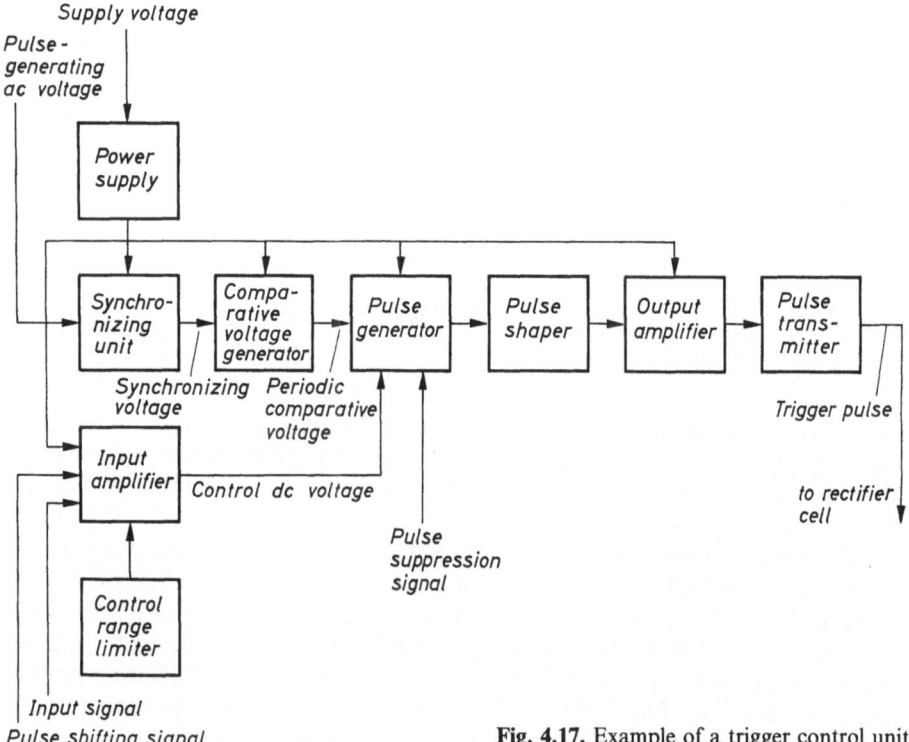

Fig. 4.17. Example of a trigger control unit

an impedance transformer to match the trigger pulse to the values of pulse current and voltage suitable for triggering the converter cell.

The synchronizing unit aids synchronization with a pulse-generating ac voltage. The signal to the input amplifier determines the variable moment when the pulse commences i.e. it alters the delay angle α. Using the control range limiter the effect of the input signal can be limited to a certain range. The pulse displacement signal is used to inhibit triggering as is the pulse suppression signal.

Not all the functional units shown in Fig. 4.13 have to be contained in an actual control unit. On the other hand there can also be other functional units [4.16, 4.18].

4.3 Cooling

In power semiconductors losses which cause heating occur in operation. Apart from the copper losses in the incoming and outgoing leads these losses arise in the actual semiconductor system, a thin silicon wafer a few 100 μm thick i.e. concentrated in a very small volume. These losses are made up from on-state, off-state, switching, and gate losses. The corresponding heat due to the losses must be dissipated to the surroundings via the case and the heat sink [4.5, 4.6].

4.3.1 Operating and Limiting Temperatures

Power semiconductors must be stored even in the unloaded state only within a permissible temperature range called the storage temperature range. Minimum values of approximately $-65°C$ and maximum values of $150°C$ to $200°C$ apply for thyristors. When the storage temperature range is exceeded changes can occur on the surface of the silicon wafer which permanently impair its blocking properties.

The operating temperature range is the name given to the range between two limiting values of coolant temperature, within which a thyristor must be operated.

The junction temperature of a thyristor in electrical operation must not exceed a certain maximum value, because otherwise deterioration occurs in the thyristor's properties. In particular the forward blocking voltage the critical rate of rise of voltage and the circuit-commutated turn-off time will be degraded. For thyristors the maximum value of junction temperature is generally stated as $125°C$. For silicon diodes it is less than $200°C$. As this maximum value ought not to be exceeded even on overload the continuous current occurring in operation must be so reduced that even an overcurrent does not cause a higher junction temperature than for example $125°C$.

In the event of a fault the junction temperature may temporarily rise to a higher value without the thyristor being permanently impaired. It can nevertheless temporarily lose its blocking capability. The lower limit of the operating temperature range for thyristors lies between $0°C$ and $-65°C$. It is amongst other things, produced by too great a gate power demand (see Fig. 4.10).

Damage occurs to a thyristor in operation when temperatures high enough to cause intrinsic conduction in the silicon are reached. This occurs according to the operating state in the range between $200°C$ and $400°C$. It is sufficient for only one point of the silicon wafer to reach such temperatures. As a result of its rising conductivity it assumes a greater part of the anode current and is further heated up. This pinch effect leads to destruction of the thyristor.

4.3.2 Losses

Electrical losses occur when a thyristor is operating. The curve of the power losses with time

$$p(t) = u_A i_A \tag{4.7}$$

is the product of thyristor voltage u_A and thyristor current i_A. For periodic operation mean power losses are defined as

$$P = \frac{1}{T} \int_0^T p(t)\,dt = \frac{1}{T} \int_0^T u_A i_A\,dt \tag{4.8}$$

where T is the period.

The energy losses which are equivalent to a quantity of heat are found by integrating Eqs. (4.6) and (4.7) over time. During periodic operation, a thyristor passes cyclically through operating states on the three branches of its characteristic curve. The losses have therefore to be made up from various components: on-state losses, off-state losses, switching losses, and additional gate losses.

On-state losses occur when the thyristor is carrying on-state current. According to Fig. 4.18 the on-state characteristic of a thyristor (also that of a semiconductor diode) can be approximated by means of the on-state threshold voltage $U_{(TO)}$ and a constant equivalent resistance r_T. The on-state voltage is then

$$u_T = U_{(TO)} + r_T i_T. \tag{4.9}$$

According to Eqs. (4.8) and (4.9) the on-state power losses are thus:

$$P_T = \frac{1}{T} \int_0^T (U_{(TO)} + r_T i_T) i_T\,dt = U_{(TO)} \frac{1}{T} \int_0^T i_T\,dt + r_T \frac{1}{T} \int_0^T i_T^2\,dt$$

$$= U_{(TO)} I_{Aav} + r_T I_{Aeff}^2. \tag{4.10}$$

The on-state losses therefore vary not only with the mean value of the anode current I_{Aav}, but also with its rms value I_{Aeff}. Therefore, the continuous limiting current of semiconductor diodes and thyristors must always be stated in

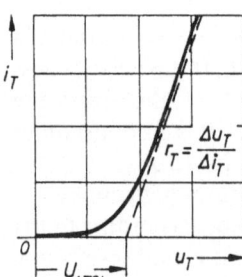

Fig. 4.18. Threshold voltage and equivalent resistance

conjunction with the shape of the current curve. For applications in the mains frequency range (50 Hz or 60 Hz) the on-state losses represent the main portion of the total losses.

Off-state and reverse blocking losses occur during the times when a thyristor (or semiconductor diode) is operated on the off-state and reverse blocking characteristics. As the off-state and reverse blocking currents are only a few mA up to the rated blocking voltage the off-state and reverse blocking losses i.e. the product of off-state or reverse blocking voltage and off-state or reverse blocking current are small. The reverse blocking current i_R is to a great extent independent of the momentary value of reverse blocking voltage so long as operation is not above the break in the reverse blocking current curve. For a sinusoidal reverse voltage (one half-cycle) reverse power losses

$$P_R = \frac{1}{T} \int_0^{T/2} p_R dt = \frac{1}{T} \int_0^{T/2} I_R \hat{u}_R \sin\ \omega t dt = \frac{1}{\pi} I_R \hat{u}_R \qquad (4.11)$$

are obtained.

On transition from the off-state characteristic to the on-state characteristic, turn-on losses occur. On transition from the on-state to the blocking state, turn-off losses occur (see Figs. 3.8 and 3.9). The momentary values of the turn-on and turn-off power losses can be very high and with power thyristors in circuits with small leakage inductances reach values of up to several kW. As, however, they only occur for a few microseconds in applications at mains frequency they represent only a small portion of the total losses. They can then be neglected in the total loss calculation but must be taken into account when designing the RC circuit (see Sect. 4.1).

In applications at higher switching frequencies in the medium-frequency band of several 100 Hz to over 10 kHz, however, the switching losses come increasingly to the fore [4.7]. The local turn-on and turn-off losses occurring in the vicinity of the gate terminal are then what determine the loading capacity of thyristors. The current carrying capacity must be considerably reduced compared to 50 Hz applications. The portion of the total losses can be considerable depending upon the switching frequency and the rate of rise of current. The loading capacity of thyristors at higher frequencies is generally given in diagrams.

With controlled rectifier cells gate losses occur so long as the gate current flows from the gate terminal to the cathode. The momentary value of the gate power losses is

$$p_G = u_G i_G. \qquad (4.12)$$

The mean gate power losses P_G can be determined from the control characteristics of a thyristor when the shape of the gate current pulse is known.

4.3.3 Thermal Equivalent Circuit

The heat rise occurring in the power semiconductor due to the losses can be calculated approximately using a thermal equivalent circuit. Figure 4.19 shows schematically the construction of a thyristor in the flat-base cell form of construction on a heat sink.

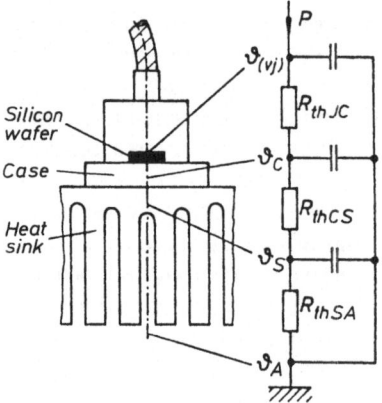

Fig. 4.19. Temperatures ϑ and thermal resistances R_{th} for a thyristor with heat sink

Table 4.1. Comparison of thermal and electrical quantities

Thermal quantities		Electrical quantities	
Heat quantity (energy)	Q in Ws	Charge	Q in As
Heat current (power)	P in W	Current	I in A
Temperature difference	$\Delta\vartheta$ in K	Voltage	U in V
Heat resistance	R_{th} in K/W	Resistance	R in V/A
Heat capacity	C_{th} in Ws/K	Capacity	C in As/V
Thermal time constant	$\tau_{th} = R_{th}C_{th}$ in s	Time constant	$t = RC$ in s

The heat generated in the semiconductor system flows via the base of the case to the heat sink and from there to the surroundings.

To simplify the calculation of such a system, temperature values ϑ are allocated to the silicon wafer, the case, heat sink, and ambient environment: junction temperature is denoted by ϑ_{vj}, case temperature by ϑ_C, heat sink temperature by ϑ_S and ambient temperature by ϑ_A. These are nominal values. In fact the temperature in the various components varies. Between the individual components of such a cooling system there are thermal resistances which are generally complex as besides the thermal conductivity all components also have a heat retention capacity. These appear as heat capacities in the equivalent circuit.

In such a thermal equivalent circuit the relationship between the power losses and the temperature is analogue to that between current and voltage in an electrical network (see Table 4.1). A heat source supplies the power losses P in thermal resistances, across which temperature differences $\Delta\vartheta$ occur.

A greatly simplified thermal equivalent circuit of a semiconductor cell with heat sink is obtained for continuous operation, that is operation with constant thermal power without taking the heat capacities into consideration (Fig. 4.20). The internal thermal resistance R_{thJC} is a characteristic parameter of the thyristor. Its upper limiting value is stated in data sheets. The external thermal resistance R_{thCA} contains the transfer between the thyristor case and the heat sink and the thermal resistance of the heat sink including the heat transfer to the surroundings. The external thermal resistance varies with the design of the heat sink. Table 4.2

Table 4.2. Typical values for internal and external thermal resistance

Case construction	Maximum mean on-state current	Heat resistance R_{thJC}	R_{thCA} with			
			Natural cooling	Forced-air ventilation with 6 m/s	Water cooling	Heat pipe cooling
	A	K/W	K/W	K/W	K/W	K/W
Screw cells	6...30	2.5...0.8	5.5...1.2	2.0...0.4		
Flat-base cells (also screw cells)	30...400	0.8...0.08	1.2...0.5	0.4...0.15	0.08...0.06	
Disc cells	200...800	0.1...0.04	0.5...0.25	0.2...0.08	0.04...0.02	0.03

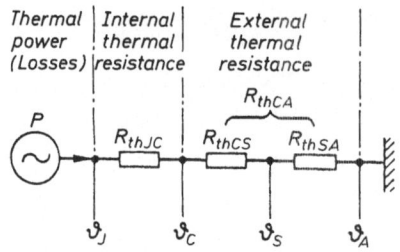

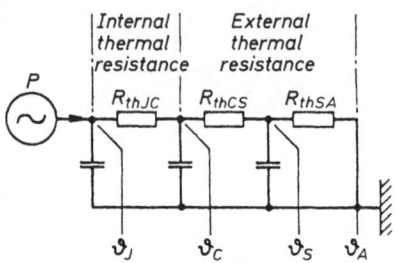

Fig. 4.20. Simplified thermal equivalent circuit of a thyristor with heat sink for steady-state operation

Fig. 4.21. Thermal equivalent circuit of a thyristor with heat sink for intermittent operation

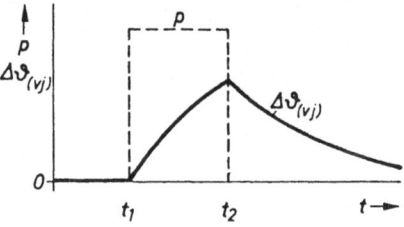

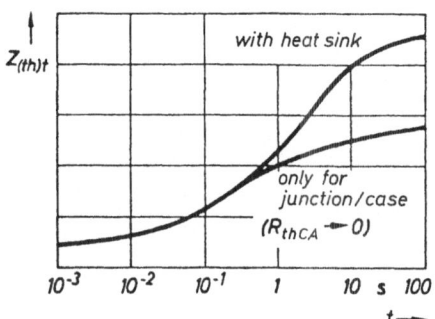

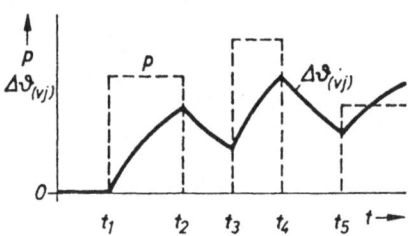

Fig. 4.22. Transient thermal impedance of a thyristor with heat sink (pulsed thermal impedance)

Fig. 4.23a,b. Curve of junction temperature. **a** With a single pulse; **b** with a chain of pulses

lists the range of values of the inner and outer heat resistance for typical constructions of semiconductor valves. The mean junction temperature

$$\vartheta_{(vj)} = P(R_{thJC} + R_{thCA}) + \vartheta_A \qquad (4.13)$$

can be calculated for given power losses P. Where the thermal power fluctuates with time the junction temperature is also subject to periodic fluctuations.

The thermal equivalent circuit in Fig. 4.20 does not take the heat retention capacity of the thermal system into account. Therefore it cannot be used for transient calculations. Figure 4.21 shows the thermal equivalent circuit for transient operation, the heat retention capacity being taken into consideration by means of heat capacities [4.13]. The result is a lattice network of thermal RC elements. Each RC element has its own thermal time constant $\tau_n = R_{(th)n}C_n$.

Such a thermal lattice network of $RC_{(th)}$ elements can be calculated in a similar way to RC lattice networks in electrical circuits. For this the thermal equivalent circuit illustrated in Fig. 4.21 is generally converted into a series circuit of thermal RC elements, because this is more readily calculated.

An analytical evaluation produces for the so-called transient thermal impedance

$$Z_{(th)t} = \sum_{n=1}^{n=m} R_{(th)n} \left(1 - e^{-\frac{t}{\tau_n}}\right). \tag{4.14}$$

The transient thermal impedance can be illustrated as a function of time (Fig. 4.22). Using the illustrated curve the heat rise of the junction can be determined if the curve of thermal power $p(t)$ against time is known. Figure 4.23 shows by way of example the temperature curve of the junction with a single impulse and with a chain of impulses.

4.3.4 Heat Sinks

Power semiconductors are generally employed with heat sinks. The thermal resistance of a heat sink varies with the material used its design, and the speed of the cooling medium. When power semiconductors and heat sinks are assembled it must be taken care that the thermal resistance of the transfer point is kept as small as possible by suitable construction. Clean contact surfaces often coated with a contact grease before assembly and correct and even contact pressure are required for this.

4.3.5 Types of Cooling

Heat is transferred in a cooling system by conduction, convection or radiation. With conduction (in gases, liquids, and solids) heat is transferred from molecule to molecule. With convection the transfer takes place by the movement of particles. One speaks of free convection if the heat is allowed to rise naturally. Forced convection can be produced by means of blowers or pumps (the cooling agent being gas or liquid). With radiation heat is transferred by electromagnetic waves (usually infra-red). Radiation is the only type of heat transfer which can take place in a vacuum.

Cooling can be classified as direct or indirect. With direct cooling the heat given off by the converter and other pieces of equipment is transfered directly to the cooling medium. With indirect cooling the heat is removed by a heat carrier or heat conductor and then transfered to the cooling medium. Air and water are common cooling media.

Natural Convection Cooling. With natural cooling the heat due to the losses is dissipated by natural air flow (natural air cooling) [4.5]. The heat is given up by the heat exchanger to the surroundings by convection and radiation. With natural air cooling the thermal resistance of a heat exchanger drops with increasing power losses, because convection and radiation vary with temperature. As unimpeded an inflow and outflow of the cooling air as possible must be strived for. The cooling fins ought therefore to be positioned at right angles.

Forced Cooling. With forced cooling differentiation is made between forced-air ventilation and liquid cooling [4.4]. With forced-air ventilation the cooling air is moved by a fan. Generally the air is drawn between the cooling fins by fans the heat being dissipated from the heat exchanger by convection. The thermal resistance of the heat exchanger depends greatly upon the speed of the cooling air flowing past the heat exchanger fins. The mean flow speed of the cooling air

$$v = \frac{Q}{A} \qquad (4.15)$$

can be calculated from the quantity of cooling medium Q per time unit and the cross-section of the air passage A of a heat exchanger: In the technical specification an air speed between the heat exchanger fins of 6 m/s is generally assumed. At higher air speeds (up to 12 m/s) higher power losses can be dissipated.

Water cooling instead of air cooling results in a heat exchanger with a lower thermal resistance (see Table 4.2). Moreover, with water cooling the entry temperature of the cooling medium can be some 10°C to 20°C lower than with air cooling. This can produce a larger temperature difference between the base of the case and the junction, resulting in operation at a higher power rating. Also with water cooling the thermal resistance of the heat exchanger drops with increasing cooling medium speed. The water is passed through cooling boxes or with parallel operation of several power semiconductors through extended hollow copper or aluminium bars to which the semiconductor diodes or thyristors are bolted.

With water cooling the heat is removed by fresh water. In the indirect case a heat carrier (air, water oil or other insulating liquids) in a closed circuit transfers the heat to the cooling medium via a heat exchanger or the case of the converter. The motion of the heat carrier arises either from natural convection or a pump or fan.

As an alternative to water oil is also used as the cooling medium. Compared to water, oil has the advantage of a higher dielectric strength. Moreover, with oil there is no danger of corrosion. Nevertheless, oil has a low specific heat and a high thermal resistance.

Important physical constants for cooling system calculations are listed in Table 4.3. The heat transfer number for metal to the cooling medium or the relative figures of merit for air, oil, and water or heat transfer media are 1 : 10 : 100, respectively.

Boiling Liquid Cooling. In special cases boiling liquid cooling can also be used with converters. Figure 4.24 illustrates the construction and operation of a heat pipe. The heat pipe transfers the heat with very little temperature drop to the base of the

Table 4.3. Physical constants of various cooling media

Cooling medium		Air	Oil	Water
Heat conductivity λ	W/mK	0,028	0,12	0,624
Specific heat c	J/kgK	1000	1900	4200
Density ϱ	kg/m³	1,09	859	988
Kinematic viscosity v	m²/s	$18 \cdot 10^{-6}$	$9,3 \cdot 10^{-4}$	$0,55 \cdot 10^{-6}$
Heat transfer number	W/m²K	35	350	3500
Metal to cooling medium		for v = 6 m/s		

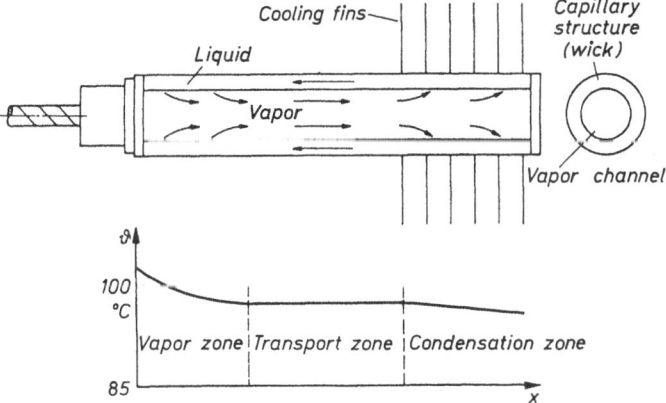

Fig. 4.24a,b. Heat pipe. **a** Construction; **b** temperature curve

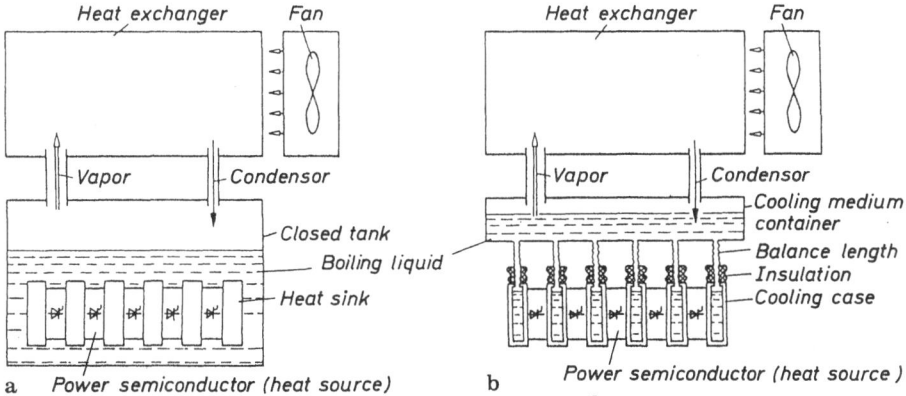

Fig. 4.25a,b. Boiling liquid cooling system. **a** Elements immersed in the liquid; **b** liquid in the cooling vessels

cooling fins which in turn transfer the heat uniformly to the cooling air. The heat pipe is evaluated and filled with a transfer medium usually water. The principal of cooling with evaporation is used in the heat transfer. The transfer medium is evaporated by the heat from the semiconductor valve. The vapor condenses in the condensation zone. The resulting condensate flows back to the heated evaporation zone through the capillary structure and the cycle begins again. The operation of heat pipes is site-dependent since the circulation is gravity-assisted. They are best used vertically [4.10].

The heat losses from power semiconductors and other elements can also be removed by boiling liquids. Figure 4.25 shows one type of boiling cooling system. The heat generating elements are either immersed in the boiling liquid (a) or the liquid boils in hollow cooling boxes (b). The liquid is in a closed system and at room temperature is usually under low pressure. The boiling point is between 47°C (Freon) and 50°C. Under nominal operation the pressure is approximately one atmosphere or slightly more.

Boiling cooling systems permit a very compact construction of converter power parts and moreover are insensitive to dirt. Boiling cooling systems shown in Fig. 4.25 were first used with converters in a railway system [4.12].

4.4 Protection Devices

The purpose of protection devices is to protect semiconductor valves and other converter components against non-permissible voltages and currents.

Overvoltage Protection. Overvoltage protection devices restrict overvoltages to voltages permissible for semiconductor valves. Overvoltages arise from switching processes in the mains supply, atmospheric influences, the switching of converter transformers or other inductances and the switching processes of the semi-conductor valve itself (see Chap. 3 and Sect. 5.1).

Silicon diodes, thyristors, and power transistors can be destroyed by overvoltages in nanoseconds. Limit values of the maximum permissible line overvoltages are prescribed as functions of the overvoltage duration for connecting converter devices to dc and ac supplies. Figure 4.26 shows as an example the permissible overvoltage ratio $(U_{dN} + \Delta U)/U_{dN}$ for the dc side input of a converter according to German regulations VDE 0160. ΔU is a non-periodic overvoltage as a deviation from the nominal input dc voltage U_{dN}. For short-term overvoltages the resulting limit curve holds for nominal input dc voltages up to 50 V. For higher output voltages the short-term permissible overvoltage is limited to a value obtained from the equation

$$\left(\frac{U_{dN} + \Delta U}{U_{dN}}\right)_{limit} = \frac{1400 \text{ V}}{U_{dN}} + 2.3. \tag{4.16}$$

The dashed line holds for $U_{dN} = 110$ V. A general standard for single-phase and three-phase systems is that electronic power equipment has to withstand at most 2 ms at twice the rated voltage. Since it is often not economical to design semiconductor valves to cope with twice the rated voltage, filters can also be connected in series with the valves especially with low power converters. These restrict the short-term overvoltage to values not dangerous to the semiconductor valve.

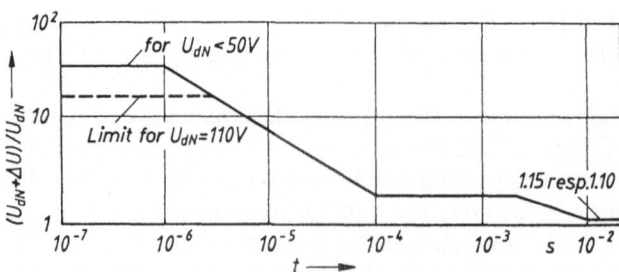

Fig. 4.26. Maximum permissible line overvoltage ratio as a function of the overvoltage duration

The most important device for restricting overvoltages is the snubber circuit applied to the semiconductor valve and other converter components as explained in Sect. 4.1. The next most important method is to install overvoltage diverters and limiters which limit overvoltages produced by atmospheric disturbances to at least 2.5 times the rated voltage. Overvoltage diverters are devices with non-linear current and voltage characteristics which, if an overvoltage occurs, switch to a low resistance (protective spark gaps, valve arrestors). Their response delay can go up to a few μs. With an overvoltage limiter, the current increases sharply when a particular voltage (the breakdown voltage) is exceeded. Devices exhibiting such behavior include selenium varistors, zener diodes, metallic paper overvoltage limiter, and metal oxid varistors. They operate almost instantaneously but their energy consumption is low.

Overcurrent Protection. This restricts overcurrents to those permissible for semiconductor elements and other devices. Fusible cut-outs, quick-break switches, power switches, and in certain cases short-circuiting devices are used. In addition protective impedances can be used to limit short-circuit currents arising

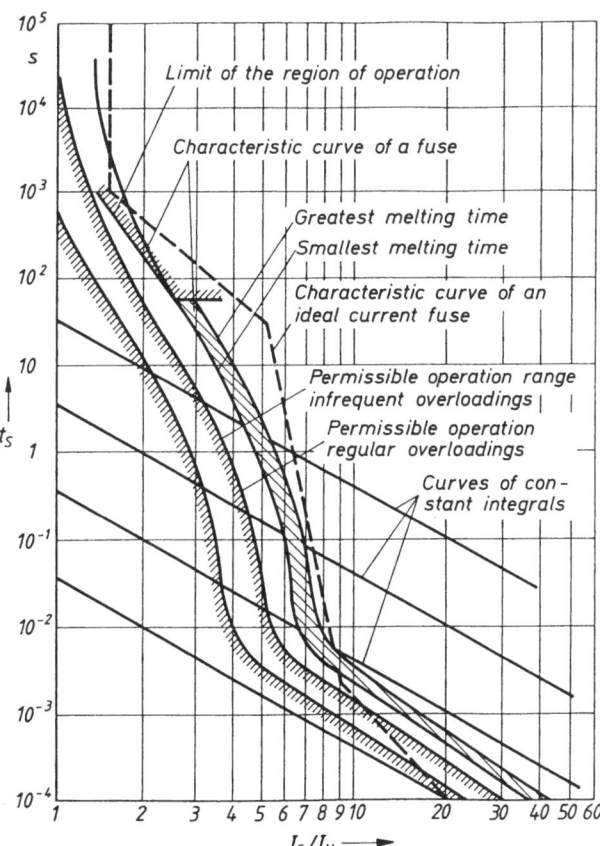

Fig. 4.27. Fusing time t_s of a super quick-break semiconductor fuse as a function of the rms value of the fuse current as a multiple of the rated fuse current

from fault conditions. Quick-break switches interrupt the short-circuit current as it increases and thereby limit its amplitude. They are usually applied on the dc side (dc quick-break switches). Power switches restrict the short-circuit current temporarily and must be able to switch off the full short-circuit current. They are usually used on the ac side.

In designing the protection one must consider the kinds of faults which are possible and the need for a high degree of operational reliability whereby the effects of overcurrent on the operation are limited. Fast-action protection restricts the duration of an overcurrent produced by a short-circuit to half a cycle (short-circuit protectin). Slow-action protection operates on the basis of thermal and magnetic cut-outs or fusible cut-outs (overload protection).

Fusible cut-outs are put in the desired break-points of the power line (wires or metal bands in a removable cartridge- or tongue-shaped unit). If an overcurrent of non-permissible size or duration occurs the current melts the fusing conductor. One or more electric arcs occur whose voltage drop relative to the voltage in the circuit reduces the current and finally breaks it. Current interruption process occurs in two stages: the fusing stage and the quenching stage. Three essential characteristic values of a fuse are the fusing integral (the integral overtime of the square of the current during the fusing stage), the quenching integral (the integral over time of the square of the current during the quenching stage), and the cut-off integral which is the sum of both.

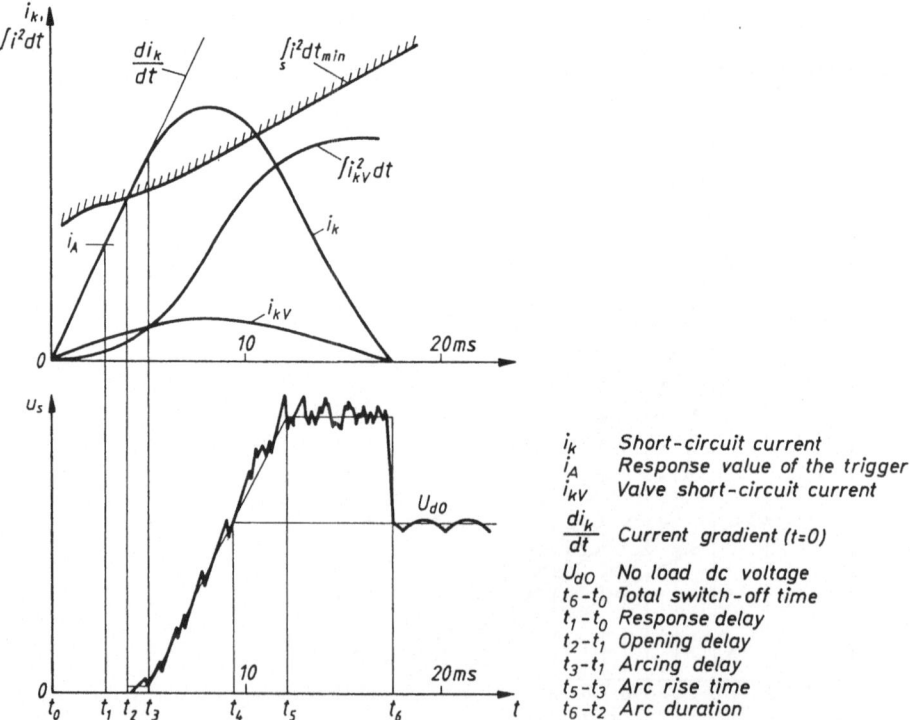

i_k	Short-circuit current
i_A	Response value of the trigger
i_{kV}	Valve short-circuit current
$\dfrac{di_k}{dt}$	Current gradient $(t=0)$
U_{do}	No load dc voltage
t_6-t_0	Total switch-off time
t_1-t_0	Response delay
t_2-t_1	Opening delay
t_3-t_1	Arcing delay
t_5-t_3	Arc rise time
t_6-t_2	Arc duration

Fig. 4.28. Current and voltage shape when switching off a short circuit with a dc quick-break switch

Super quick-break fuses have been developed for the protection of semi-conductor valves. Figure 4.27 shows the dependence of the fusing time on the value of the fusing current. Fusible cut-outs are those most used to selectively switch out defective diodes and thyristors. Their response values in this case are greater than the permissible diode or thyristor loads.

It follows that fusible cut-outs can be used not only for limiting the amplitude and time of short-circuit currents but also for overload protection. In the latter case, however, it should be remembered that the current at which the cut-out operates must be less than the maximum permissible currents of the diodes and thyristors.

DC quick-break switches operate on a principal similar to current-limiting fuses. To break the current they produce a high arc voltage which even over a long time is maintained at an almost constant value (1.7 to 2.5) times the rated switch voltage).

Figure 4.28 shows the clearance of a short circuit with a dc quick-break switch. With backlash voltage of the load the resulting voltage across the switch is the same of no-load direct voltage plus load voltage. Selectivity with respect to fuses is given if the integral $\int i_{kv}^2$ dt of the partial short circuit current per valve stays below the minimal melting integral of the fuse.

5 Switching Operations and Commutation

In Chap. 2 the system components required for a general description of the operation of a converter connection were defined. Assuming ideal conditions these are

— Voltage sources whose time response is given by $u(t)$,
— Transformers with the transformation ratio w_1/w_2 and in the case of multi-phase systems a phase dieplacement between primary and secondary electrical quantities dependent upon the transformer connection,
— Resistors R as energy converters as well as inductances L and capacitors C as magnetic and electrical energy stores.

To these can be added the converter valves performing the cyclic switching operations i.e. power semiconductors (diodes, thyristors, and power transistors). According to the requirements of the actual semiconductor switch with current conduction in one or two directions capable of being turned on only or also turned off appropriate combinations of power semiconductors are employed (see Figs. 2.2 and 2.3).

Before the various converters and their connections are described in greater detail in the following chapters the switching behavior of electrical networks will be dealt with generally. Moreover the commutation determining the internal mode of operation of converters will be studied. Depending upon the type of commutation converters can be divided into three different types as described in Chaps. 6, 7, and 8.

5.1 Switching Behavior of Electrical Networks

Any converter circuit consists of a combination of voltage sources, transformers, magnetic and electrical energy stores, energy converters (resistors), and cyclically actuated semiconductor switches. Because of their current and voltage curves, these semiconductor switches have non-linear characteristics. Under ideal conditions these characteristics have two regions or "states": the forward region characterized by a zero forward voltage drop independent of the forward current, and the reverse region characterized by a zero reverse current independent of the reverse voltage. A switching operation is defined by a transition from the reverse into the forward state or forward into the reverse state.

In such networks two switching conditions must not be violated. Neither the current in inductances L nor the voltage across capacitors C can change suddenly at instants of switching.

Both conditions result from the principles of energy. In addition mechanical switches can only interrupt currents in inductances by converting the stored magnetic energy $Li^2/2$ into heat in a switching arc. If capacitors with different voltages are switched on by mechanical switches, then high equalizing currents occur which are limited only by line inductances and contact resistances whereby the energy difference $C(u_1^2 - u_2^2)/2$ is converted into heat.

5.1.1 Switching an Inductance

Figure 5.1 shows the process of switching on an inductance L at instant t_0. The equivalent switch illustrated in the figure can be realized by a thyristor triggered at instant t_0. Assuming a voltage constant with time $u(t) = U_d$ i.e. a direct voltage the current and voltage curves illustrated are obtained. The current

$$i = \frac{U_d}{R}\left(1 - e^{-\frac{t-t_0}{\tau}}\right)$$ (5.1)

rises exponentially with the time constant $\tau = L/R$.

The voltage

$$u_L = L\frac{di}{dt} = U_d e^{-\frac{t-t_0}{\tau}}$$ (5.2)

across the inductance jumps to the value U_d at instant t_0 and thereafter decays exponentially with the same time constant τ.

The maximum rate of change of current occurs at instant t_0. By differentiation of Eq. (5.1) one obtains the value

$$\frac{di}{dt_{max}} = \frac{U_d}{L}.$$ (5.3)

So it is characteristic of the switching of inductances L that the voltage u_L changes suddenly at instants of switching and the current i kinks i.e. alters its rate of change, but retains its instantaneous value.

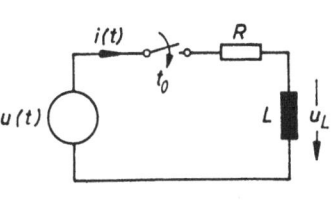

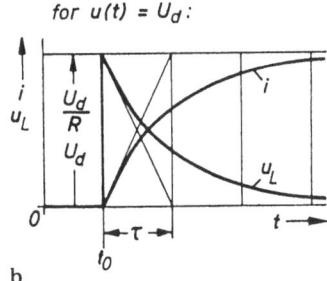

a   b

Fig. 5.1a,b. Switching an inductance

5.1.2 Switching a Capacitor

Figure 5.2 shows the process of switching on a capacitor C at instant t_0. The equivalent switch illustrated in the figure can again be realized by a thyristor triggered at instant t_0. Assuming a direct-voltage source U_d and a capacitor C discharged before the instant of switching t_0 the curves of current and voltage illustrated are obtained. After t_0, the capacitor voltage

$$u_C = U_d \left(1 - e^{-\frac{t-t_0}{\tau}} \right) \tag{5.4}$$

rises exponentially with the time constant $\tau = RC$.

The current

$$i = \frac{U_d}{R} e^{-\frac{t-t_0}{\tau}} \tag{5.5}$$

jumps from 0 to the value U_d/R at t_0 and likewise decays exponentially with the same time constant, τ.

The maximum rate of change of voltage occurs at the instant of switching t_0. By differentiation of Eq. (5.4) one obtains

$$\frac{du_C}{dt_{max}} = \frac{U_d}{RC}. \tag{5.6}$$

So it is characteristic of the switching of capacitors C that the current changes suddenly at instants of switching and the voltage u_C kinks i.e. alters its rate of change but retains its instantaneous value. Between the current and voltage of an inductance L on the one hand and the voltage and current of a capacitor C on the other there is a dualism as Figs. 5.1 and 5.2 and the associated current and voltage equations show.

These results of the switching operations for two simple networks can also be used in more extensive switching circuits. In the case of sources with a voltage varying with time $u(t)$ the momentary value of the voltage at the instant of switching t_0 should be taken into consideration.

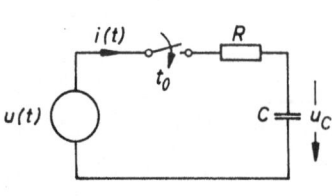

a

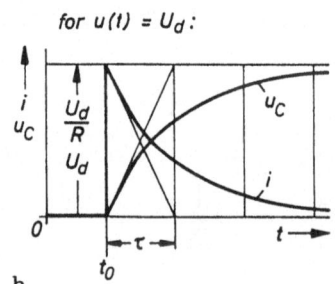

b

Fig. 5.2a,b. Switching a capacitor

5.2 Definition of Commutation

In electrical engineering one generally understands commutation to mean the transfer of current from one branch circuit to another, whereby both branches carry current during the commutating time. A basic distinction can be made between two mechanisms of commutation, namely commutation using mechanical switches and commutation using electronic switches i.e. genuine converter valves.

Mechanical switches are used for commutation in, amongst other things, mechanical controllers and commutator machines in which the commutator switches over the current from one phase winding to the following one. In electrical machines the commutator a rotary cyclically actuated mechanical switch acts as a rectifier in the generator mode and as an inverter in the motor mode. In the simplest form the current then commutates under the influence of the changing resistances between the current then commutates under the influence of the changing resistances between the current collectors (carbon brushes) and the commutator segments. By the introduction of interpoles which enforce commutation by means of commutating voltages induced in the phase windings, it also became possible to construct commutator machines of high output rating.

In power electronics the switching operations at the start and finish of each commutation process are realized not with mechanical switches but with genuine converter valves whose valve action is founded upon physical properties. This includes mercury-arc valve and the power semiconductors dealt with in Chap. 3.

The main features of a commutation process that cause the current I to be transferred from branch circuit 1 to branch circuit 2 are illustrated in Fig. 5.3. let the current I first of all flow in branch circuit 1 through the closed equivalent switch S1. Commutation is initiated by closing switch S2. Under the influence of the commutating voltage

$$u_k = u_2 - u_1 \tag{5.7}$$

a commutating current i_k begins to flow between branch circuits 1 and 2 which reduces the current I in branch circuit 1 and boosts it in branch circuit 2. After

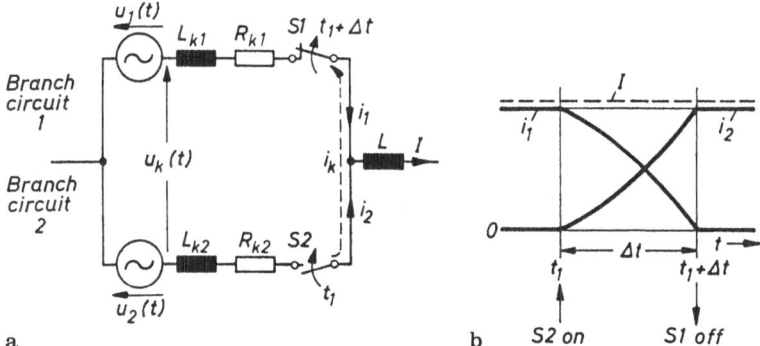

Fig. 5.3a,b. Commutation: Transfer of a current from one branch circuit to another.
a Commutating circuit; **b** current waveform

successful current transfer i.e. when current i_2 has reached the value I and hence current i_1 has become zero, the commutation process is completed by opening switch S1.

For correct commutation it is necessary for a suitable commutating voltage u_k to be present in the commutating circuit. If the natural voltages present in the single-phase or three-phase ac network are used as the commutating voltage, one talks of *natural commutation*. In place of supply voltages ac voltages generating by the load can also be used for natural commutation. It is then called load commutation.

When no natural commutating voltages are present in the commutating circuit or when these have the wrong polarity at the desired moment of commutation, commutation must be forced by providing an auxiliary voltage. For this the commutating voltage is made available either by an energy store (generally a quenching capacitor) or by increasing the resistance of the branch circuit to be quenched (using, for example, a power transistor or a thyristor capable of being turned off). This type of commutation is called *forced commutation* [5.1, 5.2].

In power electronics the switches S1 and S2 shown in Fig. 5.3 are realized by power semiconductors e.g. thyristors.

5.3 Natural Commutation

With natural commutation transfer of the current from one arm 1 of the converter to another arm 2 occurs under the influence of supply or load voltages. The current curve during the commutation process is determined not only by the commutating voltage u_k, but also by the resistances R_k and inductances L_k present in the commutating circuit (Fig. 5.4).

With sinusoidal phase voltages u_1 and u_2 in a multi-phase system the commutating voltage u_k is obtained as the difference of the voltages of two mutually commutating phases i.e. u_k is likewise a sinusoidal alternating voltage. The commutating time during which two valves relieving one another participate in carrying the current simultaneously as a result of the effective impedances in the commutating circuit is called the overlap time

$$t_u = t_2 - t_1. \tag{5.8}$$

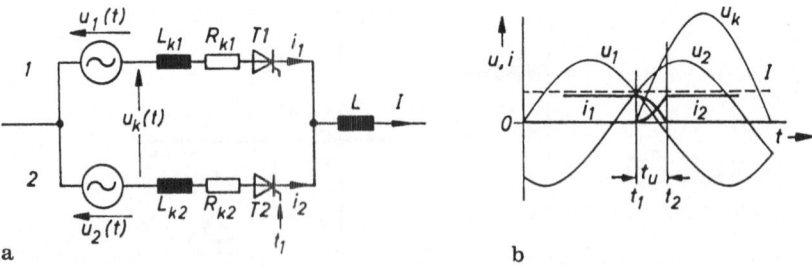

a   b

Fig. 5.4a,b. Natural commutation. **a** Commutating curcuit; **b** current and voltage waveforms (L large, I≈const)

With the commutating inductances and resistances drawn into the commutating circuit in Fig. 5.4 commutation occurs in compliance with the voltage equation

$$u_1 - L_{k1}\frac{di_1}{dt} - R_{k1}i_1 = u_2 - L_{k2}\frac{di_2}{dt} - R_{k2}i_2. \tag{5.9}$$

Using Eq. (5.7) this equation can also be written in the form

$$u_k = L_{k2}\frac{di_2}{dt} - L_{k1}\frac{di_1}{dt} + R_{k2}i_2 - R_{k1}i_1. \tag{5.10}$$

During the commutation process the sum of the currents in branch circuits 1 and 2 equals the current

$$i_1 + i_2 = I. \tag{5.11}$$

The curve of the commutation process is clearly determined by these equations. The curve of the commutating current with natural commutation is calculated in Chap. 7.

5.4 Forced Commutation

Forced commutation is the name given to commutation that is carried out either with the aid of energy stores, generally capacitive, within the converter or by increasing the resistance of the converter valve to be quenched.

Unlike power transistors, thyristors (apart from the special design of thyristors capable of being turned off), once successfully triggered, can no longer be quenched via the control circuit, so one has to resort to a supplementary quenching circuit in order to interrupt the current in a thyristor at any time independent of the presence of a suitable commutating voltage either in the supply or in the load.

Figure 5.5 shows a typical quenching circuit with quenching capacitor C. First let the main thyristor T1 carry the current I which is maintained in an external load circuit by an inductance L. When the quenching capacitor C is precharged with the polarity shown the current can be commutated from the main thyristor T1 to the quenching thyristor T1' by triggering the quenching thyristor at instant t_1. This

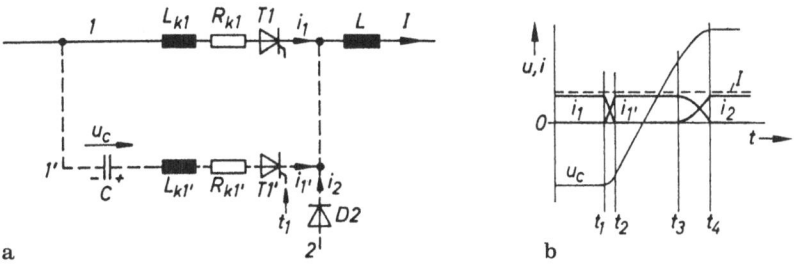

a   b

Fig. 5.5a,b. Forced commutation. **a** Commutating circuit; **b** current and voltage waveforms

first part of the commutating process is completed at instant t_2. The quenching capacitor then recharges with the opposite polarity under the influence of the current I which is assumed to be constant until at instant t_3 the current I is taken over by the auxiliary circuit 2. This is often a freewheeling circuit with a freewheeling diode D2. Commutation from the quenching circuit 1' to the freewheeling circuit 2 is completed at instant t_4. After that the current I flows via the freewheeling circuit.

With forced commutation as in Fig. 5.5 current transfer almost always occurs in several stages, generally two. The curves of current and voltage during the individual sections of the commutating process can be calculated exactly from the commutating impedance and the quenching capacitor in the same way as with natural commutation [5.3].

5.5 Types of Converters

Converters can be classified not only according to the basic functions performed (rectification, inversion, dc conversion and ac conversion) which describes the external mode of operation and was dealt with in Chap. 1, but also according to their internal mode of operation. By the 'internal mode of operation' one means the type of commutation and the origin of the elementary frequency. The elementary frequency is the frequency with which a converter arm is cyclically triggered. It can either be taken from a separate ac voltage source or be generated by an elementary frequency generator contained in the converter. This does not need to be discussed further at this point. Converters will be classified here according to the type and origin of the commutating voltage, and a distinction will be made between natural commutation and forced commutation. Converters can be classified into three different types using this scheme:

1. Converters in which no commutation processes occur,
2. Converters with natural commutation which draw their commutating voltage from the incoming supply or in special cases from the load,
3. Converters with forced commutation.

Converters without commutation are semiconductor switches and controllers for single-phase and three-phase ac dealt with in Chap. 6.

Converters with natural commutation by the supply or the load are called *externally commutated* converters. For these, external commutation can be caused either by the supply or by the load. In the case of line-commutated converters the commutating voltage required for commutation is made available by the ac supply in the case of load-commutated converters by the load. Externally commutated converters can satisfy the basic functions of rectification, inversion, and ac conversion.

Converters with forced commutation are called *self-commutated* converters. Self-commutated converters are employed mainly for inversion and dc conversion.

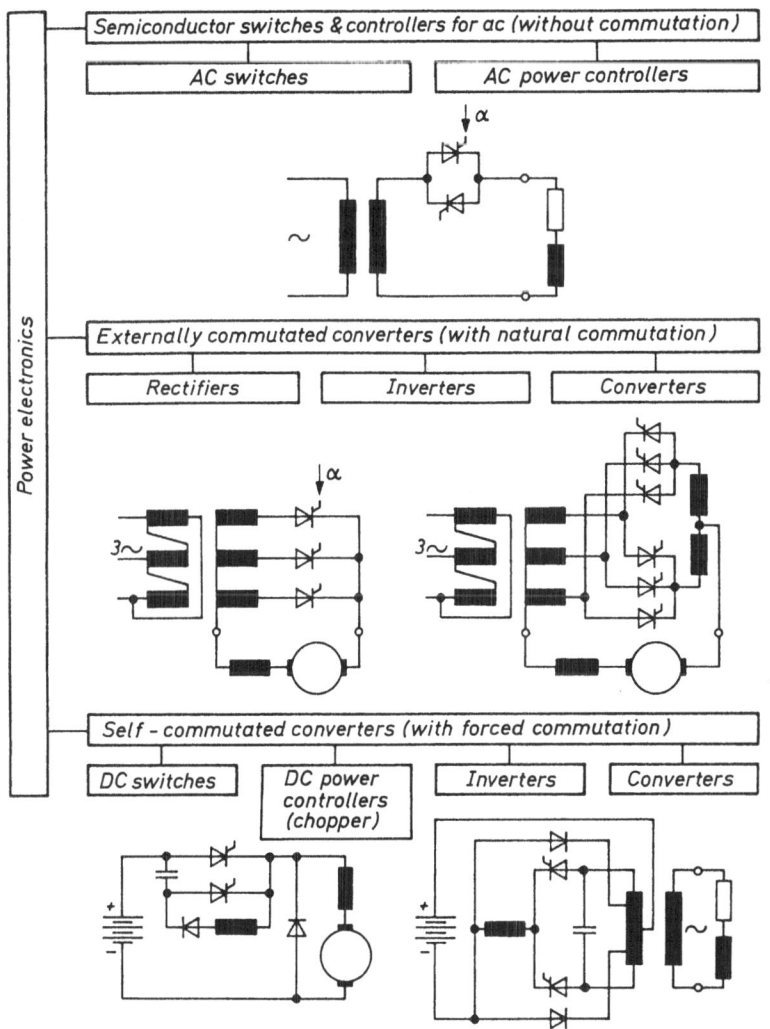

Fig. 5.6. Classification of converters according to origin of the commutating voltage

Figure 5.6 shows the classification of converters according to the origin of the commutating voltage. Characteristic circuits for each of the three types of converter are given as examples and are described in greater detail in Chaps. 6, 7, and 8.

6 Semiconductor Switches and Power Controllers for AC

In single-phase and three-phase ac circuits the current passes through zero and thereby changes its polarity after each half-wave of the mains cycle. If one introduces a semiconductor switch into a single-phase ac circuits, it must be designed to carry current in both directions; for instance it could consist of two thyristors in antiparallel (see Fig. 2.2). In such a semiconductor the current begins to flow as soon as the thyristors are triggered. The current half-cycles of different polarity are carried alternately by the two thyristors. So that a continuous alternating current can flow the thyristor taking over the current must be retriggered after each current half-cycle. If the thyristor is not triggered, the alternating current is extinguished as it passes naturally through zero.

Commutating processes as strictly defined with current being carried simultaneously by two rectifier arms relieving each other do not occur in this case. Semiconductor switches and power controllers are for this reason treated as a particular type of converter without commutation [6.7, 6.8].

6.1 Semiconductor Switches for Single-phase and Three-phase AC

Bidirectional switches can therefore be used to switch single-phase and three-phase ac circuits. Compared to mechanical switches for single-phase ac in the low-voltage field, they have both advantages and disadvantages [6.3, 6.4]. The advantages are their practically unlimited switching life, their freedom from wear, the possibility of accurately setting the instant of switching on by means of the trigger pulse, and switching off without arc at the natural passage through zero.

The disadvantages are the on-state voltage drop when conducting which often makes supplementary cooling necessary insufficient insulation capacity in the off-state with a reverse current of several milliamperes and the higher price.

Despite these disadvantages, semiconductor switches are employed in the low-voltage field where high numbers of switching cycles are required without the need for maintenance work. Figure 6.1 shows the characteristic curve of a thyristor

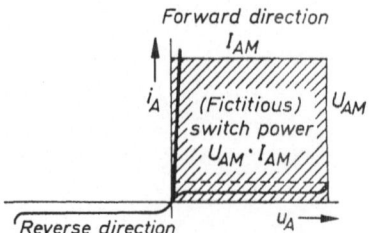

Fig. 6.1. Switching capacity of a thyristor

already dealt with in Chap. 3. The thyristor is a bistable semiconductor component, which on the reverse blocking or off-state characteristic blocks voltages with a small reverse current and on the on-state characteristic allows current to flow with a small on-state voltage drop (1 V to 2 V). It can thus be used as a switch in electrical circuits. From the product of maximum permissible peak off-state voltage U_{AM} and maximum permissible mean on-state current I_{AM} a fictitious switching capacity (hatched area in Fig. 6.1) can be defined. For power thyristors with peak off-state voltages of two and more kV and maximum average on-state currents of up to 1000 A the fictitious switching capacity, thus defined, of a thyristor is already about 1 MVA. Because of the need to maintain factors (transient voltages, overcurrents), usually only a fraction of it is utilized.

6.1.1 Semiconductor Switches

A semiconductor switch for single-phase ac constructed from thyristors in antiparallel is illustrated in Fig. 6.2. Thyristors T1 and T2 must be triggered at the beginning of the appropriate current half-cycle i_{A1} or i_{A2}. This can be done by trigger pulses which are passed through trigger pulse transmitters and then rectified by diodes. A snubber circuit should be provided for damping.
Antiparallel pair. The basic element illustrated in Fig. 6.2 of two rectifier valves in antiparallel is also designated a pair of antiparallel arms. With such a pair of antiparallel arms one connection leads into one single-phase ac system and the second connection leads into the other ac system.

With a sinusoidal current waveform $i = \hat{i} \sin \omega t$ the linear mean value

$$I_{Aav} = \frac{1}{2\pi} \int_0^\pi \hat{i} \sin \omega t \, d(\omega t) = \frac{\hat{i}}{\pi} \tag{6.1}$$

of the current in each thyristor can be calculated by integrating a current half-cycle over the cycle time t.

Accordingly the rms value

$$I_{Aeff} = \sqrt{\frac{1}{2\pi} \int_0^\pi \hat{i}^2 \sin^2 \omega t \, dt(\omega t)} = \frac{\hat{i}}{2} \tag{6.2}$$

of the current of a thyristor can be calculated.

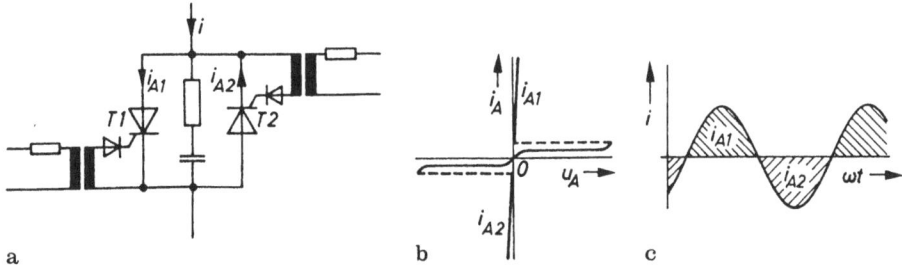

Fig. 6.2a – c. Semiconductor switch for single-phase ac. **a** Thyristors in antiparallel with suppression circuit; **b** characteristic; **c** current waveform

The rms value of the thyristor current referred to the rms value of the ac current is

$$\frac{I_{Aeff}}{I_{eff}} = \frac{\hat{\imath}/2}{\hat{\imath}/\sqrt{2}} = \frac{1}{\sqrt{2}} = 0{,}707. \tag{6.3}$$

Because the mean value is formed as the square, the rms value of the thyristor current is smaller than the rms value of the ac current only by a factor of $1/\sqrt{2}$ (not by a factor of $1/2$).

Bidirectional Thyristors. Instead of using thyristors in antiparallel, semiconductor switches for single-phase ac can also be constructed with the bidirectional thyristors (triacs) described in Chap. 3 (Fig. 6.3).

The bidirectional thyristor can be triggered by a trigger pulse of any polarity and carry current in both directions [6.1]. The characteristic curves of a bidirectional thyristor correspond to those of thyristors in antiparallel. As the triggering voltage can be of either polarity the triggering circuits are particularly simple with bidirectional thyristors. Once successfully triggered a bidirectional thyristor remains conducting so long as the current does not drop below the holding current I_H. $U_{(BO)}$ is the breakover voltage above which a bidirectional thyristor turns on even without a trigger pulse.

Bidirectional thyristors are more difficult to produce than standard thyristors because they combine two $p-n-p-n$ zone sequences in one silicon slice. Therefore they exhibit voltages and currents ratings less than those of standard thyristors. Bidirectional thyristors are, however, available for direct connection to three-phase ac systems with up to 380 V line voltage at currents of 10 to 100 A [6.5].

On-state Power Losses. The on-state power losses $P_T = u_T i_T$ of a semiconductor switch are obtained from the product of the on-state voltage u_T and on-state current i_T. They can be calculated approximately by replacing the on-state characteristic of a semiconductor valve with the on-state threshold voltage $U_{(TO)}$ and a constant equivalent resistance r_T (see Fig. 4.14). The on-state voltage is then

$$u_T = U_{(TO)} + r_T i_T. \tag{6.4}$$

The mean on-state power losses

$$P_T = \frac{1}{T} \int_0^T (U_{(TO)} + r_T i_T) i_T dt = U_{(TO)} I_{Aav} + r_T I_{Aeff}^2 \tag{6.5}$$

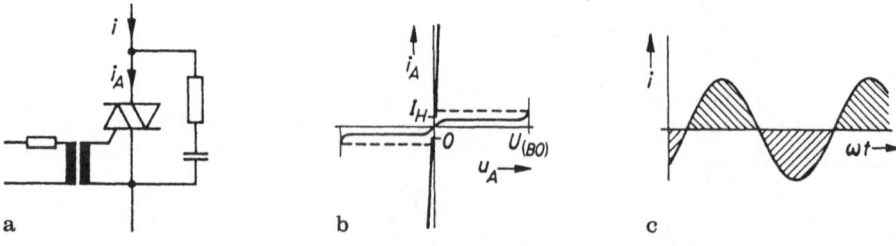

a   b   c

Fig. 6.3a–c. Semiconductor switch for single-phase ac. **a** Bidirectional thyristor (triac) with suppression circuit and trigger pulse transmitter; **b** characteristic; **c** current waveform

are obtained by integration. The result shows variation of the on-state power losses not only with the mean value I_{Aav} but also with the rms value I_{Aeff} of the valve current. Hence, the permissible maximum mean on-state current of a power semiconductor varies with the shape of the current curve. In the data sheets it is often stated for sinusoidal half-cycles. This corresponds to the loading in the case of a semiconductor switch for single-phase ac.

AC controller Connections. Table 6.1 shows the characteristics of ac controller connections. Bridge connections with a controllable short-circuit arm can like ac controller connections be used for ac conversion.

Polygon Connections. These comprise several converter principal arms in series in a ring. All connection points between the arms represent ac terminals. The load is switched between the ac terminals of the polygon connections and the ac system (characteristics of the three-phase polygon connections are shown in Table 6.1 below).

Table 6.1. AC controller and polygon connections

a The load is connected between 1 and 1′, 2 and 2′ and 3 and 3′

6.1.2 Switching Single-phase AC

Figure 6.4 shows the circuit and the voltage and current waveforms for a single-phase ac switch. A resistive-inductive load is assumed which when switched on causes the current i to lag the alternating voltage u by the load angle $\varphi = \text{arc tan}(\omega L/R)$.

When the semiconductor switch as assumed in Fig. 6.4 is triggered as the on-state current passes naturally through zero, a turn-on transient is avoided. For this the instant of switching on, namely ωt_0, must lag behind the passage of the alternating voltage through zero by the load angle φ.

If it is switched on at any other instant, then generally with a resistive-inductive load a transient component of the current exists which decays in a few cycles depending upon the damping. In this general case the current

$$i = \frac{u}{\sqrt{R^2 + (\omega L)^2}} \left[\sin(\omega t - \varphi) - e^{-\frac{R}{\omega L}(\omega t - \omega t_0)} \cdot \sin(\omega t_0 - \varphi) \right] \qquad (6.6)$$

can be calculated. After switching on it is made up of two portions, the sinusoidal on-state current and a transient component decaying with the time constant $\tau = L/R$ which becomes zero when $\omega t_0 = \varphi$ i.e. when it is switched on at the instant when the on-state current passes naturally through zero. The maximum transient

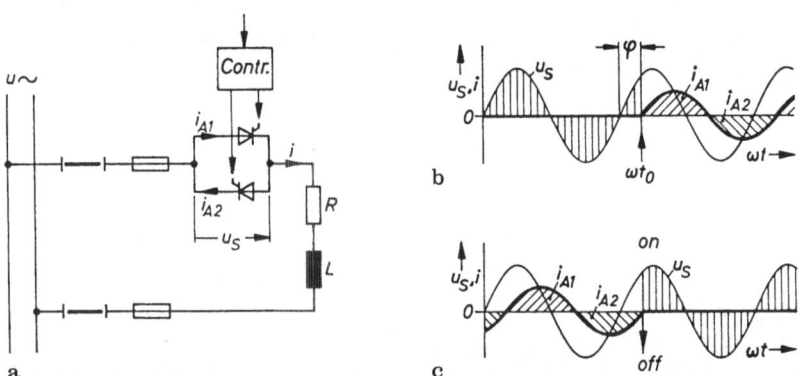

a

b

c

Fig. 6.4a—c. Switching single-phase ac on and off by semiconductor switch. **a** Circuit; **b** switching on; **c** switching off

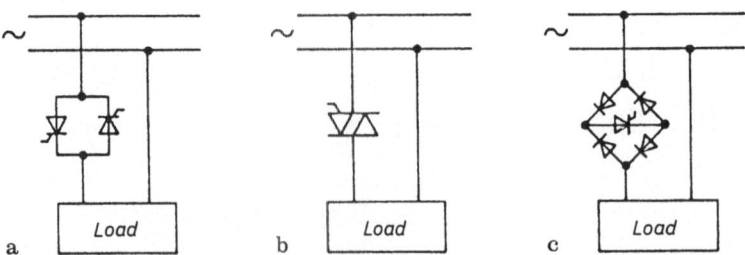

a

b

c

Fig. 6.5a—c. Switching single-phase ac. **a** Thyristors in antiparallel; **b** bidirectional thyristor (triac); **c** thyristors with diode bridge

component occurs at $\omega t_0 = \varphi + \pi/2$ i.e. when the switch is turned on at the peak of the on-state current.

To switch off the alternating current with the semiconductor switch it is sufficient to suppress further trigger pulses. The ac current continues to flow until it passes naturally through zero. The following current half-cycle no longer appears owing to the missing trigger pulse. The switch voltage u_S appears across both thyristors.

Figure 6.5 shows the most common forms of construction of semiconductor switches for single-phase ac thyristors in antiparallel, bidirectional thyristor (triac), and a thyristor in the dc arm of a diode bridge. With the latter circuit only a controlled power semiconductor for one direction of current is needed. Its current consists of rectified half-cycles.

Generally a mechanical isolator must be provided in series with semiconductor switches so that the series-connected load can be isolated during operational shutdowns. The semiconductor switch alone in the blocking state still allows a reverse current of several mA which is aggravated by the necessary RC snubber circuits. Sometime short-circuiting switches parallel to the semiconductor are also provided to avoid the losses caused by the on-state voltage drop of the power semiconductor when switched on for long periods.

6.1.3 Switching Three-phase AC

Fundamentally three-phase ac circuits can be switched on and off in the same way as single-phase ac circuits. Figure 6.6 shows the switching-on process of a three-phase resistive-inductive load with a semiconductor switch constructed from thyristors in antiparallel.

In the case of the single-phase ac circuit (see Sect. 6.1.2) it was shown that a transient response when switching on can be prevented if the semiconductor switch is triggered as the current passes naturally through zero. The conditions under which it is also possible to switch on a three-phase ac circuit without a transient response should also be examined. It appears that with simultaneous switching on of all three phases a transient response always occurs in at least two phases, as in all cases it is only possible to switch on one phase when the current passes naturally through zero. If, however, as shown in Fig. 6.6 the instants of switching on are displaced in time such that two phases (phases 1 and 2 in Fig. 6.6) are first switched on and 90° later the third phase is switched on as the current in this phase passes naturally through zero, a transient response in the current of a 3-phase resistive-inductive load can be avoided.

The thyristors for three-phase ac switches must be sized for the peak value of the line voltage although with even voltage distribution before switching on only the peak value of the phase voltage appears as the maximum off-state voltage. During switching on and off, however, higher voltage stresses can occur temporarily.

Switching off with semiconductor switches for three-phase ac is also accomplished by suppressing further trigger pulses. The current is first of all interrupted when it next passes through zero and afterwards still continues to flow in the other two phases as a single-phase current for 90° before it is completely extinguished.

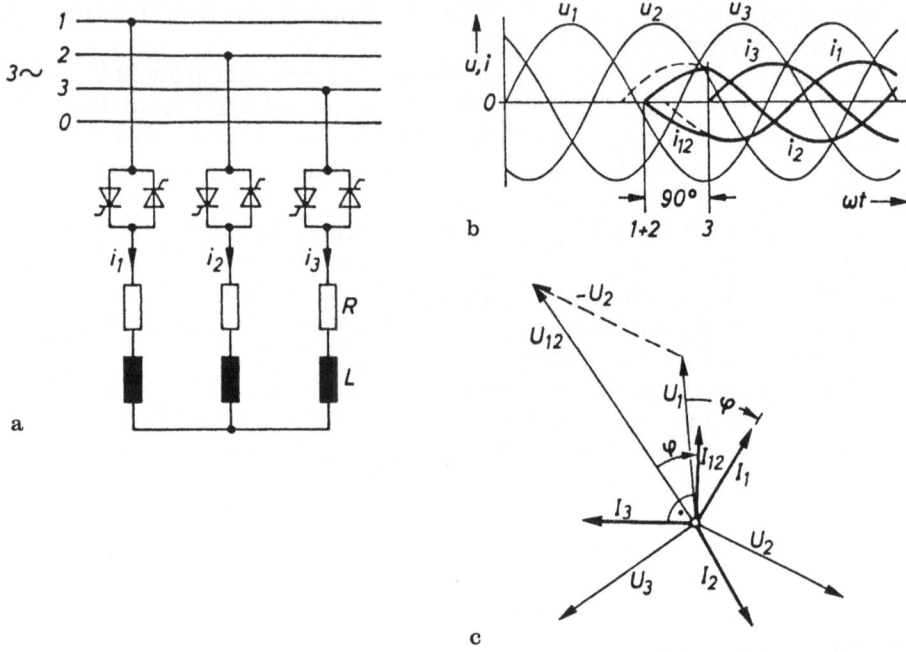

Fig. 6.6a–c. Switching on three-phase ac by semiconductor switch without transient response. **a** Circuit; **b** voltage and current waveforms; **c** phasor diagram

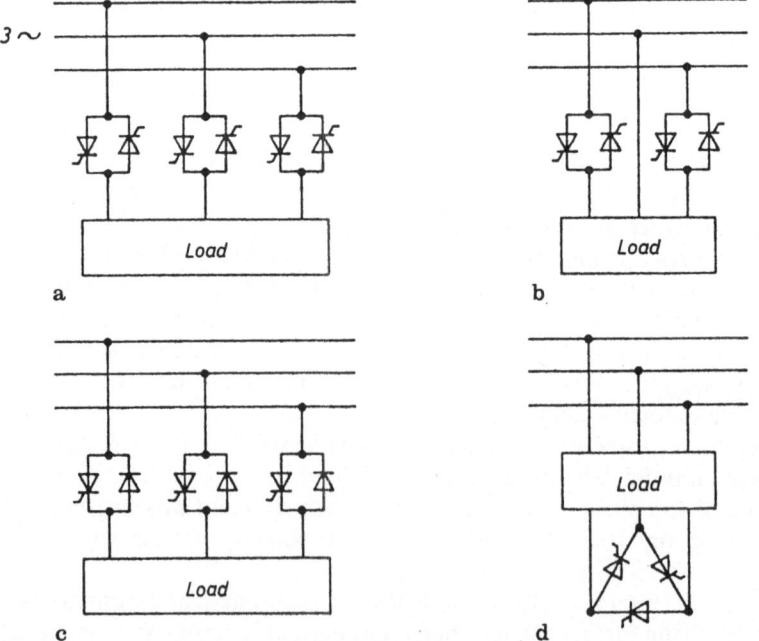

Fig. 6.7a–d. Switching three-phase ac. **a** Thyristors in antiparallel; **b** economy connection; **c** semi-controlled connection; **d** polygon connection

Figure 6.7 shows usual arrangements of semiconductor switches for three-phase ac. Instead of thyristors in antiparallel, bidirectional thyristors are often employed. With a three-phase ac system without neutral conductor switches in two phases are sufficient. In the blocking state the line voltage then lies across each thyristor switch. A three-phase load can also be switched on and off by thyristors and diodes in antiparallel, but again only if there is no neutral. The polygonal connection manages with three thyristors in the star point of a three-phase load, but the current loading of the thyristors is greater than with the other connections quoted by a factor of 3/2.

Reversing Connections. Semiconductor switches are often employed instead of mechanical contactors where high switching frequencies are required e.g. in reversing connections where rotating-field machines are switched from one direction of rotation to the other by reversing the rotating field. It is necessary to change the phase sequence for this. Figure 6.8 shows reversing connections with triac contactors. It should be ensured by taking appropriate measures in the control and RC circuits that short circuit between the mains phases does not occur by simultaneously turning on the triacs for more than one direction of rotation. For only occasional reversing processes can the rotating field also be reversed by mechanical contactors.

Comparison of Electronic and Electro-mechanical Contactors. Table 6.2 compares the properties of electronic and electro-mechanical contactors. With respect to lifetime, switching frequency, and activation electronic contactors compare favourably with electro-mechanical contactors.

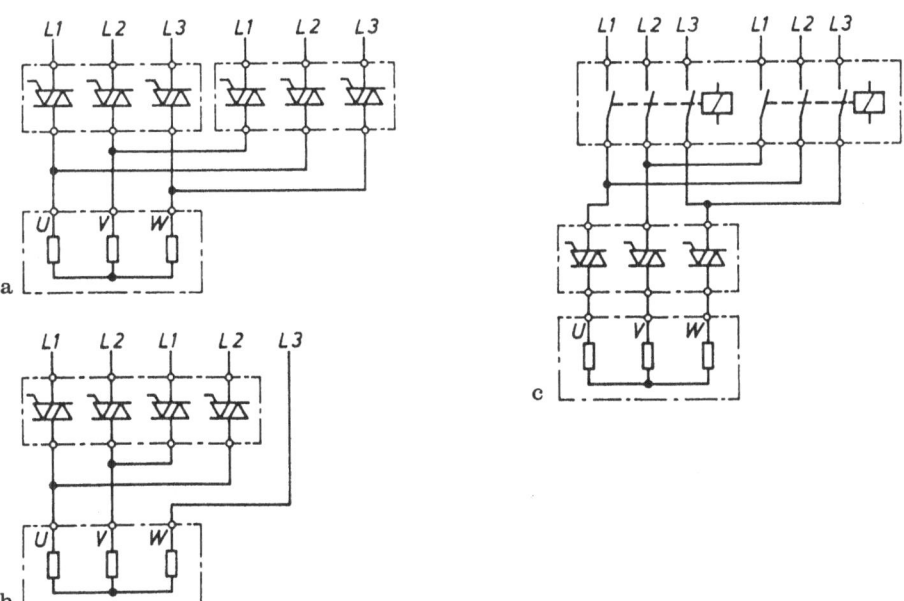

Fig. 6.8a – c. Reversing connections (reversal of rotating field) with triac contactors. **a** Two three-pole triac contactors; **b** one four-pole triac contactor; **c** using mechanical reversing contactors

Table 6.2. Comparison of electronic and electro-mechanical contactors

	Electro-mechanical contactor	Electronic contactor
Lifetime	Mechanical lifetime up to approx. 10^6 switchings; switch contact lifetime $10^5 - 10^6$ switchings	Almost unlimited switching number
Switching frequency	Up to approx. 10^4 switchings/h (max.)	Up to several kHz; max. depends on line frequency (18×10^4 switchings/h at 50 Hz)
Operation	With contacts; ac or dc current operation; increased power supply when switched on	With contacts or electronics; trigger voltage 1 ... 3 ... 5 V; trigger current 10 ... 100 ... 500 mA for each power semiconductor
Switch-on delay	10 ... 50 ms	A few μs
Switch-off delay	10 ... 50 ms	At most one half period (< 10 ms at 50 Hz)
Switch-on behaviour	Switch-on time inexact; danger of rebound; generates noise; contact bounce	Switch-on time exact; very low noise factor; no bounce
Switch-off behaviour	Switch-off time inexact; extinguishing accompanied by electric arc; generates over-voltages with low currents; generates noise; restrikes	Quenches by natural zero current flow without electric arc; very low noise generation; no restrikes
Insulation capacity	Air gap with several kV; no off-state current	Highest permissable periodic peak off-state voltage 1 ... > 2kV; a few mA off-state current; sensitive to overvoltage
Overload-capability	Very high	Limited by the limiting current of the power semiconductor
Short-circuit stability	Danger of welding; danger of lift-off	Peak impulse current limit of the power semiconductor; protection by fuse or fast-action switch
Heat due to power losses	Low; closed contact voltage drop < 10mV; can be worsened during operation	High ($\approx 2\% - 4\%$ of P_N); forward anode voltage 1.2 ... 1.5 ... > 2V; cooling necessary

Table 6.2 (continued)

	Electro-mechanical contactor	Electronic contactor
Environmental influences	Greater wear when dirty, eventual breakdown; sensitive to unfavourable gases; no danger of explosion only when encapsulated	Heating up of the environment; insensitive to unfavourable gases
Typical power range	2 − 1000A; 600V	1 − 1000A; 600V
Maintenance	Replacement of switch contacts according to their loading	No maintenance; eventually greasing of the fan
Size	Compact	2 − 8 times larger
Price	Very cheap	5 − 25 times more expensive
Supply	Extensive, mature technology	Steadily increasing, further developments possible

On the other hand they have higher heat losses when in an on-state condition and a low insulation capability when off. They do not compare favorably in size or cost to mechanical contactors.

Applications. Electronic contactors are useful when high switching frequencies and electronic adaptability to the switching process are required. Triac contactors can be built to take voltages up to 500 V. They are used for example for reversable drives. In welding control thyristors in antiparallel are increasingly taking over from ignitron contactors.

6.1.4 Switching Inductances and Capacitors

The switching on and off of a mixed resistive-inductive load is dealt with in Sect. 6.1.2. The particular conditions on switching pure reactances (inductances and capacitors) with semiconductor switches will now be examined once again [6.9, 6.14, 6.15].

Switching an Inductance. Figure 6.9 shows the switching of an inductance L with a semiconductor switch both without a transient response and with the maximum transient response.

As a consequence of the ever-present winding capacitance C_L of the inductance L, even when switching on without a transient response i.e. triggering the semiconductor switch as the ac current i passes naturally through zero, a medium-frequency transient oscillation occurs which on switching on is superimposed upon the voltage u_L across the inductance. The angular frequency of this oscillation is approximately $1/\sqrt{LC}$; the maximum amplitude reached is the peak value of the alternating voltage. On switching off the switch voltage u_S has a

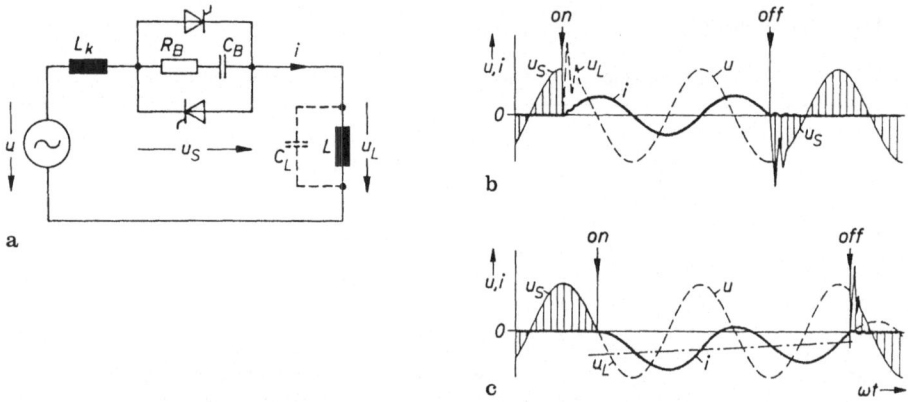

Fig. 6.9a – c. Switching an inductance by semiconductor switch. **a** Circuit; **b** switching on without transient response; **c** switching on with transient response

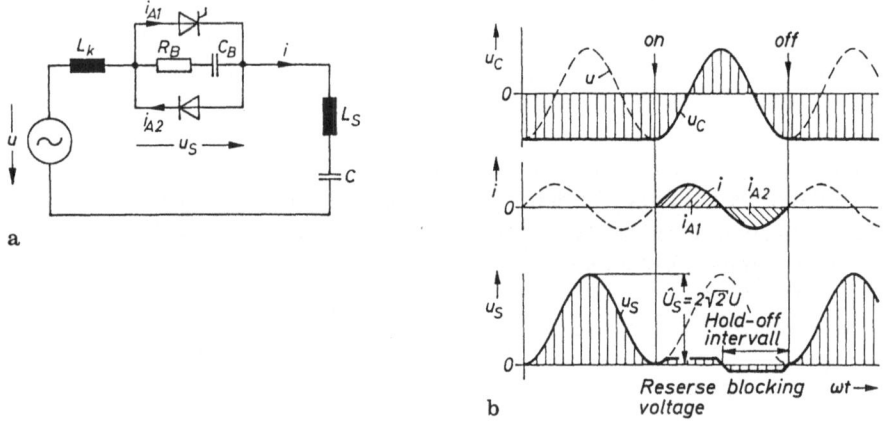

Fig. 6.10a,b. Switching a capacitor by semiconductor switch (thyristor with diode in antiparallel) without transient response. **a** Circuit; **b** voltage and current waveforms

medium-frequency oscillation superimposed upon it, the angular frequency of which is nearly $1/\sqrt{(L+L_k)C_B}$; its amplitude is damped by the RC circuit resistor R_B. The maximum transient response occurs when the semiconductor switch is triggered on as the ac voltage u passes through zero i.e. at the peak value of the on-state current i. It decays exponentially with a characteristic time determined by the losses in the single-phase ac circuit.

Switching a Capacitor. When switching on a capacitor C a transient response when switching as the current passes through zero can only be prevented if the capacitor has previously been charged to the peak value of the alternating voltage u. Figure 6.10 shows the circuit under consideration and the waveforms of voltage and current on switching on and off without transients. In this case the semiconductor switch consists of a thyristor with diode connected in antiparallel. Before switching on it should be ensured that the capacitor C is charged via a protective resistor to the peak value of the alternating voltage. Owing to the charging up of the capacitor the maximum blocking voltage across the semiconductor switch in the off-state is twice the peak value of the mains voltage. To limit transient currents due to fluctuations in the supply voltage or faults a protective inductance L_S is required, unless the supply inductance L_k is sufficiently large.

6.2 Semiconductor Power Controllers for Single-phase and Three-phase AC

Semiconductor switches allow not only the switching on and off of single-phase and three-phase ac circuits dealt with up to now, but also the switching on repeated in each half-cycle whereby the current in each case flows from the instant of triggering until its natural passage through zero. With this technique the power of single-phase and multi-phase ac loads can be contactlessly and continuously varied, a process called phase-angle control. Converters employed for this purpose using antiparallel pairs are therefore called single-phase ac power controllers or three-phase power controllers [6.2].

The control angle α, also called delay angle, at which the semiconductor power controllers are cyclically triggered is defined as the anlge between passage of the phase voltage through zero i.e. passage of the uncontrolled on-state current with resistive load through zero and the instant of triggering. By altering the control angle α the flow of power between a single-phase or three-phase source and the load can be continuously adjusted [6.10, 6.11, 6.12, 6.13].

6.2.1 Controlling Single-phase AC

Figure 6.11 shows the basic circuit of a single-phase ac power controller with which first of all a resistive load R is assumed. The cyclic instants of triggering of the two thyristors connected in antiparallel are delayed relative to passage of the alternating voltage u through zero by the control angle α.

Because of this delay the semiconductor switch blocks the hatched portion of the voltage. With resistive load R the current jumps after triggering of the thyristors in each case to the instantaneous value of the on-state current and then varies sinusoidally from the instant of triggering until passage through zero. By altering the control angle α the current of the load R can be continuously adjusted between the maximum value U/R at $\alpha = 0$ and zero at $\alpha = 180°$.

Current Curve. Figure 6.12 shows the current curve with a single-phase ac power controller as a function of the control angle with three different loads: resistive load, resistive-inductive load, and inductive load. In each case the current waveform is drawn for several discrete control angles.

With resistive load the current with the semiconductor switch triggered can easily be calculated in accordance with

$$i = \frac{\hat{u}}{R} \sin \omega t. \tag{6.7}$$

With the semiconductor blocked (from $\omega t = 0$ to α or from $\omega t = \pi$ to $\pi + \alpha$), $i = 0$.

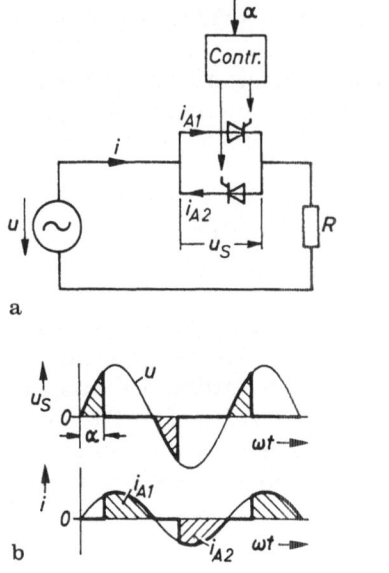

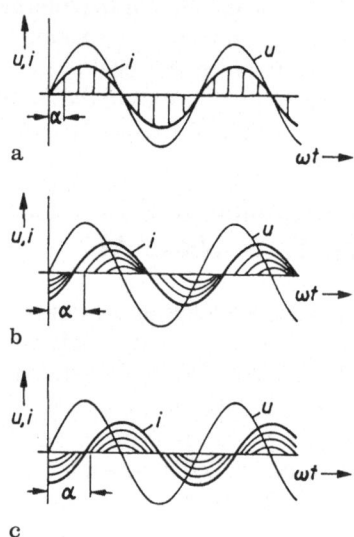

Fig. 6.11a,b. Controlling single-phase ac power with semiconductor power controller. **a** Circuit with thyristors in antiparallel; **b** voltage and current waveforms

Fig. 6.12a–c. Current waveform with single-phase ac power controller as function of control angle α with different loads. **a** Ohmic load; **b** ohmic-inductive load; **c** inductive load

With inductive load L the load current i is obtained from equation

$$i = \frac{\hat{u}}{\omega L}\left[\sin\left(\omega t - \frac{\pi}{2}\right) - \sin\left(\alpha - \frac{\pi}{2}\right)\right]. \tag{6.8}$$

The inductive current has a phase displacement of 90° with respect to the alternating voltage u. The control angle can therefore alter the load current only within the range from 90° to 180°. Equation (6.8) applies only in the range from $\omega t = \alpha$ to $\omega t = 2\pi - \alpha$ for the positive current half-cycle. The rest of the time the semiconductor switch blocks, and the current $i = 0$. So the load current consists of sinusoidal current cusps which are displaced with respect to the base line by the amount $\hat{u}\sin(\alpha - \pi/2)/\omega L$ varying with the control angle α. Intermediate regions of current zero occur. With a resistive-inductive load, the current is no longer sinusoidal. Instead the current is made up of a sinusodial component and a transient component decaying with the time constant $\tau = L/R$.

$$i = \frac{\hat{u}}{\sqrt{R^2 + (\omega L)^2}}\left[\sin(\omega t - \varphi) - e^{-\frac{R}{\omega L}(\omega t - \alpha)}\cdot\sin(\alpha - \varphi)\right] \tag{6.9}$$

This current waveform corresponds to that which occurs upon switching-on a semiconductor switch when the instant of switching on ωt_0 is replaced by the control angle α (see Eq. (6.6)).

Control characteristic. From Eqs. (6.7), (6.8), and (6.9) the control characteristics of a single-phase ac power controller can be calculated as a function of the

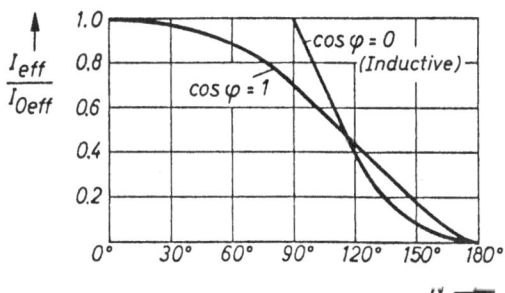

Fig. 6.13. Control characteristic of single-phase ac power controller (rms value of load current)

Table 6.3. Formulae for single-phase ac power controller

Current waveform:
for resistive load R:

$$i = \frac{u}{R} \sin \omega t \quad \text{for } \alpha \leqq \omega t \leqq \pi \text{ or } \pi + \alpha \leqq \omega t \leqq 2\pi, \qquad \text{otherwise } i = 0$$

for inductive load L:

$$i = \frac{\hat{u}}{\omega L} \left[\sin\left(\omega t - \frac{\pi}{2} \right) - \sin\left(\alpha - \frac{\pi}{2} \right) \right] \qquad \begin{array}{l} \text{for } \alpha \leqq \omega t \leqq 2\pi - \alpha \text{ or } \pi + \alpha \leqq \omega t \leqq \pi - \alpha, \\ \text{otherwise } i = 0 \end{array}$$

for resistive-inductive load R and L:

$$i = \frac{\hat{u}}{\sqrt{R^2 + (\omega L)^2}} \left[\sin(\omega t - \varphi) - e^{-\frac{R}{\omega L}(\omega t - \alpha)} \sin(\alpha - \varphi) \right] \quad \text{where } \varphi = \arctan \frac{\omega L}{R}$$

for $\alpha \leqq \omega t$ until current passes through zero or $\pi + \alpha \leqq \omega t$ until current passes through zero, otherwise $i = 0$

Control characteristic:
for resistive load R:
Mean value of load current $\dfrac{I_{av}}{I_{0av}} = \dfrac{1 + \cos \alpha}{2} \ (\alpha \text{ from } 0 \ldots \pi)$

rms value of load current $\dfrac{I_{eff}}{I_{0eff}} = \sqrt{\dfrac{1}{\pi} \left(\pi - \alpha + \dfrac{1}{2} \sin 2\alpha \right)} \ (\alpha \text{ from } 0 \ldots \pi)$

for inductive load L:
Mean value of load current $\dfrac{I_{av}}{I_{0av}} = \sin \alpha + (\pi - \alpha) \cos \alpha \ \left(\alpha \text{ from } \dfrac{\pi}{2} \ldots \pi \right)$

rms value of load current $\dfrac{I_{eff}}{I_{0eff}} =$

$$\sqrt{\dfrac{4}{\pi} \left[(\pi - \alpha) \left(\cos^2\alpha + \dfrac{1}{2} \right) + \dfrac{3}{2} \sin \alpha \cos \alpha \right]} \ \left(\alpha \text{ from } \dfrac{\pi}{2} \ldots \pi \right)$$

---

control angle α. For this the control characteristic is understood to be the rms value of the load current I_{eff} (as reference value usually the maximum rms value I_{0eff} — no phase control is taken) as a function of the control angle α. Figure 6.13 shows this control characteristic of a single-phase ac power controller. The parameter is the power factor cos φ of the load ($\varphi = \arctan^{-1}\omega L/R$). Control characteristics can also be calculated for the mean value of the load current I_{av} (the reference value is I_{0av}) instead of for rms value. Table 6.3 contains the equations for calculating the current waveforms and the control characteristics.

6.2.2 Controlling Three-phase AC

The same conditions as apply to controlling single-phase systems with semi-conductor switches apply to the multi-phase ac power controller. The basic circuit for a three-phase ac power controller is shown in Fig. 6.6 as a three-phase ac switch with three pairs of thyristors connected in antiparallel. The control angle α again corresponds to the angle between passage of a phase voltage through zero i.e. passage of the uncontrolled resistive on-state current of this phase through zero and the initiation of the associated trigger pulse. Owing to the linking of the three phases, however, the voltage and current conditions are no longer as transparent as with the single-phase ac power controller.

By enlarging the control angle α from 0° to 150° with resistive load and from 90° to 150° with inductive load the power of a symmetrical three-phase load is continuously variable between the maximum value and zero. Figure 6.14 shows the control characteristic of a three-phase power controller i.e. the referred effective control angle α. Here again the power factor cos φ, of the load is the parameter [6.16, 6.17].

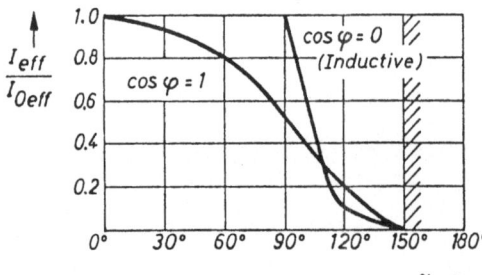

Fig. 6.14. Control characteristic of a three-phase ac power controller (rms value of load current)

6.2.3 Reactive and Distortion Power

With phase-angle control non-sinusoidal currents arise in the load not only with the single-phase power controller but also with the three-phase power controller and hence also in the single-phase or three-phase system (see current waveform in Figs. 6.11 and 6.12 with single-phase ac power controller).

With cyclic processes a non-sinusoidal quantity (here the current i) can be split up into fundamental and harmonic frequency components by Fourier

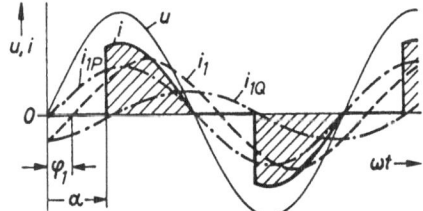

Fig. 6.15. Fundamental frequency reactive power of a single-phase ac power controller with resistive load

analysis. This analysis produces a fundamental component of defined amplitude and phase and a series of higher order harmonics, the frequencies of which are multiples of the fundamental frequency. The amplitude and phase of each harmonic can likewise be determined by Fourier analysis. Generally the amplitude of the harmonics reduces as the ordinal number increases.

In Fig. 6.15 the load current arising from a resistive load and control angle α is illustrated hatched in. This cyclic current i at constant control angle α has a fundamental oscillation i_1, the amplitude, and phase relationship φ_1 of which can be calculated by Fourier analysis. It can be recognised that the fundamental oscillation i_1 of the current lags the ac voltage u. Therefore, in single-phase ac systems with a control angle α, inductive fundamental frequency reactive power Q_1 also occurs with resistive load and is calculated from

$$Q_1 = UI_1 \sin \varphi_1. \tag{6.10}$$

The fundamental oscillation i_1 can be split up into an active component i_{1P} and a reactive component i_{1Q}. The peak value of the active component of the fundamental frequency current is obtained from

$$\hat{\imath}_{1P} = \frac{2}{\pi} \int\limits_{\alpha}^{\pi} \hat{\imath} \sin^2 \omega t d\omega t = \frac{\hat{\imath}}{\pi} (\pi - \alpha + \sin \alpha \cos \alpha) \tag{6.11}$$

and the peak value of the reactive component from

$$\hat{\imath}_{1Q} = \frac{2}{\pi} \int\limits_{\alpha}^{\pi} \hat{\imath} \sin \omega t \cos \omega t d\omega t = -\frac{\hat{\imath}}{\pi} \sin^2 \alpha. \tag{6.12}$$

From the active and reactive components one can obtain the phase displacement angle φ_1 of the fundamental frequency

$$\varphi_1 = \text{arc} \tan \frac{|\hat{\imath}_{1Q}|}{|\hat{\imath}_{1P}|} = \text{arc} \tan \frac{\sin^2 \alpha}{\pi - \alpha + \sin \alpha \cos \alpha}. \tag{6.13}$$

Besides the fundamental frequency reactive power Q_1 a distortion power D arises which with a sinusoidal voltage and non-sinusoidal current can be calculated from the defining equation

$$D = U\sqrt{I_2^2 + I_3^2 + \dots}. \tag{6.14}$$

The active power P is obtained from

$$P = UI_1 \cos \varphi_1. \tag{6.15}$$

$\cos \varphi_1$ is called the fundamental frequency power factor or displacement factor.

It might at first be thought surprising that reactive power should arise with a resistive load r, because this is linked with power pulsation between the load and the ac system, but a resistive load does not have the ability to store energy like an inductor or a capacitor. Closer examination shows that the non-linear characteristic of the semiconductor switch, namely, its capacity to block voltage or pass current with negligible on-state voltage, causes distortion and fundamental frequency reactive power. The two augment each other at any instant so that no energy flows back from the resistive load to the ac voltage source. In general it should be emphasized that reactive power and distortion power are calculated quantities that have no direct physical meaning. Physical meaning is attached to the curve of power against time (see also Sects. 7.12 and 11.2)

$$p = ui. \tag{6.16}$$

6.2.4 Control Techniques

When operating antiparallel pairs constructed of power semiconductors for switching and for power control, a distinction should be made between various

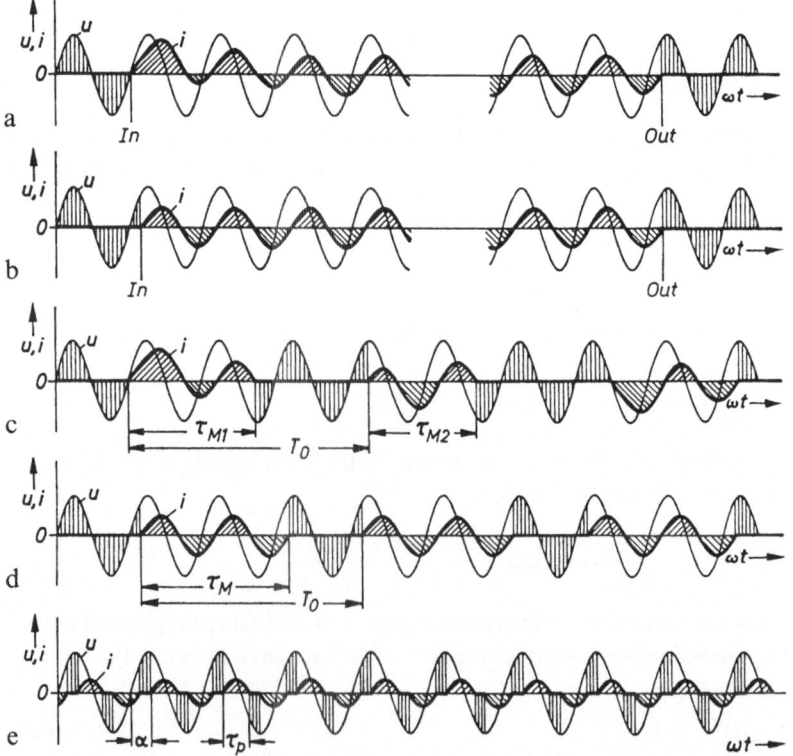

Fig. 6.16a – e. Control techniques of semiconductor switches and power controllers for single-phase ac. **a** Switching operation (non-periodic), not synchronized; **b** switching operation (non-periodic), synchronized; **c** multi-cycle control (periodic), asynchronous; **d** multi-cycle control (periodic), synchronous; **e** phase delay control

control techniques (Fig. 6.16). In one instance the semiconductor switches for two directions of current can be employed to switch single-phase and three-phase ac circuits on and off. This is called switching operation (non-periodic). Switching on can either be unsynchronized or when the transient component is eliminated synchronized. After each passage through zero a trigger pulse must be applied for the next current half-cycle. Switching off is by suppression of further trigger pulses. Figure 6.16a shows unsynchronized and Fig. 6.16b synchronized switch duty. If semiconductor switches for two directions of current are used not only to switch on and off but also to control the flow of power between an ac current source and the load, this can be by phase control with control angle α. This control technique is illustrated in Fig. 16.6e. With this technique the duration of current flow τ_p is less than 180°. The shape of the current waveform depends upon the type of load (see Fig. 6.12).

Another technique for controlling the power flow is called multi-cycle control. A distinction can be made between the asynchronous case in which the repetition frequency $1/T_0$ is not synchronized with the supply system frequency (Fig. 6.16c) and the synchronous case in which this is so (Fig. 6.16d). Multi-cycle control generates subharmonics of the supply system frequency, in the load or supply system current.

7 Externally Commutated Converters

Externally commutated converters need a separate source of ac voltage not belonging to the converter to provide it with the commutating voltage during the period of commutation. In the case of line-commutated converters the supply system is this source of ac voltage with load-commutated converters the load represents this ac voltage source. Externally commutated converters operate with natural commutation.

The line-commutated converter is the type of converter most often employed [9,23]. Converter engineering with mercury-arc rectifiers has almost exclusively employed line-commutated converters. With the development of power electronics the self-commutated converters to be dealt with in the following chapter also acquired considerable importance as did the semiconductor switches and power controllers for single-phase and three-phase ac. Nevertheless line-commutated converters still have the greater economic significance in power electronics.

Commutation has already been defined in Chap. 5 and the two different types natural commutation and forced commutation have been dealt with. Natural commutation determines the internal method of operation of externally commutated converters. Their main feature is that after successful triggering a converter valve with a higher momentary potential takes over the current from the converter valve previously carrying the current. The current is taken over during a finite commutating period which is also called the overlap time. A prerequisite for assumption of the current by the following valve is a suitable commutating voltage which for line-commutated converters is provided by the supply system where necessary via a transformer and for load-commutated converters by the load.

7.1 Line-commutated Rectifiers and Inverters

Line-commutated rectifiers and inverters perform the basic functions of rectification or inversion and draw their commutating voltage from the single-phase or three-phase ac supply system i.e. they use the voltages available in the supply system for commutation. With a single-phase ac voltage the commutating voltage only has the correct polarity during a half cycle i.e. the possible commutating zone in the case of converters with natural commutation is limited to this half cycle.

7.1.1 Operation in the Rectifier Mode

Formation of a dc voltage with a line-commutated rectifier will first be examined without considering commutation i.e. ignoring commutating reactances and resistances. Figure 7.1a shows the connection of a simple line-commutated converter. This is the so-called three-pulse center tap connection (Code sign M3), also designated 'star connection'. This typical converter connection is used below to examine the characteristic properties of line-commutated converters. These properties can readily be extended to other converter connections.

No-load Direct Voltage. If it is assumed that thyristors T1, T2, and T3 are replaced by uncontrolled diodes, a dc voltage u_d that follows the cusps of the phase voltages u_1, u_2, and u_3 (Fig. 7.1b) appears across the load. By integration the mean value of u_d for uncontrolled rectifier operation is obtained from

$$U_{di} = \frac{1}{(2\pi)/3} \int_{-\pi/3}^{+\pi/3} \sqrt{2}U \cos \omega t \, d\omega t = \frac{1}{(2\pi)/3} \sqrt{2}U \sin \omega t \Big|_{-\pi/3}^{+\pi/3}$$

$$= \frac{3}{\pi} \sqrt{2}U \sin \frac{\pi}{3}. \tag{7.1}$$

U_{di} is designated the ideal no-load direct voltage resulting from the phase voltage U on the secondary side of the converter transformer ignoring resistive and inductive voltage drops.

Commutating and Pulse Numbers. Equation (7.1) can be extended to converters with any number of pulses by introducing q as the commutating number of a commutating group. The commutating number q then indicates the number of commutating events occurring during one cycle of the supply system within a group of mutually commutating valves.

The pulse number p is defined as the total number of non-simultaneous commutations of a converter connection during one cycle of the ac supply. Thus

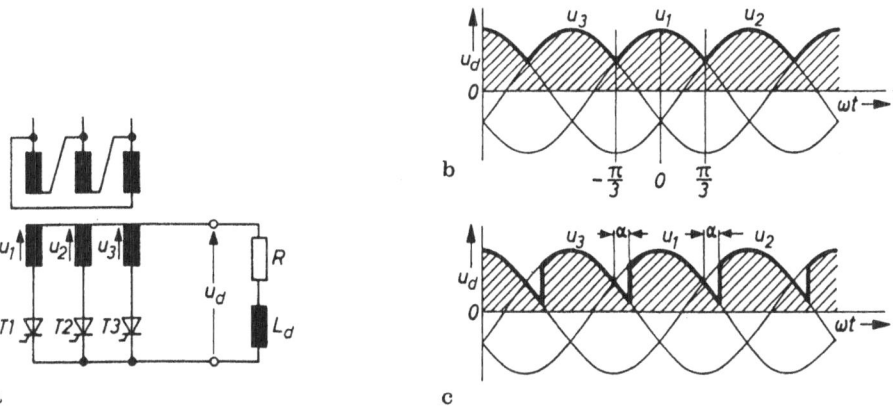

a b c

Fig. 7.1a–c. Formation of dc voltage by line-commutated rectifier. **a** Three-pulse center tap connection; **b** full modulation; **c** phase angle control with control angle α

the ideal no-load direct voltage U_{di} becomes

$$U_{di} = \frac{1}{(2\pi)/q} \int_{-\pi/q}^{+\pi/q} \sqrt{2}U \cos \omega t \, d\omega t = \frac{1}{(2\pi)/q} \sqrt{2}U \sin \omega t \Big|_{-\pi/q}^{+\pi/q}$$

$$= \frac{q}{\pi} \sqrt{2}U \sin \frac{\pi}{q}, \tag{7.2}$$

which can be generalized by the introduction of a factor s. Factor s is 1 for center tap connections and 2 for the bridge connections dealt with later. Thus, the general equation for the ideal no-load direct voltage of a line-commutated converter is

$$U_{di} = s\frac{q}{\pi} \sqrt{2}U \sin \frac{\pi}{q}. \tag{7.3}$$

Control angle. With controlled converter valves e.g. thyristors transfer of the current to the next arm occurs only after this triggering. Transfer can therefore be delayed with respect to the natural intersection of the phase voltages. The control angle α is defined as the time span by which the instant of triggering is retarded compared to that with full modulation. The control angle is generally stated in electrical degrees. The converter is then said to have phase angle control.

With a control angle α the curve of dc voltage u_d shown in Fig. 7.1c is obtained. The mean value of the dc voltage with a control angle α can be calculated by integrating within limits differing by the control

$$U_{di\alpha} = \frac{1}{(2\pi)/q} \int_{-\frac{\pi}{q}+\alpha}^{+\frac{\pi}{q}+\alpha} \sqrt{2}U \cos\omega t \, d\omega t$$

$$= \frac{1}{(2\pi)/q} \sqrt{2}U \sin\omega t \Big|_{-\frac{\pi}{q}+\alpha}^{+\frac{\pi}{q}+\alpha}$$

$$= \frac{q}{\pi} \sqrt{2}U \sin\frac{\pi}{q} \cos\alpha. \tag{7.4}$$

Introduction of the factor s produces

$$U_{di\alpha} = s\frac{q}{\pi} \sqrt{2}U \sin\frac{\pi}{q} \cos \alpha. \tag{7.5}$$

From Eqs. (7.3) and (7.5) the important relationship for the ideal no-load direct voltage $U_{di\alpha}$ is obtained occurring at control angle α

$$U_{di\alpha} = U_{di} \cos \alpha \tag{7.6}$$

which says that the mean value of the dc voltage of line-commutated converters varies as the cosine of the control angle α.

7.1.2 Operation in the Inverter Mode

The control angle α can be continuously increased from full modulation at $\alpha = 0$ (maximum value of direct voltage U_{di}) whereby the output dc voltage varies in accordance with Eq. (7.6). At $\alpha = 90°$ the mean value of the dc voltage is zero. On further increasing the control angle beyond $90°$ the mean value of the dc voltage, becomes negative and continues to rise with negative polarity as the control angle is increased. At $\alpha = 180 - \gamma$ where γ is the margin angle defined below it reaches the maximum possible negative mean value.

The range with control angles of $\alpha = 90°$ to $180° - \gamma$ with negative mean value of dc voltage is called operation in the inverter mode because the direction of energy flow is opposite to that when operating in the rectifier mode. On operation in the inverter mode energy from the dc load is passed back into the operating state the converter operates as a line-commutated inverter.

Figure 7.2 shows the process of controlling a converter in three-pulse center tap connection from the rectifier into the inverter mode whereby on account of the preset valve action the dc current I_d maintains its direction while the polarity of the mean value of the dc voltage U_d reverses. In the rectifier mode the load shown as a dc machine absorbs energy from the three-phase ac supply. In the inverter mode it injects energy back into the three-phase ac supply.

The commutating voltage, that is the difference in the phase voltage of mutually commutating valves, has the correct polarity in the range from $\alpha = 0$ to $180°$ for in this range the potential of the relieving phase is always higher than that of the previous phase. On increasing the control angle beyond $180°$ the polarity of

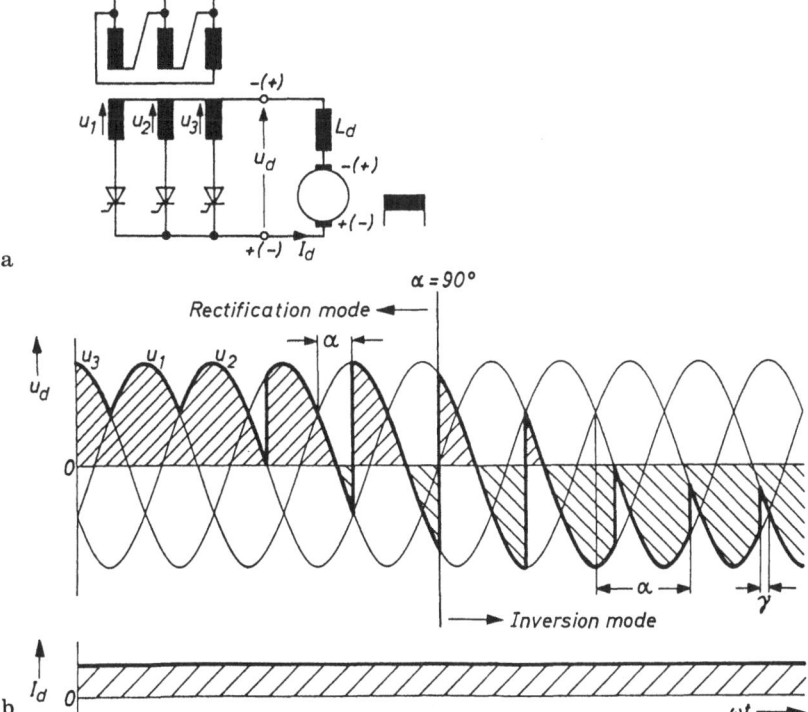

Fig. 7.2a,b. Modulation from operation in the rectifier mode into operation in the inverter mode. **a** Three-pulse center tap connection; **b** voltage and current waveforms

the commutating voltage would reverse. The potential of the relieving phase is then lower than that of the phase carrying the current. This range is forbidden for natural commutation because it leads to short circuits in the commutating circuit.
Margin Angle. To insure that the device turning off has adequate time to regain its forward blocking capability the control angle α must not be increased to 180°. Rather for operation in the inverter mode a safety clearance must be maintained from the intersection of the phase voltages. This clearance measured in electrical degrees is called the margin angle γ.

7.1.3 Line Commutation

The waveform of the commutating current with line-commutated converters will now be examined more closely.
Commutating Voltage. The commutating voltage u_k is a sinusoidal alternating voltage arising in a multi-phase system as a result of the difference between the voltages of two mutually commutating phases (Fig. 7.3). From the phasor diagram of the voltages U_1 and U_2 of two commutating phases separated by an angle $2\pi/q$ the commutating voltage can be calculated from

$$U_k = 2\ U\ \sin\frac{\pi}{q}. \tag{7.7}$$

In a three-phase system $U_k = \sqrt{3}U$ and is therefore equal to the line voltage.
Commutating Current. The waveform of current during a natural commutation was dealt with in Chap. 5 (see Fig. 5.4). The resistances in the commutating circuit were ignored and it was assumed that the commutating inductances L_k are the same size; so the commutating Eqs. (5.9) or (5.10) derived in Chap. 5 simplifies to

$$u_k = 2L_k\frac{di_k}{dt}. \tag{7.8}$$

Here, i_k is the short-circuit current flowing in the commutating circuit and is obtained from

$$i_k = i_2 = I_d - i_1. \tag{7.9}$$

With a sinusoidal commutating voltage $u_k = \sqrt{2}U_k\sin\ \omega t$ one obtains from Eq. (7.8) the waveform of the short-circuit current i_k in the commutating circuit:

$$i_k = \frac{1}{2L_k}\int\sqrt{2}U_k\sin\ \omega t\ dt. \tag{7.10}$$

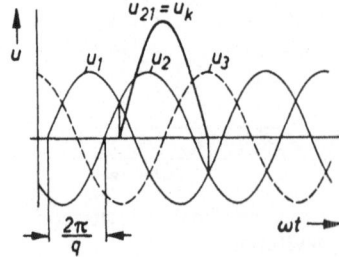

a

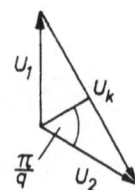

b

Fig. 7.3a,b. Commutating voltage U_k. **a** Transient response; **b** phasor diagram

With the initial condition $t_0 = 0$, $i_k = 0$,

$$i_k = \frac{\sqrt{2}U_k}{2\omega L_k} (1 - \cos \omega t).$$

(7.11)

This equation reproduces the waveform of the short-circuit current which is illustrated in Fig. 7.4. With phase angle control and a control angle α the initial condition $\omega t_0 = \alpha$, $i_k = 0$ applies i.e.

$$i_k = \frac{\sqrt{2}U_k}{2\omega L_k} (\cos \alpha - \cos \omega t).$$

(7.12)

If the phase short circuit occurring when two valves are carrying current simultaneously were still to exist after the end of the commutation process, the short-circuit current would rise again to its maximum value with $\alpha = 0$ to

$$2\sqrt{2}I_k = \frac{\sqrt{2}U_k}{\omega L_k}.$$

(7.13)

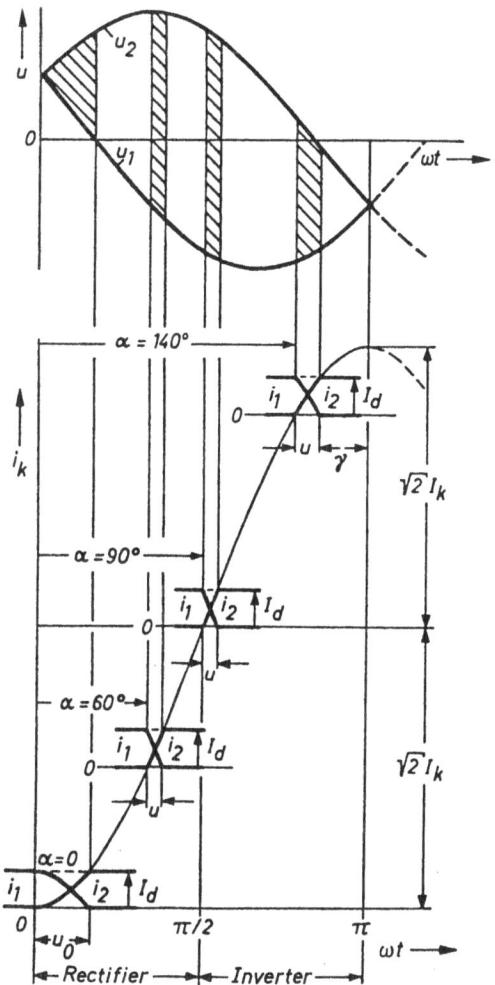

Fig. 7.4. Waveforms of commutating currents

In fact, in fault-free converter operation the phase short circuit is removed at the end of the commutating time because the current in the relieved valve becomes zero and the valve blocks. Depending upon the control angle α the transient response of the commutating currents consists of the corresponding sections of the short-circuit current waveform. When $\alpha=0$ commutation is relatively slow because the commutating voltage $u_k=u_2-u_1$ is small (hatched in at the top of Fig. 7.4).

When $\alpha=90°$ commutation occurs below the maximum value of the commutating voltage u_k with the greatest rate of rise; thereafter, the rate slows on further increasing the control angle α for operation in the inverter mode.

Overlap Time. The commutating time t_u, i.e. the time span during which two commutating converter arms carry current simultaneously, is called the overlap time or overlap angle u, generally stated in electrical degrees. It can be calculated from Eq. (7.8) by integrating this equation over the commutating time

$$\int\limits^{t_u} u_k dt = \int\limits^{t_u} 2L_k \frac{di_k}{dt} dt = 2L_k \int\limits^{t_u} di_k. \tag{7.14}$$

Since the commutating current i_k changes from zero to I_d during the overlap time t_u,

$$\int\limits^{t_u} u_k dt = 2L_k I_d. \tag{7.15}$$

With control angle α this becomes

$$\int\limits_{\alpha/\omega}^{\alpha/\omega+t_u} \sqrt{2}U_k \sin\,\omega t\,\,dt = \frac{1}{\omega}\sqrt{2}U_k[-\cos\,\omega t|_{\alpha/\omega}^{\alpha/\omega+t_u}] = 2L_k I_d \tag{7.16}$$

producing

$$\cos(\alpha+u) = \cos\,\alpha - \frac{2\omega L_k I_d}{\sqrt{2}U_k} = \cos\,\alpha - \frac{I_d}{\sqrt{2}I_k}. \tag{7.17}$$

The initial overlap angle with $\alpha=0$ is obtained from

$$\cos\,u_0 = 1 - \frac{I_d}{\sqrt{2}I_k}. \tag{7.18}$$

Figure 7.5 illustrates the waveforms of voltage and current at three different control angles ($\alpha=0$, $\alpha=30°$, and $\alpha=140°$, i.e. operation in the inverter mode) taking commutation into consideration. Only the inductances L_k in the commutating circuit are considered. The resistances R_k drawn in broken lines are ignored. It is assumed that there is a large smoothing choke L_d on the dc side.

During the overlap time t_u both commutating valves carry current. The commutating voltage u_k appears across the two commutating inductances L_K. When the inductances L_k are identical it is divided equally between them. The waveform of dc voltage u_d during the commutating time t_u is the average of the two commutating phase voltages. After completion of commutation until the beginning of the next one u_d is the phase voltage of the valve carrying the current. Due to the inductive voltage drop across the commutating inductances the mean value U_d of the dc voltage is reduced by the hatched areas below the voltage-time curve. This so-called dc voltage regulation D_x is calculated later.

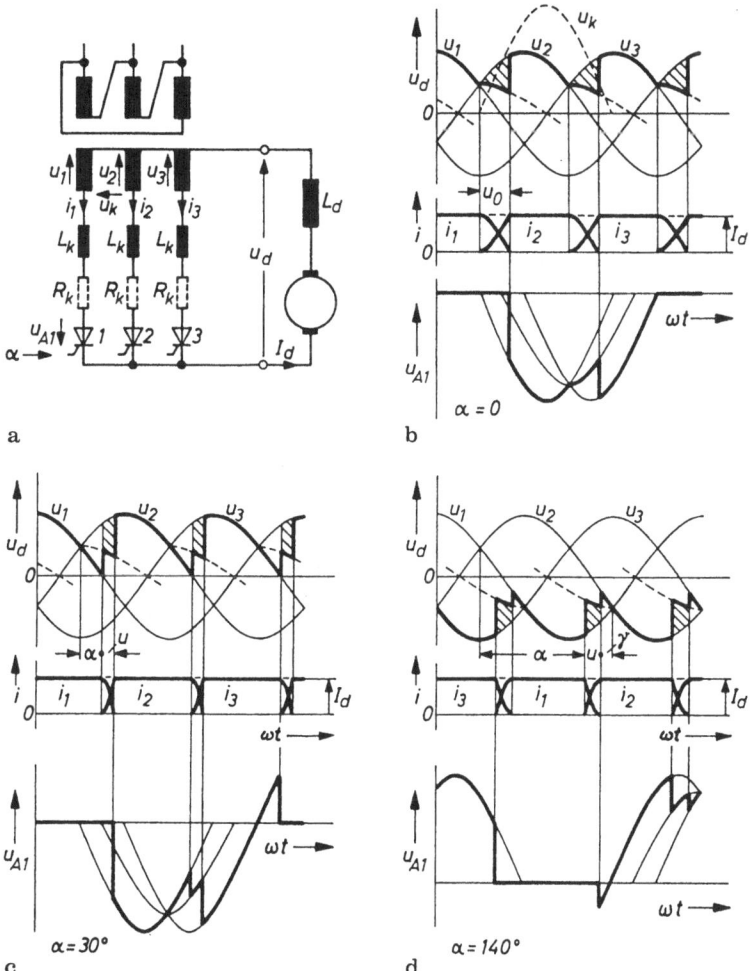

Fig. 7.5a−d. Commutation with three-pulse center tap connection (star connection).
a Connection; **b** full modulation; **c** operation in rectifier mode with phase angle control;
d operation in inverter mode

The reverse blocking voltage in the converter valves is the line voltage between the phase in which the valve lies and the phase actually carrying current.

Effect of Commutating Resistance. It will now be shown how the waveform of the commutating currents can be calculated taking into consideration the resistances R_k in the commutating circuit. Assuming constant and identical commutating inductances L_k and commutating resistances R_k and with complete smoothing of the dc current I_d by a large smoothing inductance L according to Fig. 7.6 the following equations are obtained

$$i_1 + i_2 = I_d \tag{7.19}$$

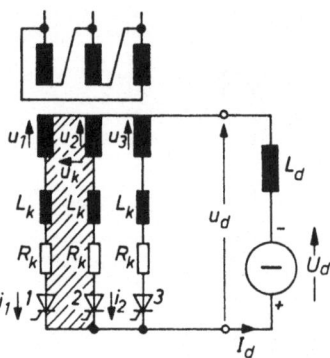

Fig. 7.6. Calculation of commutation taking resistances into account

and

$$u_k - L_k \frac{di_2}{dt} - R_k i_2 + L_k \frac{di_1}{dt} + R_k i_1 = 0 \tag{7.20}$$

which describe the commutation process.

Substituting Eq. (7.19) in (7.20) produces

$$u_k - 2L_k \frac{di_2}{dt} - 2R_k i_2 + R_k I_d = 0. \tag{7.21}$$

The general assumed solution in (7.21) is made up of a particular solution and a homogeneous solution

$$i_2 = i'_2 + i''_2. \tag{7.22}$$

Substituting this assumed solution in (7.21) produces

$$u_k - 2L_k \frac{di'_2}{dt} - 2L_k \frac{di''_2}{dt} - 2R_k i'_2 - 2R_k i''_2 + R_k I_d = 0. \tag{7.23}$$

The particular solution is therefore to be sought in accordance with

$$u_k - 2L_k \frac{di'_2}{dt} - 2R_k i'_2 = 0 \tag{7.24}$$

and the homogeneous solution in accordance with

$$L_k \frac{di''_2}{dt} + R_k i''_2 - \frac{1}{2} R_k I_d = 0. \tag{7.25}$$

For Eq. (7.24) where $u_k = \sqrt{2} U_k \sin \omega t$ the particular solution is obtained from the assumption

$$i'_2 = \hat{i}_k \sin(\omega t - \varphi) \tag{7.26}$$

whereby

$$\hat{i}_k = \frac{\sqrt{2} U_k}{\sqrt{(2R_k)^2 + (2\omega L_k)^2}} \tag{7.27}$$

and

$$\tan \varphi = \frac{\omega L_k}{R_k}.$$ (7.28)

The homogeneous solution of Eq. (7.25) is obtained from the assumption

$$i''_2 = Ae^{-\frac{t}{\tau}} + B.$$ (7.29)

The unknown constants τ and B can be determined by substituting Eq. (7.29) in Eq. (7.25). $\tau = L_k/R_k$ and $B = 0.5\ I_d$ are obtained. The constant A can be obtained from the initial conditions of the equation

$$i''_2 = Ae^{-\frac{t}{\tau}} + 0.5\ I_d.$$ (7.30)

At the instant t_0 of starting commutation $i_2 = i'_2 + i''_2 = 0$. The relationship

$$i_2 = i'_2 + i''_2 = \hat{i}_k \sin(\omega t_0 - \varphi) + Ae^{-\frac{t_0}{\tau}} + 0.5\ I_d = 0$$ (7.31)

is obtained as the equation for determining the unknown constant A. So

$$A = -\frac{\hat{i}_k \sin(\omega t_0 - \varphi) + 0.5\ I_d}{e^{-\frac{t_0}{\tau}}}.$$ (7.32)

The homogeneous solution i''_2 thus becomes

$$i''_2 = -[\hat{i}_k \sin(\omega t_0 - \varphi) + 0.5\ I_d]e^{-\frac{t - t_0}{\tau}} + 0.5\ I_d$$ (7.33)

and the current i_2 during commutation

$$i_2 = i'_2 + i''_2 = \sqrt{2}I_k\left[\sin(\omega t - \varphi) - \sin(\omega t_0 - \varphi)e^{-\frac{t - t_0}{\tau}}\right]$$
$$+ 0.5\ I_d\left(1 - e^{-\frac{t - t_0}{\tau}}\right).$$ (7.34)

Inserting $\omega t_0 = \alpha$ in this equation produces

$$i_2 = \sqrt{2}I_k\left[\sin(\omega t - \varphi) - \sin(\alpha - \varphi)e^{-\frac{\omega t - \alpha}{\omega \tau}}\right] + 0.5\ I_d\left(1 - e^{-\frac{\omega t - \alpha}{\omega \tau}}\right).$$ (7.35)

From this equation one can calculate the exact waveform of the commutating current even taking the commutating resistances R_k into account. For $R_k = 0$ Eq. (7.35) is simplified to

$$i_2 = -\hat{i}_k \cos \omega t + \hat{i}_k \cos \alpha = \frac{\sqrt{2}U_k}{2\omega L_k}(\cos \alpha - \cos \omega t).$$ (7.36)

This equation corresponds to Eq. (7.12) derived earlier for the commutating current i_k, ignoring R_k.

7.1.4 Load Characteristic

The ideal no-load dc voltage U_{di} of a line-commutated rectifier can be calculated in accordance with Eq. (7.3). If such a rectifier is loaded with dc current I_d, a mean value of the dc voltage U_d is produced at the output which as a result of voltage drops is smaller than U_{di}. The voltage drop is made up of three components; the inductive direct voltage regulation D_x, the resistive direct voltage regulation D_r, and the on-state voltage U_F of the converter valves.

Inductive Direct Voltage Regulation. The inductive direct voltage regulation D_x is caused by the inductances L_k in the commutating circuit. Its magnitude should be calculated with the aid of Fig. 7.5. During the overlap time u the dc voltage u_d follows the dotted line between two phase voltages. So in the commutating inductances the dotted areas under the voltage-time curve are lost. From Eq. (7.15) an area under the voltage-time curve of

$$\frac{1}{2}\int_0^{t_u} u_k dt = L_k I_d \tag{7.37}$$

can be calculated. The number of commutations occuring per second is $f \cdot s \cdot q$ where f is supply frequency, s is a factor depending upon the connection (1 for center tap connection and 2 for bridge connection), and q the commutating number of a commutating group. The inductive direct voltage regulation D_x thus becomes

$$D_x = fsqL_k I_d. \tag{7.38}$$

If the inductive direct voltage regulation D_x is referred to the ideal no-load direct voltage U_{di}, the relative inductive ac voltage regulation

$$d_x = \frac{D_x}{U_{di}} = \frac{fsqL_k I_d}{U_{di}} \tag{7.39}$$

is obtained.

With the aid of Eqs. (7.2), (7.7), and (7.11), this equation can be converted into

$$d_x = \frac{I_d}{2\sqrt{2}I_k}. \tag{7.40}$$

Thus the relationship

$$\cos(\alpha+u) = \cos \alpha - 2d_x \tag{7.41}$$

is obtained for the overlap anlge u from Eq. (7.17) and

$$\cos u_0 = 1 - 2d_x \tag{7.42}$$

is obtained from Eq. (7.18) for the initial overlap angle u_0.

The mean value of the direct voltage U_d for a control angle $\alpha=0$ can be expressed as

$$U_d = U_{di} - D_x = U_{di}(1-d_x) = U_{di}\frac{1+\cos u_0}{2}. \tag{7.43}$$

With a non-zero control angle α the mean value of the dc voltage $U_{d\alpha}$ is calculated from

$$U_{d\alpha} = U_{di}\cos \alpha - D_x = U_{di}(\cos \alpha - d_x) = U_{di}\frac{\cos \alpha + \cos(\alpha+u)}{2}. \tag{7.44}$$

Resistive Direct Voltage Regulation. The resistive direct voltage regulation D_r can be calculated from the resistances R_k in the commutating circuit. The following applies

$$D_r = s R_k I_d \tag{7.45}$$

with converters of high power ratings, the inductive direct voltage regulation dominates owing to the comparatively large short-circuit voltage of several percent with such installations.

On-state Voltage. A further voltage drop is caused by the on-state voltage of the converter valves. Whereas with mercury-arc rectifiers a voltage drop of 20 V to 50 V occurs due to the arc the on-state voltage drop with semiconductor diodes and thyristors is only 1 V to 3 V per device.

Figure 7.7 shows the load characteristic of a fully modulated line-commutated rectifier. The on-state voltage U_F of the valves can be assumed to be approximately independent of the current. The resistive direct voltage regulation D_r which can also contain the resistive valve portion and the inductive direct voltage regulation D_x vary directly with the current.

If the output dc voltage with phase angle control is reduced by increasing the control angle α, the load characteristic is displaced downwards dependent upon the control angle α (Fig. 7.8).

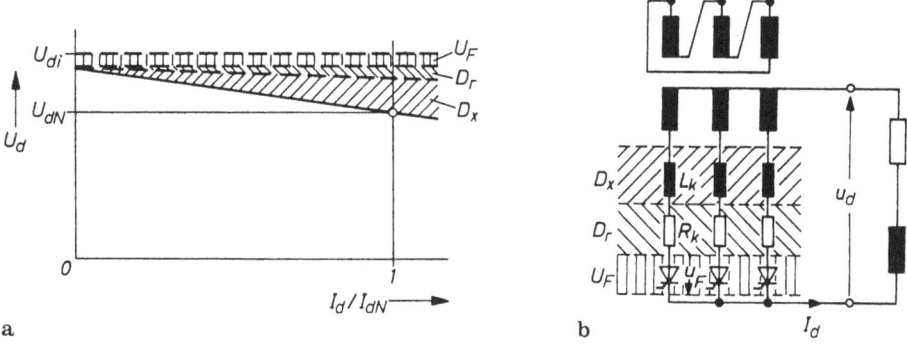

Fig. 7.7a,b. Load characteristic of fully modulated line-commutated rectifier. **a** Linearized load characteristic for $\alpha = 0$; **b** connection

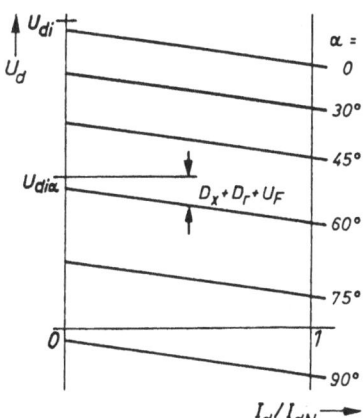

Fig. 7.8. Load characteristic of a line-commutated rectifier with control angle α as parameter

During operation in the rectifier mode the voltage drops cause a reduction in the output voltage U_d. An increase in U_d is produced during operation in the inverter mode (see Fig. 10.4).

Margin Angle and Angle of Advance.

The margin angle γ necessary during operation in the inverter mode still has to be calculated taking the commutation overlap time into consideration. In Fig. 7.5 operation in the inverter mode is illustrated with $\alpha = 140°$.

From Fig. 7.4 the relationship

$$I_d = \sqrt{2}I_k[\cos \gamma - \cos(u+\gamma)] \tag{7.46}$$

can be derived which can be converted into

$$\cos(u+\gamma) = \cos \gamma - \frac{I_d}{\sqrt{2}I_k} = \cos \gamma - 2d_x. \tag{7.47}$$

The angle $u+\gamma$ is also designated the angle of advance β. Taking overloads I_d/I_{dN} and supply voltage dips U/U_N into account the required angle of advance β can be calculated from

$$\cos \beta = \cos(u+\gamma) = \cos \gamma - 2d_x \frac{I_d}{I_{dN}} \frac{U_N}{U}. \tag{7.48}$$

During operation in the inverter mode the margin angle γ indicates the period of reverse blocking voltage across the valve. This is also called the hold-off time (see Sect. 8.1.2). The margin angle γ must be larger than the circuit-commutated recovery time t_q of the converter valves. If it falls below this value, a short circuit occurs between the relieving ac current phases. This is called commutation failure.

7.1.5 Converter Connections

Converter connections are standardized (in Germany in the DIN standards 41761). Table 7.1 lists the nomenclature and characteristics of basic converter connections. Two types of connections are distinguished: single-way and double-way connections.

With single-way connections the terminals on the ac side of the converter assembly and hence the converter transformer windings on the valve side or, if a converter transformer is not present, then the connections of the ac system carry a uni-directional current. They are each connected to only one principal arm.

With double-way connections the terminals on the ac side of the converter assembly and hence the converter transformer windings on the valve side (if there is no converter transformer, then the terminals of the ac system) carry an ac current.

It is feature of all center tap connections that the terminals of the converter arms with the same polarity are connected together and form one dc terminal while the second dc terminal is formed by the center tap of the ac system. Center tap connections have the code letter M. The commutation number q and pulse number p equal the number of converter arms.

Table 7.1. Nomenclature and notation for fundamental converter connections

Connection Type	Connection family		Individual connection		Example
	Nomenclature	Code letter	Characteristic number	Nomenclature	
Single-way connection	Center tap connection	M		p-pulse center tap connection	
Double-way connection	Bridge connection	B	Pulse number p^a	p-pulse bridge connection	

a If the pulse numbers is variable for the type of operation (e.g. asymmetric control), it holds for full modulation.

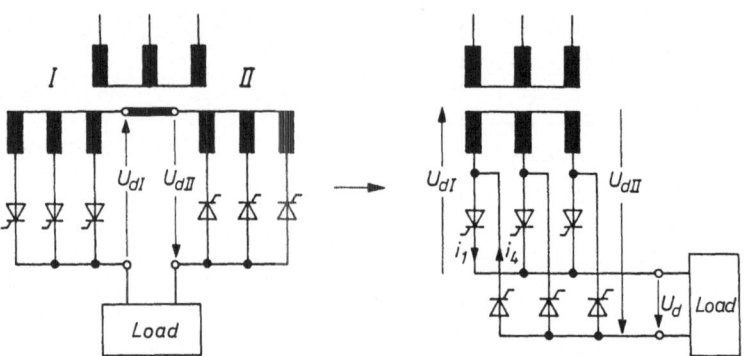

Fig. 7.9. Bridge connection as series connection of two M3 converters

Figure 7.9 illustrates how a bridge connection can be considered two converters in center tap connection connected in series. By connecting in series the two partial converters I and II in three-pulse center tap connection (M3) shown in the figure where converter II has a negative dc voltage U_{dII} due to reversal of the polarity of the thyristors, the six-pulse bridge connection illustrated on the right of the figure is created. With this connection the secondary winding of the converter transformer can be used for both parts of the converter. The resulting dc voltage is

$$u_d = u_{dI} + u_{dII}. \tag{7.49}$$

A six-pulse ripple is produced because the three-pulse partial voltages u_{dI} and u_{dII} are mutually displaced by 180°.

Figure 7.10 shows the voltage and current waveforms with the six-pulse bridge connection (also three-phase ac bridge connection) when fully modulated (control angle $\alpha = 0$). The dc voltage u_d is produced by adding the partial voltages u_{dI} and u_{dII}. In the converter transformer as well as in the secondary winding the current i is a pure ac current. Ignoring the overlap angle it consists of 120° rectangular blocks of alternating polarity with 60° vacancies.

Figure 7.11 shows the voltage and current waveforms with a control angle $\alpha = 90°$. The mean value of the partial dc voltages U_{dI} and U_{dII} is zero in this operating condition as is the mean value of the total dc voltage U_d. At $\alpha = 90°$ the dc voltage u_d has the maximum ripple w_i which is defined as

$$w_i = \frac{\sqrt{\sum U_{vi}^2}}{U_{di}} \tag{7.50}$$

where U_{vi} is the rms value of the vth harmonic.

The six-pulse bridge connection is the connection most frequently employed for line-commutated converters. Its principal advantage over the center tap connection is better utilization of the converter transformer. Whereas with center tap connections the converter transformer secondary has current blocks of only one direction flowing in it (single-way) with bridge connections there is no dc current component in the secondary windings of the transformer i.e. current flows in both directions (double-way).

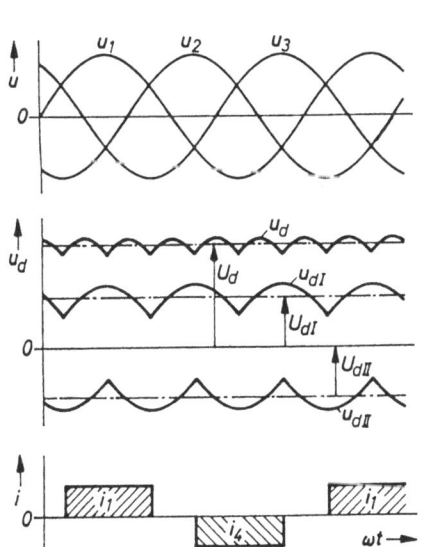

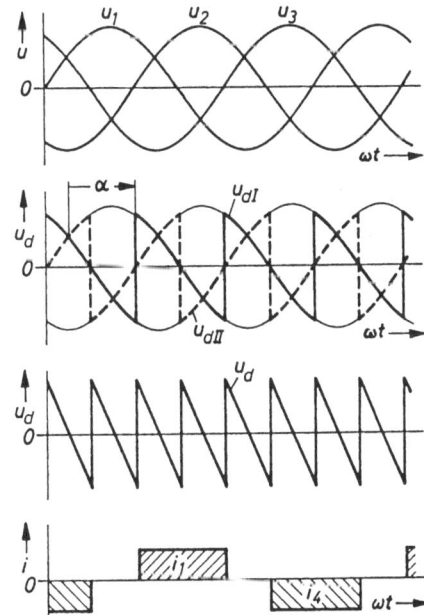

Fig. 7.10. Voltage and current waveforms with six-pulse bridge connection (three-phase ac bridge connection). Control angle $\alpha=0$

Fig. 7.11. Voltage and current waveforms with six-pulse bridge connection (three-phase ac bridge connection). Control angle $\alpha=90°$

Valve Voltage. The voltage stress on the converter valves is equal to the line voltage between mutually commutating phases. The maximum voltage occurring across the valve is

$$\hat{u}_A = \sqrt{2}U_k. \tag{7.51}$$

Using Eq. (7.3) or (7.7) this can be converted into

$$\hat{u}_A = \sqrt{2}\cdot 2U \sin\frac{\pi}{q} = \frac{2\pi}{s\,q}U_{di}. \tag{7.52}$$

For bridge connections $s=2$, while for center tap connections $s=1$. With bridge connections the converter valves are stressed with only half the reverse voltage for the same no-load dc voltage U_{di}. Nevertheless, owing to the two partial converters being connected in series, twice the number of valves are needed.

With non-zero phase angle control the voltage across the converter valve jumps, after the valve current is extinguished, to an instantaneous value that is called the initial inverse voltage and can be calculated according to

$$u_{ii} = \sqrt{2}U_k\sin(\alpha+u). \tag{7.53}$$

The peak value of the initial inverse voltage occurs when $\sin(\alpha+u)=1$, $u_{ii} = \sqrt{2}U_k$. Since the valve voltage approaches this calculated value of the initial inverse voltage with a damped oscillation, a snubber circuit is necessary (see Sect. 4.1).

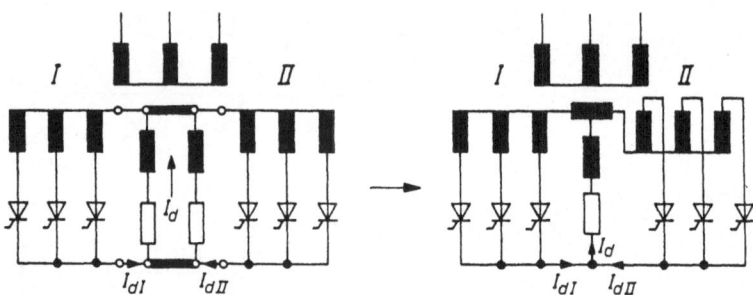

Fig. 7.12. Interphase transformer connection as parallel connection of two M3 converters

Interphase Transformer Connection. The parallel connection of two converters in three-pulse center tap connection is illustrated in Fig. 7.12. An effective six-pulse connection is desired. To achieve this one partial converter must operate with a phase displacement of 180° with respect to the other, which can be achieved by two secondary windings of the converter transformer being phase-displaced by 180°. Parallel operation of the two partial converters I and II is then, however, only possible via chokes which absorb the differential voltage and limit the equalizing current of three times the supply frequency caused by the different momentary values of the partial voltages u_{dI} and U_{dII}. As the chokes are magnetized by the dc current they have to be sized accordingly. If the two chokes are coupled in what is called an interphase transformer, the dc component of the magnetizing current is eliminated and the size of the magnetic component reduced.

The dc load current is the sum of the currents I_{dI} and I_{dII}:

$$I_d = I_{dI} + I_{dII}. \tag{7.54}$$

The interphase transformer connection is employed for electro-chemical rectifiers where a relatively low dc voltage and high dc current are required (see Sect. 13.1.5). With the same voltage stress on the valves this connection supplies double the dc current but only half the dc voltage of a bridge connection.

In Fig. 7.13 the most important converter connections are listed with their characteristic values. These are the two-pulse and three-pulse center tap connections M 2 and M 3, the double-star connection with interphase transformer M 3.2, and the two-pulse and six-pulse bridge connections B 2 and B 6.

Center Tap Connections. The top of Table 7.2 shows in detail the characteristics of simple center tap connections (M1, M2, M3, M6). The table contains nomenclature, code letters, symbols, circuit details, voltage-current waveforms, and electrical values. These values are ideal values for sinusoidal input voltages, neglecting the voltage drops in the converter and with ideally smoothed dc current (except for the single-pulse center tap connection M1). The terminal notations help with the coordination of the converter transformer and converter assembly. The symbols + and − indicate the dc terminals. These are valid for the conducting direction of the valve. In the rectifier mode the + has a positive potential with respect to the −.

	Converter connection	Transformer connection	Direct voltage (at α = 0)		Valve voltage		Valve current			Transformer	AC supply
			$\dfrac{U_{di}}{U_2}$	$w_i = \dfrac{\sqrt{\Sigma U_{vi}^2}}{U_{di}}$	$\dfrac{\hat{u}_A}{U_2}$	$\dfrac{\hat{u}_A}{U_{di}}$	Angle of current flow	$\dfrac{I_{Aov}}{I_d}$	$\dfrac{I_{Aeff}}{I_d}$	$\dfrac{S_{Tr}}{U_{di}\cdot I_d}$	$\dfrac{S_{Li}}{U_{di}\cdot I_d}$
Centertap connections	Two pulse centertap connection M2 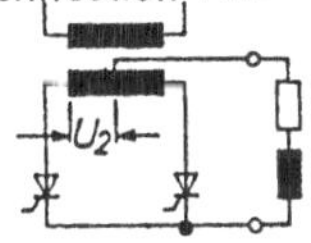	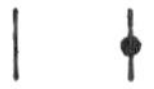	$\dfrac{2}{\pi}\sqrt{2}$ 0.90	2-pulse 0.482	$2\sqrt{2}$ 2.83	π 3.14	180	$\dfrac{1}{2}$ 0.5	$\dfrac{1}{\sqrt{2}}$ 0.707	1.34	1.11
	Three-pulse centertap connection M3 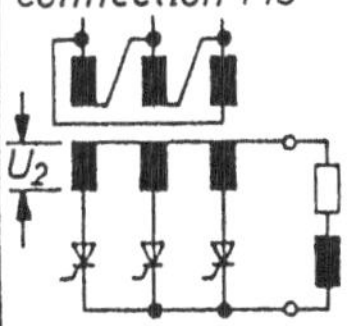	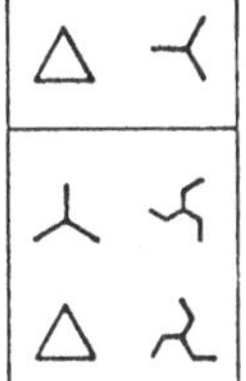	$\dfrac{3\sqrt{6}}{2\pi}$ 1.17	3-pulse 0.183	$\sqrt{3}\cdot\sqrt{2}$ 2.45	$\dfrac{2\pi}{3}$ 2.09	120	$\dfrac{1}{3}$ 0.333	$\dfrac{1}{\sqrt{3}}$ 0.577	1.35 1.46	1.21
	Double star connection with interphase transformer M3.2 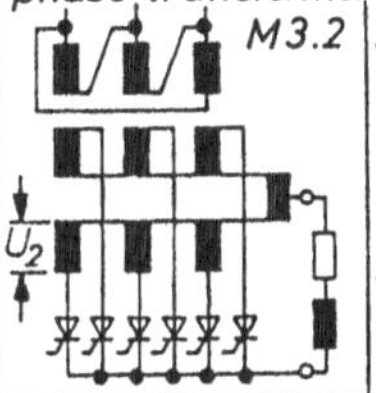	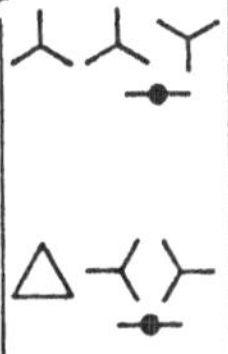	$\dfrac{3\sqrt{6}}{2\pi}$ 1.17	6-pulse 0.042	$\sqrt{3}\cdot\sqrt{2}$ 2.45	$\dfrac{2\pi}{3}$ 2.09	120	$\dfrac{1}{6}$ 0.167	$\dfrac{1}{2\sqrt{3}}$ 0.289	1.26	1.05
Bridge connections	Two-pulse bridge connection B2 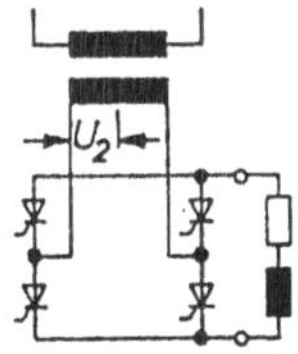	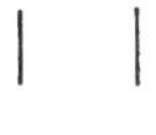	$\dfrac{4}{\pi}\sqrt{2}$ 1.80	2-pulse 0.482	$2\sqrt{2}$ 2.83	$\dfrac{\pi}{2}$ 1.57	180	$\dfrac{1}{2}$ 0.5	$\dfrac{1}{\sqrt{2}}$ 0.707	1.11	1.11
	Six-pulse bridge connection B6 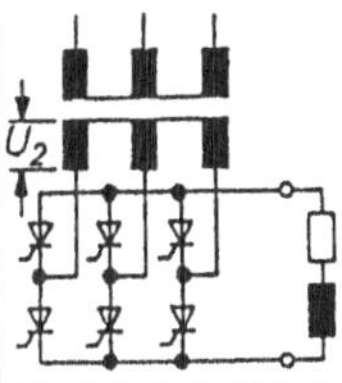	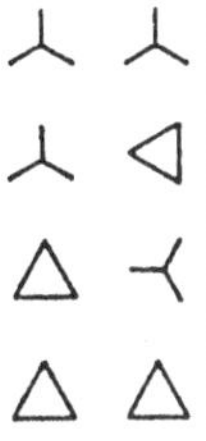	$\dfrac{3\sqrt{6}}{\pi}$ 2.34	6-pulse 0.042	$\sqrt{3}\cdot\sqrt{2}$ 2.45	$\dfrac{\pi}{3}$ 1.05	120	$\dfrac{1}{3}$ 0.333	$\dfrac{1}{\sqrt{3}}$ 0.577	1.05	1.05

Fig. 7.13. Converter connections

Table 7.2. Simple and compound center tap connections

Converter connection		Phasor diagram of the transformer voltages		Principal circuit of the converter assembly	Pulse number and commutation					DC voltage U_{di}			
Nomenclature DIN 41761	Code letter	line-side	valve-side		p	q	s	g	δ	Waveform	$\dfrac{U_{di}}{U_{v0}}$	$\dfrac{d_{xt}}{u_{xt}}$	w_{Ui} in %
Single-pulse center tap connection (single-way connection)	M1				1	1	1	1	1		0.450 $\dfrac{\sqrt{2}}{\pi}$	—	121
Two-pulse center tap connection (single-phase center tap connection)	M2				2	2	1	1	1		0.450 $\dfrac{\sqrt{2}}{\pi}$	0.707 $\dfrac{1}{\sqrt{2}}$	48.3
Three-pulse center tap connection (star connection)	M3				3	3	1	1	1		0.675 $\dfrac{3}{\pi\sqrt{2}}$	0.866 $\dfrac{\sqrt{3}}{2}$	18.3
Six-pulse center tap connection (double star connection)	M6 or M6/0 or M6/30				6	6	1	1	1		1.350 $\dfrac{3\sqrt{2}}{\pi}$	1.500 : 0.500	4.2
Double three-phase center tap connection (parallel) (double star connection with interphase transformer)	M 3.2/0 or M 3.2/30				6	3	1	2	1		0.675 $\dfrac{3}{\pi\sqrt{2}}$	0.500 $\dfrac{1}{2}$	4.2
Threefold two-pulse center tap connection (parallel)	M 2.3/30				6	2	1	3	1		0.450 $\dfrac{\sqrt{2}}{\pi}$	0.750 : 0.500	4.2

a DC premagnetization in the transformer.

The meaning of the symbols in Table 7.2 (and Tables 7.3, 7.5, and 7.6, which follow) is as follows (see also Publication 146 of the IEC)

p   pulse number
q   commutating number
s   number of series connected commutating groups
g   number of sets of commutating groups between which I_{dN} is divided
δ number of commutating groups commutating simultaneously per reactor
U_{di0} ideal no-load direct voltage (at $\alpha=0$)
U_{v0} rms phase-to-phase voltage on cellside of dc transformer at no load
d_{xt} inductive direct voltage regulation due to converter transformer in percent of U_{di0}

Voltage at the converter arm		Arm current $I_p =$ valve-side conductor current I_v					Line-side conductor current			Line side apparent power
$\dfrac{U_{im}}{U_{di}}$	$\dfrac{U_{i0m}}{U_{di}}$	Waveform	$\dfrac{I_{peff}}{I_d}$	$\dfrac{I_{pmax}}{I_d}$	$\dfrac{I_{pav}}{I_d}$	$\dfrac{I_v}{I_d}$	in $\curlywedge$-connection Waveform	in $\triangle$-connection Waveform	$\dfrac{I_{Li}}{I_d}$	$\dfrac{S_{Li}}{U_{di}I_d}$
3.142 π	3.142 π		1.571 $\dfrac{\pi}{2}$	3.142 π	1.000	1.571 $\dfrac{\pi}{2}$			1.211 $\sqrt{\dfrac{\pi^2}{4}-1}$	2.69 $\dfrac{\pi}{2}\sqrt{\dfrac{\pi^2}{2}-2}$
3.142 π	3.142 π		0.707 $\dfrac{1}{\sqrt{2}}$	1.000	0.500 $\dfrac{1}{2}$	0.707 $\dfrac{1}{\sqrt{2}}$			0.500 $\dfrac{1}{2}$	1.11 $\dfrac{\pi}{2\sqrt{2}}$
2.094 $\dfrac{2\pi}{3}$	2.094 $\dfrac{2\pi}{3}$		0.577 $\dfrac{1}{\sqrt{3}}$	1.000	0.333 $\dfrac{1}{3}$	0.577 $\dfrac{1}{\sqrt{3}}$			0.472 $\dfrac{\sqrt{2}}{3}$	1.21 $\dfrac{2\pi}{3\sqrt{3}}$
2.094 $\dfrac{2\pi}{3}$	2.094 $\dfrac{2\pi}{3}$		0.408 $\dfrac{1}{\sqrt{6}}$	1.000	0.167 $\dfrac{1}{6}$	0.408 $\dfrac{1}{\sqrt{6}}$			0.816 $\sqrt{\dfrac{2}{3}}$	1.05 $\dfrac{\pi}{3}$
2.094 $\dfrac{2\pi}{3}$	2.418 $\dfrac{4\pi}{3\sqrt{3}}$		0.289 $\dfrac{1}{2\sqrt{3}}$	0.500 $\dfrac{1}{2}$	0.167 $\dfrac{1}{6}$	0.289 $\dfrac{1}{2\sqrt{3}}$			0.408 $\dfrac{1}{\sqrt{6}}$	1.05 $\dfrac{\pi}{3}$
3.142 π	3.142 π		0.236 $\dfrac{1}{3\sqrt{2}}$	0.333 $\dfrac{1}{3}$	0.167 $\dfrac{1}{6}$	0.236 $\dfrac{1}{3\sqrt{2}}$			0.272 $\dfrac{1}{3}\sqrt{\dfrac{2}{3}}$	1.05 $\dfrac{\pi}{3}$

u_{xt}	inductive component of the relative short-circuit voltage u_{kt} of the converter transformer
w_{Ui}	ideal ac voltage content of the dc voltage (at $\alpha = 0°$)
U_{im}	ideal crest voltage of an arm
U_{i0m}	ideal crest no-load voltage of an arm with non-operating interphase transformer
I_d	direct current (mean)
I_p	arm current (rms)
I_{pmax}	arm current (peak)
I_{pav}	arm current (mean)
I_v	current on cell side of transformer (rms)
I_{Li}	ideal current on line side (rms)
S_{Li}	ideal apparent power on line side

The given values are ideal values. They hold for a sinusoidal terminal voltage in the absence of voltage drops in the converter with inductive load (connection M1 with resistive load). It is assumed that there is full modulation ($\alpha=0$) and a unit turns ratio of the converter transformer.

Center tap connections utilize poorly the converter transformer. The single-pulse center tap connection M1 is used only at low powers (because of inefficient copper use, the presence of a dc magnetizing current in the transformer and the high ripple content of the dc voltage). With the three-pulse center tap connection the transformer secondary is usually connected in zig-zag to prevent residual ampere turns for every transformer leg. The six-pulse center tap connection M6 used to be applied with mercury arc valves.

Parallel Connection. Basic converter connections may be connected in parallel to increase the dc output. If the pulse number is also to be increased, the partial

Table 7.3. Simple and compound bridge connections

Converter connection Nomenclature DIN 41761	Code letter	Phasor diagram of the transformer voltages line-side	valve-side	Principal connection layout of the converter assembly	Pulse number and commutation p q s g δ	D C voltage U_{di} Waveform	$\dfrac{U_{di}}{U_{v0}}$	$\dfrac{d_{xt}}{u_{xt}}$	W_{ui} in %
Two-pulse bridge connection (single-phase bridge connection)	B 2				2 2 2 1 2		0.900 $\dfrac{2\sqrt{2}}{\pi}$	0.707 $\dfrac{1}{\sqrt{2}}$	48.3
Six-pulse bridge connection (three-phase bridge connection)	B 6				6 3 2 1 1		1.350 $\dfrac{3\sqrt{2}}{\pi}$	0.500 $\dfrac{1}{2}$	4.2
Double six-pulse bridge connection (parallel) with full modulation. Connection angle of the twelve-pulse unit=15°	B 6.2/15				12 3 2 2 1		0.52 ⋮ 0.26 0.518 1.350 $\dfrac{3\sqrt{2}}{\pi}$ 0.52 ⋮ 0.26		1.03
Double six-pulse bridge connection (in series). Connection angle of the twelve-pulse unit=15°	B 6.2S15				12 3 4 1 1		0.52 ⋮ 0.26 2.701 $\dfrac{6\sqrt{2}}{\pi}$ 0.518		1.03

connections must have different relative phase angles. The commutating groups of the partial connection can be decoupled by means of interphase transformers.

If the partial connections connected in parallel on the dc side have the same relative phase angles, then the number of parallel connections is stated in front of the code letter of the basic converter connection e.g. 2M3. With different relative phase angles the number of parallel connections is stated after the pulse number separated by a point, so that the product of these numbers is the resultant pulse number of the composite converter connection e.g. M3.2. The bottom of Table 7.2 shows characteristics of composite center tap parallel connections.

The double three-pulse center tap connection M3.2 is used to feed dc current into galvanic and electrolysis plants with a power range from several hundred kW to over 50 MW where high dc currents with moderate dc voltages (less than 700 V) are necessary. The triple two-pulse center tap connection M2.3 connected in

Voltage at the converter arm		Arm current I_p				Valve-side conductor current I_v		Line-side conductor current		Line-side apparent power
$\frac{U_{im}}{U_{di}}$	$\frac{U_{i0m}}{U_{di}}$	Waveform	$\frac{I_{peff}}{I_d}$	$\frac{I_{pmax}}{I_d}$	$\frac{I_{pav}}{I_d}$	Waveform	$\frac{I_v}{I_d}$	Waveform	$\frac{I_{Li}}{I_d}$	$\frac{S_{Li}}{U_{di} I_d}$
1.571	1.571		0.707	1.000	0.500		1.000		1.000	1.11
$\frac{\pi}{2}$	$\frac{\pi}{2}$		$\frac{1}{\sqrt{2}}$		$\frac{1}{2}$					$\frac{\pi}{2\sqrt{2}}$
1.047	1.047		0.577	1.000	0.333		0.816		0.816	1.05
$\frac{\pi}{3}$	$\frac{\pi}{3}$		$\frac{1}{\sqrt{3}}$		$\frac{1}{3}$		$\sqrt{\frac{2}{3}}$		$\sqrt{\frac{2}{3}}$	$\frac{\pi}{3}$
1.047	1.047									
$\frac{\pi}{3}$	$\frac{\pi}{3}$									
			0.289	0.500	0.167		0.408		0.789	1.01
1.047	1.170		$\frac{1}{2\sqrt{3}}$	$\frac{1}{2}$	$\frac{1}{6}$		$\frac{1}{\sqrt{6}}$		$\frac{1+\sqrt{3}}{2\sqrt{3}}$	$\pi\frac{1+\sqrt{3}}{6\sqrt{2}}$
$\frac{\pi}{3}$										
0.524	0.524		0.577	1.000	0.333		0.816		1.578	1.01
$\frac{\pi}{6}$	$\frac{\pi}{6}$		$\frac{1}{\sqrt{3}}$		$\frac{1}{3}$		$\sqrt{\frac{2}{3}}$		$\frac{1+\sqrt{3}}{\sqrt{3}}$	$\pi\frac{1+\sqrt{3}}{6\sqrt{2}}$

zig-zag is seldom applied because of the large combined power rating of the rectifier and interphase transformers.

Bridge Connections. Bridge connections are double-way connections. They comprise only arm pairs constructed from two converter principal arms with opposite conduction directions at a center terminal. Each ac side connection is made at the center terminal of an arm pair. Terminals of common polarity of the arm pairs are connected together and thus constitute a dc terminal.

The commutation number q is equal to the number of dc terminals and hence to the number of arm pairs. Bridge connections having an even commutation number q have a pulse number p equal to q. If q is odd, then $p = 2q$.

The top of Table 7.3 lists the characteristics of simple bridge connections (B2 and B6).

The two-pulse bridge connection B2 finds application in the lower power range (because of the relatively large dc voltage ripple) and also in traction applications having single-phase ac overhead lines up to several MW. The six-pulse bridge connection B6 is the standard connection for diode and thyristor converters. It is used in the power range from several kW to several hundred MW (dc drives, rectifier subsystems, electrolysis rectifiers, high voltage dc conversion).

The middle of Table 7.3 shows the characteristics of the double six-pulse bridge connection (parallel). The double six-pulse bridge connection (parallel) B6.2/15 has a twelve-pulse ripple factor in the dc voltage and low harmonics in the ac line current.

Series Connection. Basic converter connections are connected in series in order to increase the rated dc voltage. If the pulse number also has to be increased, the partial connections must also have different relative phase circuit angles. S denotes

Table 7.4. Formulae for line commutated converters

Direct voltage:

Ideal no-load direct voltage (at $\alpha = 0$): $U_{di} = s \dfrac{q}{\pi} \sqrt{2} U_s \sin \dfrac{\pi}{q} = s \dfrac{q}{\pi} \sqrt{2} \dfrac{U_{v0}}{2}$

U_s = valve-side star voltage (phase voltage), U_{v0} = communication voltage controlled ideal no-load direct voltage at delay angle α: $U_{di\alpha} = U_{di} \cos\alpha$ for resistive load:

$$U_{di} = U_{di} \cos\alpha \quad \text{für} \quad 0 \leq \alpha \leq \dfrac{\pi}{2} - \dfrac{\pi}{p}$$

$$U_{di\alpha} = U_{di} \dfrac{1 - \sin\left(\alpha - \dfrac{\pi}{p}\right)}{2 \sin \dfrac{\pi}{p}} \quad \text{for} \quad \dfrac{\pi}{2} - \dfrac{\pi}{p} \leq \alpha \leq \dfrac{\pi}{2} + \dfrac{\pi}{p}$$

for half-controllable connections: $U_{di\alpha} = \dfrac{U_{di}}{2}(1 + \cos\alpha)$

Ripple factor of the dc voltage:

ideal ac voltage component of the ordinal number v: $U_{vi} = \dfrac{\sqrt{2}}{v^2 - 1} U_{di}$

ideal ac voltage superimposedon the dc voltage: $U_{\ddot{u}i} = \sqrt{\Sigma U_{vi}^2}$

ideal ac voltage content (ideal ripple factor): $w_i = \dfrac{\sqrt{\Sigma U_{vi}^2}}{U_{di}}$

Table 7.4. (continued)

Commutation:

Commutation voltage: $U_{v0} = 2U_s \sin \dfrac{\pi}{q}$.

Commutation current: $i_k = \dfrac{1}{2L_k} \int \sqrt{2}U_{v0} \sin \omega t\, dt = \dfrac{\sqrt{2}U_{v0}}{2\omega L_k}(1 - \cos \omega t)$.

Angle of overlap U: $\cos(\alpha + u) = \cos\alpha - \dfrac{2\omega L_k I_d}{\sqrt{2}U_{v0}} = \cos\alpha - 2d_x$.

Angle of advance β with inverter mode: $\cos\beta = \cos(\gamma + u) = \cos\gamma - 2d_x$.

Initial overlap u_0: $\cos(\alpha + u_0) = 1 - 2d_x$.

Regulation:

Total direct voltage regulation (with $\alpha = $ const):

$U_{d0\alpha} - U_{d\alpha} = U_{dxL} + U_{dxt} + U_{dxb} + U_{dt} + n(U_{dv} - U_{dvo})$.

Total relative inductive direct voltage regulation:

$d_x = \dfrac{U_{dx}}{U_{di}} = \dfrac{\delta \cdot s \cdot q \cdot f \cdot L_k I_d}{g U_{di}} = \dfrac{I_d}{2\sqrt{2}I_k}$.

Nominal values: $\dfrac{d_x}{d_{xN}} = \dfrac{I_d}{I_{dN}}$.

Total relative resistive direkt voltage regulation: $d_r = \dfrac{R_k I_d}{U_{di}}$.

Valve voltage:

Ideal peak blocking voltage at the converter arm: $U_{im} = 2U_{v0} = \dfrac{2}{s}\dfrac{\pi}{q} U_{di}$.

U_{im} for non-operative interphase transformer (almost no-load): U_{i0m}.

Valve current:

Arm current: I_p

with good smoothing one has

for simple center tap connections: $I_{peff} = \dfrac{I_d}{\sqrt{p}}$　　$I_{pav} = \dfrac{I_d}{q}$　　$I_{pmax} = I_d$

for compound center tap connections: $I_{peff} = \dfrac{I_d}{g\sqrt{q}}$　　$I_{pav} = \dfrac{I_d}{gq}$　　$I_{pmax} = \dfrac{I_d}{g}$.

for bridge connections: $I_{peff} = \dfrac{I_d}{\sqrt{q}}$　　$I_{pav} = \dfrac{I_d}{q}$　　$I_{pmax} = I_d$.

$p = $ pulse number, $Q = $ commutating number, $g = $ number of commutating groups into which the dc current is divided

---

the dc side partial connections connected in series and is stated behind the code for the basic converter connections. The number of series-connections is also given.

The bottom of Table 7.3 shows for example the characteristics of the series double six-pulse bridge connection. The double six-pulse bridge connection (in series) B6.2S15 is used for high power dc drives and for HVDC transmission (twelve-pulse).

Table 7.4 lists the equations for line-commutated converters.

The direct voltage U_{di} can be calculated for all connections in accordance with Eq. (7.3). For the two-pulse bridge connection it should be noted that the phase voltage U_2 is half the secondary ac voltage. The valve voltage is obtained from Eq. (7.52).

The conduction angle of the valves is 180° with single-phase connections and 120° with three-phase connections. As the on-state losses in power semiconductors vary not only with the mean value I_{Aav} but also with the rms value I_{Arms} as large a conducting period as possible is expedient for economic utilization.

7.1.6 Converter Transformer

With most converters a transformer is placed between the single-phase or three-phase ac system and the converter valves. This transformer is called the converter transformer. Its connection is matched to the specified converter connection and the windings must be sized for the stresses and current waveforms during converter operation.

For a given connection the secondary voltage of the converter transformer determines the output dc voltage (see Eq. (7.3)). Moreover, the winding insulation of the primary and secondary windings acts as galvanic isolation. As a consequence of the switching functions of the valves the windings of the converter transformer carry non-sinusoidal currents. Thus, the rating of a converter transformer is increased compared to that of a transformer processing the same power with sinusoidal currents. With bridge connections pure ac currents also flow in the secondary winding of the converter transformer whereas with center tap connections the secondary windings are badly utilized as they carry current blocks in one direction only, i.e. are loaded with a dc component.

To increase the number of pulses and hence reduce the harmonics on the dc and ac sides a converter transformer can be used to increase the number of phases of the secondary winding or perform a phase displacement. Like mains transformers converter transformers can also be equipped with tap changers to match the voltage. Owing to the generally smaller currents the tap changers are generally applied on the primary side. Converter transformers are analyzed and rated according to the known laws of transformer construction. Its short circuit voltage u_{kt} depending on leakage reatances to a great extent determines the inductive direct voltage regulation D_x (see Eq. (7.38)) and hence the load characteristic of the converter. With interphase transformers both partial systems must have the same short circuit impedance voltage so that the ac current is evenly dividet between the two partial converters.

Transformer Rating. The rating of a converter transformer will now be calculated for two typical connections. This is done by determining the apparent power on the primary and secondary sides of the transformer. Half the sum of the primary and secondary apparent powers produces the transformer rating S_{Tr}. When this is referred to the ideal dc power $U_{di}I_d$ of the converter the determining equation is

$$\frac{S_{Tr}}{U_{di}I_d} = \frac{\frac{1}{2}\Sigma U_{Tr}\cdot I_{Tr}}{U_{di}I_d}. \qquad (7.55)$$

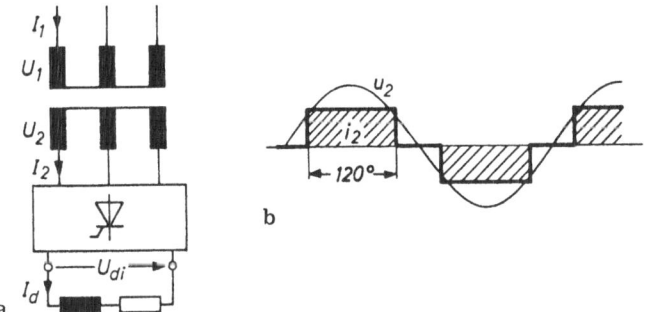

Fig. 7.14a,b. Calculation of rating of converter transformer (six-pulse bridge connection).
a Connection; **b** transformer current

With the six-pulse bridge connection the current illustrated in Fig. 7.14 flows in the secondary and primary windings of the converter transformer. The referred rating is obtained according to Eq. (7.55) as

$$\frac{S_{Tr}}{U_{di}I_d} = \frac{\frac{1}{2}(3U_1I_1 + 3U_2I_2)}{U_{di}I_d}. \tag{7.56}$$

U_1I_1 equals U_2I_2 owing to power balance. According to Eq. (7.3) for the six-pulse bridge connection

$$U_{di} = \frac{3\sqrt{6}}{\pi}U_2. \tag{7.57}$$

Moreover

$$I_2 = \sqrt{\frac{2}{3}}I_d \text{ or } I_d = \sqrt{\frac{3}{2}}I_2. \tag{7.58}$$

If these values are substituted in Eq. (7.56), the referred rating of the converter transformer in the six-pulse bridge connection becomes

$$\frac{S_{Tr}}{U_{di}I_d} = \frac{3U_2I_2}{\frac{3\sqrt{6}}{\pi}U_2 \cdot \sqrt{\frac{3}{2}}I_2} = \frac{\pi}{3} = 1.05. \tag{7.59}$$

So the rating is only 5% higher than for a transformer with sinusoidal currents. With the six-pulse bridge connection the same value 1.05 is obtained for the apparent power S of the three-phase ac system.

If the same calculation is performed for a converter transformer with the interphase transformer connection (Fig. 7.15), the referred transformer rating becomes

$$\frac{S_{Tr}}{U_{di}I_d} = \frac{\frac{1}{2}(3U_1I_1 + 2 \cdot 3U_2I_2)}{U_{di}I_d}. \tag{7.60}$$

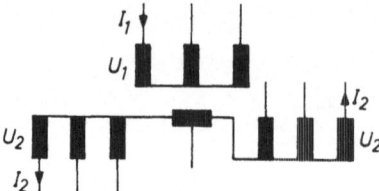

Fig. 7.15. Converter transformer with interphase transformer connection

With the interphase transformer connection

$$U_{di} = \frac{3\sqrt{6}}{2\pi} U_2 \text{ resp. } U_{di} = \frac{3\sqrt{6}}{2\pi} \cdot \frac{w_2}{w_1} U_1 \tag{7.61}$$

and

$$I_2 = \frac{I_d}{2\sqrt{3}} \text{ resp. } I_1 = \frac{w_2}{w_1} \sqrt{\frac{2}{3}} I_d \tag{7.62}$$

where w_1/w_2 is the transformation ratio of the transformer. When this is substituted Eq. (7.60) the referred transformer rating with the interphase transformer connection becomes

$$\frac{S_{Tr}}{U_{di}I_d} = \frac{1/2(3 \cdot 2\pi/3\sqrt{6} \cdot U_{di}\sqrt{2/3}I_d + 2 \cdot 3\ 2\pi/3\sqrt{6}\ U_{di} \cdot I_d/2\sqrt{3})}{U_{di}I_d} =$$

$$= \frac{1}{2}\left(\frac{\pi}{3} + \sqrt{2}\frac{\pi}{3}\right) = 1.26. \tag{7.63}$$

So the converter transformer is utilized less well with the interphase transformer connection than with the six-pulse bridge connection. The referred system apparent power becomes 1.05, as for the six-pulse bridge connection, because the same currents flow on the primary side of the transformer, and hence in the supply system, as with the six-pulse bridge connection.

Converter Transformer Connections. Table 7.5 shows converter transformer connections (terminals are denoted according to German standards VDE 0532).

When sizing converter transformers the particular operating conditions (harmonics, short-circuit capacity) must be taken into account. One assumes idealized rectangular currents. Skin effects due to harmonics are generally not considered in the layout. Also the positive effect of overlapping is ignored. In installations with closed-loop control the converter transformer must be so sized that a dc voltage reserve is available to counteract disturbances. Likewise with dual converters a voltage reserve is necessary to ensure inverter operation. The inductive component of the short-circuit voltage u_{xt} of the converter transformer proportionally determines the inductive dc voltage regulation and thereby the load characteristic of the line-commutated converter. To connect an interphase transformer there must be short-circuit voltages of equal magnitude available for both commutation groups so that the dc current is equally distributed.

With center tap connections rectified currents flow in the valveside windings (direct current, fundamental, and harmonic components). This produces a dc magnetization in each leg and results in a constant magnetic flux which coming

Table 7.5. Connections and electrical values of converter transformers

Converter connection Code letter DIN 41761	Connection group of the converter transformer (and, where applicable, of the interphase transf.)	Pulse number and commutation p	q	D.C. voltage $\frac{U_{di}}{U_{v0}}$	Valve-side conductor current $\frac{I_v}{I_d}$	Line-side apparent power $\frac{S_{Li}}{U_{di}\,I_{dN}}$	Transformer rating primary	secondary	total	Short circuit paths when measuring losses P_A	P_B	P_C	Load losses P_{vi}	Short-circuit paths when determining u_{kt}	$\frac{d_{xt}}{u_{xt}}$
M2	Iin	2	2	0.450	0.707	1.11	1.11	1.57	1.34	N-1	N-2		$\frac{P_A+P_B}{2}$	1-2	0.707
M3/0 M3/30 M3/60 M3/90	Dzn 0 / Yzn 5 / Dzn 6 / Yzn 11	3	3	0.675	0.577	1.21	1.21	1.71	1.46	1-2-3			$P_A+\frac{r_2}{3}I_d^2$	1-2-3	0.866
M3/30	Dyn 5						1.21	1.48	1.35						
M6/30	Dyn (5+11)	6	6	1.350	0.408	1.05	1.28	1.81	1.55	1-3-5	2-4-6		$1.5\,\frac{P_A+P_B}{2}$	Mean value of 1-3-5 and 2-4-6	1.5 to 0.5
M6/30 M6/0	Yzn(5+7) / Dzn(0+10)	6	6			1.05	1.05	1.79	1.42	1-2, 3-4, 5-6	2-3, 4-5, 6-1	Mean value of 1-3-5 and 2-4-6	$\frac{P_A+2P_B+3P_C}{6}$	2-4-6	0.5
M3.2/30 M3.2/0(+Sd)	Y yn0,yn6 / D yn5,yn11(In)	6	3	0.675	0.289	1.05	1.05	1.48	1.26	1-3-5	2-4-6		$\frac{P_A+P_B}{2}$	1-3-5 and 2-4-6	0.5
M2.3/30 (+Sd)	D in in (Zn)	6	2	0.450	0.236	1.05	1.11	1.57	1.34	1-3-5 N1-N2-N3	2-4-6 N1-N2-N3		$1.125\,\frac{P_A+P_B}{2}$	1-4, 2-5, 3-6	0.75 to 0.5
B2	Ii	2	2	0.900	1.000	1.11	1.11	1.11	1.11	1-2			P_A	1-2	0.707
B6/30 B6/0	Dd0 or Yy0 / Dy5 or Yd5	6	3	1.350	0.816	1.05	1.05	1.05	1.05	1-2-3			P_A	1-2-3	0.5
B6.2/15 (+Sd)	D y5 d6 or Y d5 y6(In)	12	3	1.350	0.408	1.01	1.012	1.05	1.03	1-3-5 Mean value of 1-3-5 and 2-4-6	2-4-6	1-3-5 2-4-6	$0.035(P_A+P_B)+0.930P_C$	Mean value of 1-3-5 and 2-4-6	0.52
B6.2/15 (+Sd)	Y r1(+15°) or D r0(+15°)(In In)	12	3	1.350	0.408	1.01	1.012	1.021	1.016	1-2 2-3 4-5 6-1	2-3 4-5 6-1	1-2 3-4 5-6	$1.34P_A-0.08P_B-0.27P_C$	1-3-5 and 2-4-6	0.26

from the yokes can only flow through the air or the transformer frame. Since of the dc magnetization the magnetizing current increases especially if the converter carries impulse currents greater than the nominal current. With unsmoothed dc current the dc flux can still be superimposed by an ac component. Line-side delta windings suppress this ac component since they represent a short-circuit for currents with a three-fold line frequency or multiples thereof. With three-pulse center tap connections the dc magnetization of the transformer legs can be avoided by using a valve-side zig-zag winding (see Table 7.2).

With bridge connections operating symmetrically, pure ac current also flows in the valve-side windings.

The volume and weight of a transformer are proportional to the nominal power raised to the 3/4 power. In fact,

$$V \sim W \sim (S_{Tr}/\omega)^{3/4}.$$

Converter transformers arranged separately in a converter assembly should carry name plate with the heading 'converter transformer'.

This should include the transformer data and some converter data. The power rating of the transformer need not be indicated. Converter transformers are tested according to preset test procedures (in Germany VDE 0532 or VDE 0550; also in Publication 146 of the IEC).

7.1.7 Reactive Power

Power Definitions. First, a few definitions will be given. With sinusoidal waveforms of voltage and current the conditions are simple to comprehend. The apparent power S according to

$$S = UI \tag{7.64}$$

equals the product of the rms values of voltage and current.

The phase displacement angle between the sinusoidal voltage and sinusoidal current is φ. The active power P is

$$P = UI \cos \varphi. \tag{7.65}$$

The quantity

$$Q = UI \sin \varphi \tag{7.66}$$

is called reactive power. The relationship between apparent, active, and reactive power is given by

$$S = \sqrt{P^2 + Q^2}. \tag{7.67}$$

The relationship

$$\frac{P}{S} = \cos \varphi \tag{7.68}$$

is called the active power factor or displacement power factor and the relationship

$$\frac{Q}{S} = \sin \varphi \qquad (7.69)$$

is called the reacitve power factor.

The powers so defined are calculated quantities. Only the momentary value $p(t)$ of the power has any physical meaning and according to

$$p(t) = ui = UI[\cos \varphi - \cos(2\omega t - \varphi)] \qquad (7.70)$$

oscillates about the mean value $UI \cos \varphi$ at twice the frequency of the supply system.

These definitions are not adequate when considering the reactive power of converters, because non-sinusoidal currents flow on the ac side. The definitions are therefore to accomodate the assumption of a sinusoidal voltage waveform and non-sinusoidal current waveform.

The apparent power S is again

$$S = UI = U\sqrt{I_1^2 + I_2^2 + I_3^2 + \ldots} \ . \qquad (7.71)$$

Only the fundamental frequency component of the current contributes to the active power

$$P = P_1 = UI_1 \cos \varphi_1. \qquad (7.72)$$

The reactive power Q is

$$Q = \sqrt{S^2 - P^2}. \qquad (7.73)$$

It contains two constituents, namely, the fundamental frequency reactive power

$$Q_1 = UI_1 \sin \varphi_1 \qquad (7.74)$$

and the harmonic distortion power

$$D = U\sqrt{I_2^2 + I_3^2 + \ldots} \ . \qquad (7.75)$$

The quantities thus defined can be illustrated by right-angled triangles which can be combined to form a tetrahedron (Fig. 7.16). S_i is the fundamental frequency apparent power defined by

$$S_1 = UI_1. \qquad (7.76)$$

The following definitions are important.

The power factor is

$$\lambda = \frac{P}{S} = g_i \cos \varphi_1; \qquad (7.77)$$

where

$$g_i = \frac{I_1}{I} \qquad (7.78)$$

is the fundamental frequency component of the current.

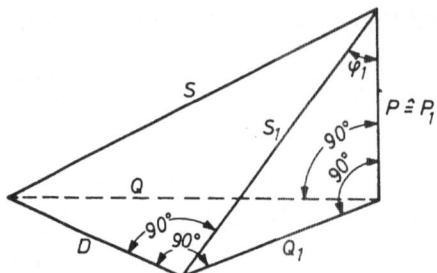

Fig. 7.16. Definition of apparent, active, and reactive power

cos φ_1 is called the fundamental frequency power factor or displacement factor.

With non-sinusoidal current the power factor λ is smaller than the displacement factor by the ratio g_i of the rms values of the fundamental component of the current to the total current. The power factor λ and fundamental frequency power factor cos φ_1 are often confused with one another. Only with sinusoidal current are they the same.

According to

$$S^2 = P^2 + Q_1^2 + D^2 \tag{7.79}$$

with sinusoidal voltage and non-sinusoidal current the apparent power S is made up of the active power P the fundamental frequency reactive power Q_1 and the distortion power D (see Fig. 7.16). The transient response of the power and other power definitions will be addressed in Sect. 11.2.

With line-commutated converters reactive power arises due to the displacement between the phase currents and the phase voltages [7.2]. There are two causes of this:

- Control reactive power arises due to displacement of the valve currents by the control angle α when phase angle control is employed and
- Commutation reactive power arises due to the overlap angle of the valve current caused by the inductances in the short-circuit path.

Control Reactive Power. The control reactive power can with idealized commutation i.e. without reactances in the commutation circuit be easily calculated. Figure 7.17 shows the supply current i_L with a six-pulse converter for example with the six-pulse bridge connection. With full modulation ($\alpha=0$) the phase voltage u and fundamental frequency component i_{1L}, of the current are in phase. The fundamental frequency reactive power Q_1 is 0. With phase angle control and control angle α the valve currents, and hence the supply currents, are retarded by the control angle α.

$$\varphi_1 = \alpha \tag{7.80}$$

and

$$\cos \varphi_1 = \cos \alpha \tag{7.81}$$

with completely smooth direct current I_d the fundamental frequency apparent power is

$$S_1 = U_{di} I_d. \tag{7.82}$$

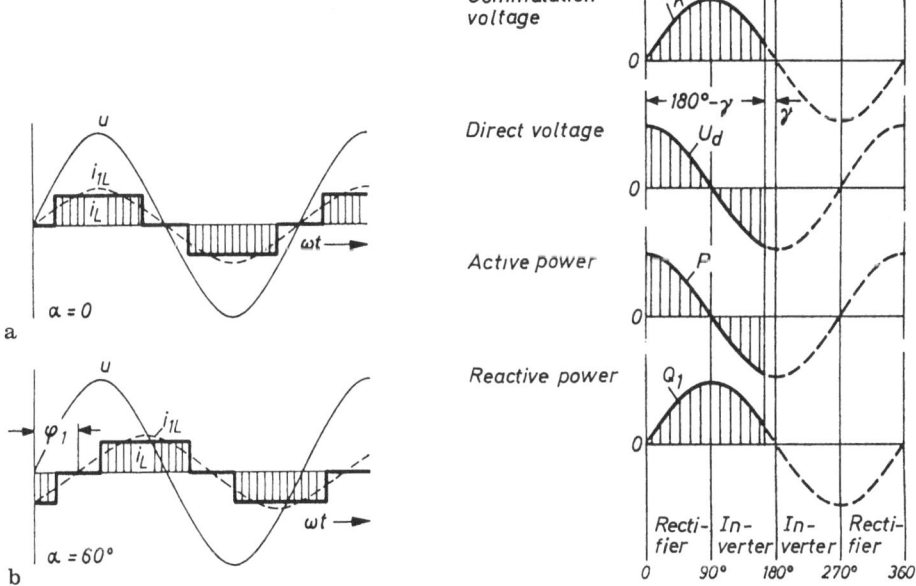

Fig. 7.17a,b. Supply current with six-pulse converter

Fig. 7.18. Control range of line-commutated converter with natural commutation

The active power P equals the fundamental frequency active power

$$P = P_1 = U_{di\alpha}I_d = U_{di}I_d\cos\alpha \qquad (7.83)$$

and the fundamental frequency reactive power is

$$Q_1 = U_{di}I_d\sin\alpha. \qquad (7.84)$$

According to these equations the displacement factor $\cos\varphi_1$ equals the cosine of the control angle α and the phase displacement angle φ_1 of the fundamental frequency component equals the control angle α.

Since according to Eq. (7.6) the direct voltage of a line-commutated converter is proportional to $\cos\alpha$, it must also be proportional to $\cos\varphi_1$. This fixed relationship between the output direct voltage U_d and the displacement factor $\cos\varphi_1$ is characteristic of line-commutated converters with natural commutation.

In Fig. 7.18 the relationships are plotted as a function of the control angle α (not as a funtion of the time ωt!). Natural commutation is only possible in the range from $\alpha = 0$ to $\alpha = 180° - \gamma$. The fundamental frequency reactive power Q_1 is inductive in this range. The converter behaves as an inductance on the single-phase or three-phase ac supply system. The range of capacitive reactive power from $\alpha = 180°$ to $\alpha = 360°$ can only be achieved with forced commutation (see Sect. 8.3.5).

Commutation Reactive Power. The commutation reactive power caused by the overlap angle u is not easy to calculate accurately. For this it would be necessary to analyse the valve current waveforms given by Eq. (7.35) or (7.36) by Fourier analysis. An approximate calculation can, however, be easily performed in the following way. Assuming that the amplitude of the fundamental frequency component of the valve current is not affected by the overlap angle u the following relationships apply

$$P = P_1 = U_{d\alpha}I_d = U_{di}I_d(\cos\alpha - d_x)$$ (7.85)

and

$$\cos\varphi_1 = \frac{P_1}{S_1} \approx \frac{U_{di}I_d(\cos\alpha - d_x)}{U_{di}I_d} = \cos\alpha - d_x.$$ (7.86)

These equations with the element $-d_x$ take into account not only the control reactive power but also the commutation reactive power. For operation in the inverter mode

$$\cos\varphi_1 \approx \cos\gamma - d_x.$$ (7.87)

Circle Diagram of the Reactive Power. The circle diagram of the referred reactive power taking Eqs. (7.84) and (7.6) into account is the semicircle illustrated in Fig. 7.19. Taking the overlap angle into account

$$\frac{Q_1}{U_{di}I_d} \approx \sqrt{1 - (\cos\alpha - d_x)^2}$$ (7.88)

and

$$\frac{U_d}{U_{di}} = \cos\alpha - d_x.$$ (7.89)

These equations also approximately describe a semicircle. The initial values achieved on operation in the rectifier or inverter mode are shown in Fig. 7.19 as a function of the initial overlap angle.

Sequence Control. As the reactive power Q is an additional load on the converter transformer and supply system but provides no contribution to the mean power transmission, it is undesirable. Figure 7.20 shows sequence control with which the two partial converters I and II are connected in series, only partial converter II being controlled.

This connection is also called a buck and boost connection. The output dc voltage $U_{d\alpha}$ is produced as the sum of the dc voltages of the two partial converters

$$U_{d\alpha} = U_{di\alpha I} + U_{di\alpha II} = \frac{U_{di}}{2}(1 + \cos\alpha_{II}).$$ (7.90)

Connection of one uncontrolled and one controlled converter in series produces a control range for the output dc voltage of $+100\%$ to 0. If two controlled partial converters are connected in series, these can be modulated one behind the other with $\alpha_I = 0$ and α_{II} between 0 and $180° - \gamma$. In this case the dc voltage $U_{d\alpha}$ is obtained from

$$U_{d\alpha} = U_{di\alpha I} + U_{di\alpha II} = \frac{U_{di}}{2}(\cos\alpha_I + \cos\alpha_{II}).$$ (7.91)

The modulation zone ranges from $+100\%$ to -100%.

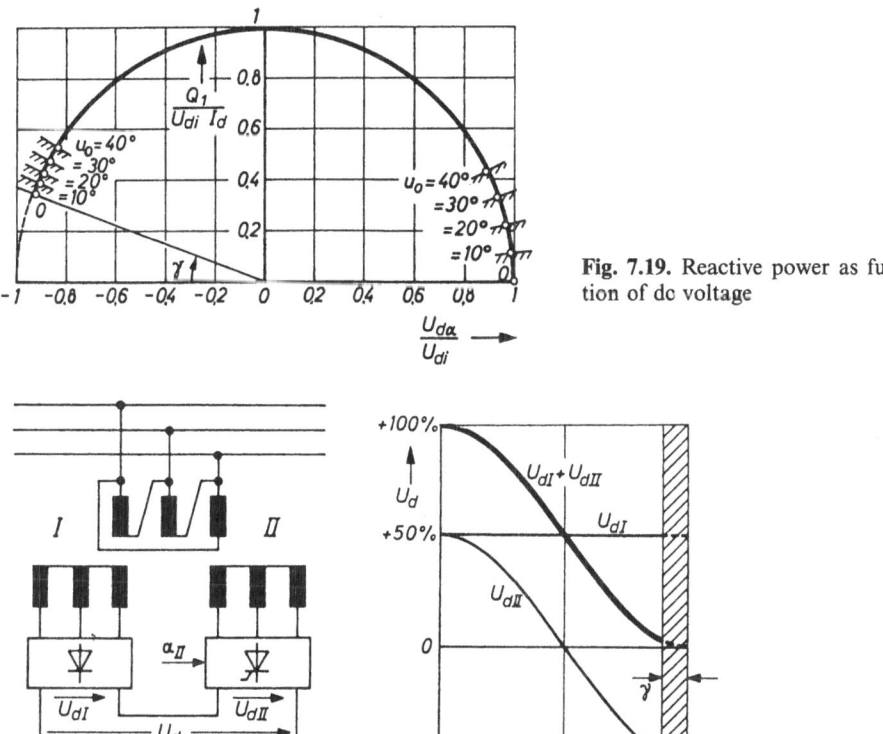

Fig. 7.19. Reactive power as function of dc voltage

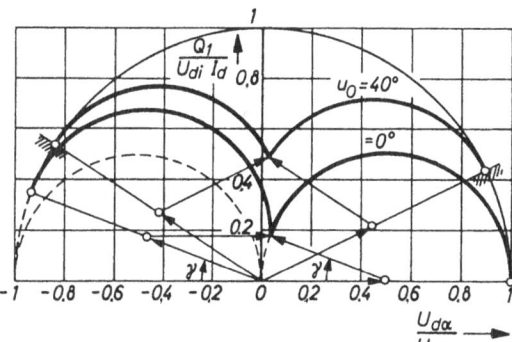

Fig. 7.20a,b. Control characteristic with sequence control. **a** Series connection of one controllable and one uncontrollable converter; **b** dc voltage across load

Fig. 7.21. Reduction in system reactive power with sequence control

Due to sequence control there is a considerable reduction in the reactive power in the single-phase or three-phase ac system. The referred fundamental frequency reactive power $Q_1/(U_{di}I_d)$ has a circle diagram composed of two pitch circles (Fig. 7.21). The construction of the pitch circles as a function of the initial overlap angle u_0 and the margin angle γ is given by auxiliary lines.

Instead of connecting two partial converters in series the same effect can also be achieved by asymmetrically controlling the valves of a bridge connection, because a bridge connection corresponds to the series connection of two partial center tap converters.

7.1.8 Half-controllable Connections

A connection is said to be half-controllable when only half the converter arms are controllable. Half-controllable bridge connections have a special significance [7.7, 7.16]. Their advantage lies for one thing in the smaller number of controlled valves needed. They create moreover a saving in reactive power Q as was described with sequence control.

Two-pulse Bridge Connection. Figure 7.22 shows the connection and voltage and current waveforms of the fully controllable two-pulse bridge connection. All converter valves carry 180° current blocks. The characteristic values of this connection are stated in Fig. 7.13 and in Table 7.3. With phase angle control the rectangular ac current i in the supply system is retarded by the control angle α.

Figure 7.23 shows the connection, voltage and current waveforms of the single-pole controllable two-pulse bridge connection which is also said to be symmetrically half-controllable. Here, valves 1' and 2' are uncontrolled diodes. The connection can be considered according to its method of operation as a series connection of one uncontrollable and one controllable partial converter in two-pulse center tap connection. With respect to the reactive power it therefore behaves like a buck and boost connection (see Figs. 7.20 and 7.21). The output direct voltage $U_{d\alpha}$ is thus given by Eq. (7.90).

In the single-phase ac supply the current blocks i are shortened by the control angle α

Figure 7.24 shows the connection, voltage and current waveforms with the so-called arm-pair half-controllable two-pulse bridge connection which is also called an asymmetrically half-controllable connection. With this connection valves 2 and 2' are uncontrollable diodes. With respect to the supply reactive power it behaves as a single-pole controllable two-pulse bridge connection. On the other hand owing to the diode freewheeling for the load the current blocks in the controllable valves 1 and 1' are shortened with phase angle control by the control angle α. Those current blocks in diodes 2 and 2' are correspondingly lengthened.

The asymmetrically half-controllable two-pulse bridge connection is employed in locomotives and motor coaches on the railways (see Sect. 13.1.6). The dc voltage U_d can be controlled between 100% and 0. Operation in the inverter mode is not possible so there is no regenerative braking for locomotives or motor coaches equipped in this way.

Six-pulse Bridge Connection. Half-controllable connections are also possible with multi-phase converters. Figure 7.25 shows the half-controllable six-pulse bridge connection with which valves 1', 2' and 3' of the upper partial converter II are uncontrollable diodes. The connection behaves as a series connection of one uncontrollable and one controllable partial converter in three-pulse center tap connection. A freewheeling branch can be provided for the load circuit. With phase angle control of the partial converter I the dc voltage u_d has three pulses.

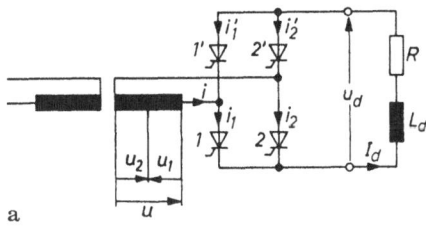

Fig. 7.22a,b. Two-pulse bridge connection (fully modulated)

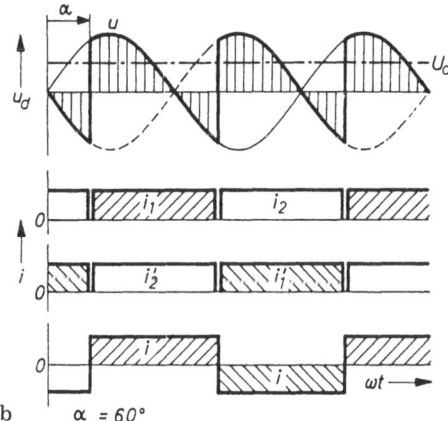

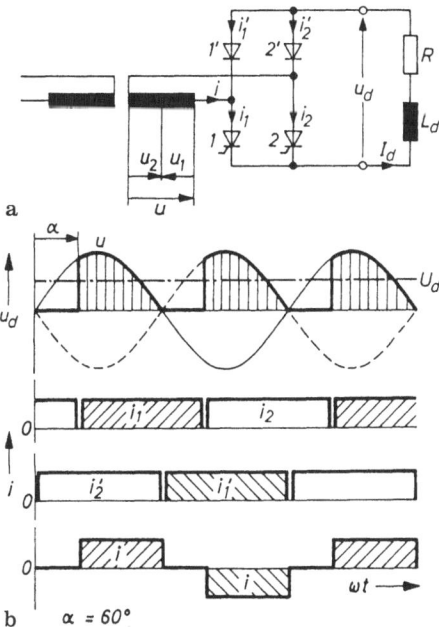

Fig. 7.23a,b. Single-pole controlled two-pulse bridge connection (symmetrically half-controllable)

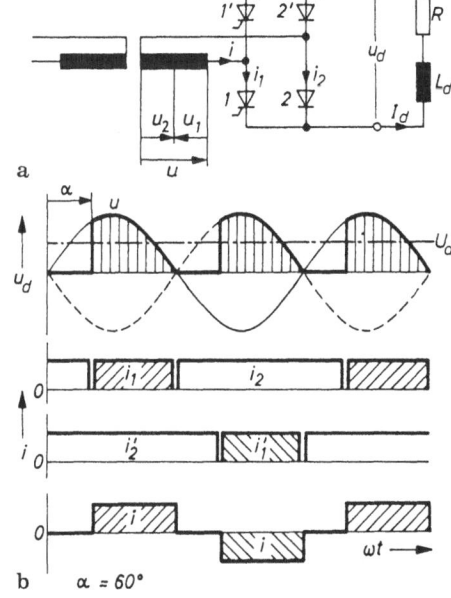

Fig. 7.2a,b. Arm-pair half-controllable two-pulse bridge connection (asymmetrically half-controllable)

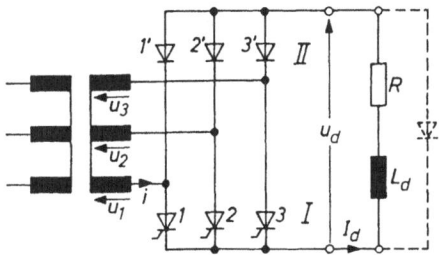

Fig. 7.25. Half-controllable six-pulse bridge connection

Table 7.6. Half-controllable bridge connections

Converter connection		Phasor diagram of the transformer voltages		Principal circuit of the converter assembly	Pulse number and commutation	D C voltage U_{di}				Voltage at the converter arm		Load type
Nomenclature DIN 41761	Code letter	line-side	valve-side		$p\ q\ s\ g\ \delta$ Waveform		$\dfrac{U_{di}}{U_{v0}}$	$\dfrac{d_{xt}}{u_{xt}}$	$\dfrac{w_{Ui}}{\text{in \%}}$	$\dfrac{U_{im}}{U_{di}}$	$\dfrac{U_{i0m}}{U_{di}}$	

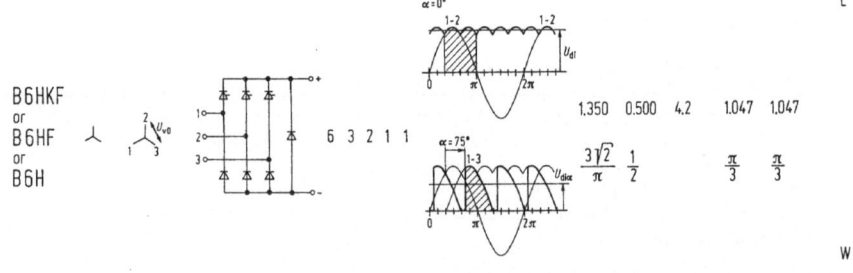

Single-pol controllable two-pulse bridge connection. The cathodes of the controllable valve form a d c terminal												L
B 2HK or **B 2H**						$\alpha=0°$						W
							0.900	0.707	48.3	1.571	1.571	
Arm-pair half controllable two-pulse bridge connection		U_{v0}			$2\ 2\ 2\ 1\ 2$	$\alpha=75°$	$\dfrac{2\sqrt{2}}{\pi}$	$\dfrac{1}{\sqrt{2}}$		$\dfrac{\pi}{2}$	$\dfrac{\pi}{2}$	L
B 2HZ or **B 2H**												W
Half controllable six-pulse bridge connection with free-wheeling arm						$\alpha=0°$						L
B6HKF or **B6HF** or **B6H**					$6\ 3\ 2\ 1\ 1$		1.350	0.500	4.2	1.047	1.047	
						$\alpha=75°$	$\dfrac{3\sqrt{2}}{\pi}$	$\dfrac{1}{2}$		$\dfrac{\pi}{3}$	$\dfrac{\pi}{3}$	
												W

Controll angle α	Arm current I_p Controllable principal arm				Non-controllable principal arm				Valve-side and line-side line current			Line-side apparent power
α	Waveform	$\dfrac{I_{peff}}{I_d}$	$\dfrac{I_{pmax}}{I_d}$	$\dfrac{I_{pav}}{I_d}$	Waveform	$\dfrac{I_{peff}}{I_d}$	$\dfrac{I_{pmax}}{I_d}$	$\dfrac{I_{pav}}{I_d}$	Waveform	$\dfrac{I_v}{I_d}$	$\dfrac{I_{Li}}{I_d}$	$\dfrac{S_{Li}}{U_{di}\,I_d}$
$0°\ldots\alpha_{max}$	$\alpha=75°$	0.707 $\dfrac{1}{\sqrt{2}}$	1.000			0.707 $\dfrac{1}{\sqrt{2}}$	1.000			$\sqrt{\dfrac{\pi-\alpha}{\pi}}$	$\sqrt{\dfrac{\pi-\alpha}{\pi}}$	
$0°$		0.785 $\dfrac{\pi}{4}$	1.571 $\dfrac{\pi}{2}$	0.500 $\dfrac{1}{2}$		0.785 $\dfrac{\pi}{4}$	1.571 $\dfrac{\pi}{2}$	0.500 $\dfrac{1}{2}$		1.111 $\dfrac{\pi}{2\sqrt{2}}$	1.111 $\dfrac{\pi}{2\sqrt{2}}$	
$90°$	$\alpha=90°$	1.111 $\dfrac{\pi}{2\sqrt{2}}$	3.142 π			1.111 $\dfrac{\pi}{2\sqrt{2}}$	3.142 π			1.571 $\dfrac{\pi}{2}$	1.571 $\dfrac{\pi}{2}$	1.11 $\dfrac{\pi}{2\sqrt{2}}$
$0°\ldots180°$	$\alpha=75°$	$\sqrt{\dfrac{\pi-\alpha}{2\pi}}$	1.000 $\dfrac{\pi-\alpha}{2\pi}$			1.000 $\dfrac{\pi+\alpha}{2\pi}$		$\sqrt{\dfrac{\pi+\alpha}{2\pi}}$		$\sqrt{\dfrac{\pi-\alpha}{\pi}}$	$\sqrt{\dfrac{\pi-\alpha}{\pi}}$	
$9°$		0.785 $\dfrac{\pi}{4}$	1.571 $\dfrac{\pi}{2}$	0.500 $\dfrac{1}{2}$		0.785 $\dfrac{\pi}{4}$	1.571 $\dfrac{\pi}{2}$	0.500 $\dfrac{1}{2}$		1.111 $\dfrac{\pi}{2\sqrt{2}}$	1.111 $\dfrac{\pi}{2\sqrt{2}}$	
$90°$	$\alpha=90°$	1.111 $\dfrac{\pi}{2\sqrt{2}}$	3.142 π			1.111 $\dfrac{\pi}{2\sqrt{2}}$	3.142 π			1.571 $\dfrac{\pi}{2}$	1.571 $\dfrac{\pi}{2}$	
$0°\ldots60°$	$\alpha=30°$	0.577 $\dfrac{1}{\sqrt{3}}$	0.333 $\dfrac{1}{3}$			0.577 $\dfrac{1}{\sqrt{3}}$	0.333 $\dfrac{1}{3}$			0.816 $\sqrt{\dfrac{2}{3}}$	0.816 $\sqrt{\dfrac{2}{3}}$	
$60°\ldots180°$	$\alpha=75°$	$\sqrt{\dfrac{\pi-\alpha}{2\pi}}$	1.000 $\dfrac{\pi-\alpha}{2\pi}$		Free-wheeling arm (non controllable) $\sqrt{\dfrac{\pi-\alpha}{2\pi}}$ $\sqrt{\dfrac{3\alpha-\pi}{2\pi}}$	1.000 $\dfrac{\pi-\alpha}{2\pi}$		$\dfrac{3\alpha-\pi}{2\pi}$		$\sqrt{\dfrac{\pi-\alpha}{\pi}}$	$\sqrt{\dfrac{\pi-\alpha}{\pi}}$	1.05 $\dfrac{\pi}{3}$
$0°$		0.578 $\sqrt{\dfrac{\pi^2}{54}+\dfrac{\pi}{12\sqrt{3}}}$	1.047 $\dfrac{\pi}{3}$	0.333		0.578 $\sqrt{\dfrac{\pi^2}{54}+\dfrac{\pi}{12\sqrt{3}}}$	1.047 $\dfrac{\pi}{3}$	0.333		0.817 $\sqrt{\dfrac{\pi^2}{27}+\dfrac{\pi}{6\sqrt{3}}}$	0.817 $\sqrt{\dfrac{\pi^2}{27}+\dfrac{\pi}{6\sqrt{3}}}$	
$0°$	$\alpha=90°$	0.740 $\dfrac{\pi}{3\sqrt{2}}$	2.094 $\dfrac{2\pi}{3}$	$\dfrac{1}{3}$		0.740 $\dfrac{\pi}{3\sqrt{2}}$	2.094 $\dfrac{2\pi}{3}$	$\dfrac{1}{3}$		1.047 $\dfrac{\pi}{3}$	1.047 $\dfrac{\pi}{3}$	

Table 7.6 lists the characteristics of half-controllable bridge connections.

The single-pole controllable two-pulse bridge connection B2HK is non-standard since the load current has no free wheeling path. The arm-pair half-controllable two-pulse bridge connection B2H2 (whose power range is up to 10 MW) is the connection most used in ac railway systems. Usually two half-controllable bridges are used in series with sequence control. The connection is also used in the lower power range for field and armature feeding of electrical machines.

The half controllable six-pulse bridge connection with free arm B6HKF is three-pulsed (second harmonic in line current) in the case of phase control.

7.1.9 Harmonics

As a result of the switching function of the valves converters generate harmonics in the voltage and current not only on the single-phase or three-phase ac side but also on the dc side. It would be desirable to have as sinusoidal a current as possible on the ac side and a completely smoothed direct current on the dc side. Since the converter valves in fact perform non-linear switching functions but have no storage effect, this can only be achieved by means of supplementary filters and smoothing devices constructed from electrical and magnetic stores (capacitors and inductances) (see Sect. 11.4).

Effect on Supply System. The effect of static converters on the ac supply system is discussed more fully in Sect. 9.3. Static converters not only load the supply with distorted currents, but can also cause considerable distortion of the voltage supply through e.g. the commutating processes which represent a temporary short circuit of the supply phases via the commutating inductances (see Figs. 9.7, 9.8, and 9.9).

Ideal Values of the Harmonics. Assuming a sinusoidal symmetrical line ac current and a completely smoothed dc current one can easily calculate the ac voltage superimposed on the dc side and the harmonics in the line-side ac current (Table 7.7). With the ideal dc voltage U_{di} the resulting superimposed ac voltages are of the order

$$v = kp, \text{ for } k = 1, 2, 3,..., \tag{7.92}$$

the rms values in the case of full modulation being given by

$$U_{vi} = \frac{\sqrt{2}}{v^2 - 1} U_{di}. \tag{7.93}$$

The ideal ripple factor w_i is the ratio of the rms value of the superimposed ideal ac voltage to the ideal dc voltage:

$$w_i = \frac{\sqrt{\sum U_{vi}^2}}{U_{di}}. \tag{7.94}$$

AC line current harmonics I_{vi} having the order

$$v = kp \pm 1, \ k = 1, 2, 3,..., \tag{7.95}$$

Table 7.7. Superimposed ac voltages on the dc side (with full modulation) and harmonics in the line current (ideal values)

P	2		3		6		12		18		24	
	$\frac{U_{vi}}{U_{di}}$ %	$\frac{I_{vi}}{I_{li}}$ %	$\frac{U_{vi}}{U_{di}}$ %	$\frac{I_{vi}}{I_{li}}$ %	$\frac{U_{vi}}{U_{di}}$ %	$\frac{I_{vi}}{I_{li}}$ %	$\frac{U_{vi}}{U_{di}}$ %	$\frac{I_{vi}}{I_{li}}$ %	$\frac{U_{vi}}{U_{di}}$ %	$\frac{I_{vi}}{I_{li}}$ %	$\frac{U_{vi}}{U_{di}}$ %	$\frac{I_{vi}}{I_{li}}$ %
2	41.14	–	–	50.00	–	–	–	–	–	–	–	–
3	–	33.33	17.68	–	–	–	–	–	–	–	–	–
4	9.43	–	–	25.00	–	–	–	–	–	–	–	–
5	–	20.00	–	20.00	–	20.00	–	–	–	–	–	–
6	4.04	–	4.04	–	4.04	–	–	–	–	–	–	–
7	–	14.29	–	14.29	–	14.29	–	–	–	–	–	–
8	2.24	–	–	12.50	–	–	–	–	–	–	–	–
9	–	11.11	1.77	–	–	–	–	–	–	–	–	–
10	1.43	–	–	10.00	–	–	–	–	–	–	–	–
11	–	9.09	–	9.09	–	9.09	–	9.09	–	–	–	–
12	0.99	–	0.99	–	0.99	–	0.99	–	–	–	–	–
13	–	7.69	–	7.69	–	7.69	–	7.69	–	–	–	–
14	0.73	–	–	7.14	–	–	–	–	–	–	–	–
15	–	6.67	0.63	–	–	–	–	–	–	–	–	–
16	0.55	–	–	6.25	–	–	–	–	–	–	–	–
17	–	5.88	–	5.88	–	5.88	–	–	–	5.88	–	–
18	0.44	–	0.44	–	0.44	–	–	–	0.44	–	–	–
19	–	5.26	–	5.26	–	5.26	–	–	–	5.26	–	–
20	0.35	–	–	5.00	–	–	–	–	–	–	–	–
21	–	4.76	0.32	–	–	–	–	–	–	–	–	–
22	0.29	–	–	4.55	–	–	–	–	–	–	–	–
23	–	4.35	–	4.35	–	4.35	–	4.35	–	–	–	4.35
24	0.25	–	0.25	–	0.25	–	0.25	–	–	–	0.25	–
25	–	4.00	–	4.00	–	4.00	–	4.00	–	–	–	4.00
w_i	48.34	–	18.27	–	4.20	–	1.03	–	0.46	–	0.25	–
$\frac{I_{Li}}{I_{li}}$	–	111.07	–	120.92	–	104.72	–	101.15	–	100.51	–	100.29
g_{Li}	–	0.900	–	0.827	–	0.955	–	0.989	–	0.995	–	0.997

have rms values (assuming ideal values independent of the degree of modulation)
given by

$$I_{vi} = \frac{1}{v} I_{1i} \qquad (7.96)$$

where I_{1i} is the rms value of the fundamental oscillation of the ideal line current.
The ideal ac line current has the rms value

$$I_{Li} = \sqrt{I_{1i}^2 + \Sigma I_{vi}^2} = I_{1i}\sqrt{1 + \Sigma \frac{1}{v^2}}. \qquad (7.97)$$

The ideal fundamental frequency component of the ac line current is

$$g_{1i} = I_{1i}/I_{Li}. \qquad (7.98)$$

The pulse number of the converter connection therefore determines the ordinal
number of the harmonics which occur. The magnitude of these harmonics then
depends only on the ordinal number v (and no longer on the pulse number p of the
converter connection).

With phase control the harmonics increase.

Line-side Line Current. Reactances in the commutation circuit reduce the
harmonic currents of high ordinal number since the current increase during the
commutation interval is limited. Figure 7.26 illustrates an example of the time
diagram of the line-side currents with full modultion and with phase control for
three- and six-pulse converters.

The harmonics with ordinal numbers from 5 to 25 can be determined from Fig.
7.27 as a function of the control angle α and the relative inductive direct voltage
regulation. These calculations assume an ideally smoothed dc current. The ordinal
numbers v are obtained from Table 7.7 or Eq. (7.95) using the pulse number of the
converter connections.

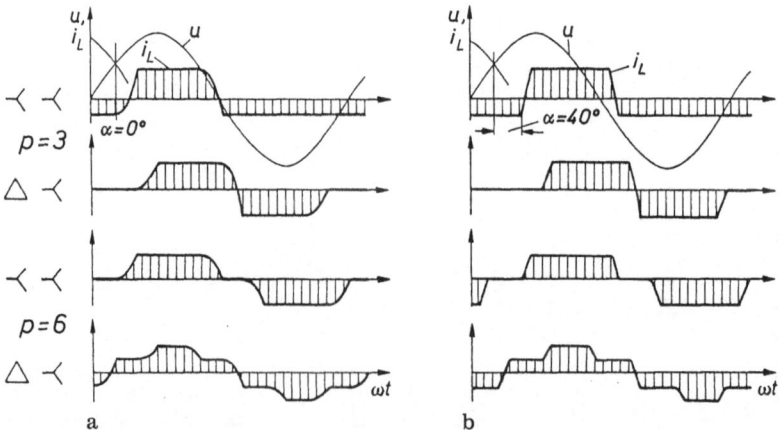

Fig. 7.26a,b. Line-side line current waveform for three- and six-pulse line commutated converters
(initial overlapping $u_0 = 30°$). **a** With full phase control ($\alpha = 0$); **b** for termination control
($\alpha = 40°$)

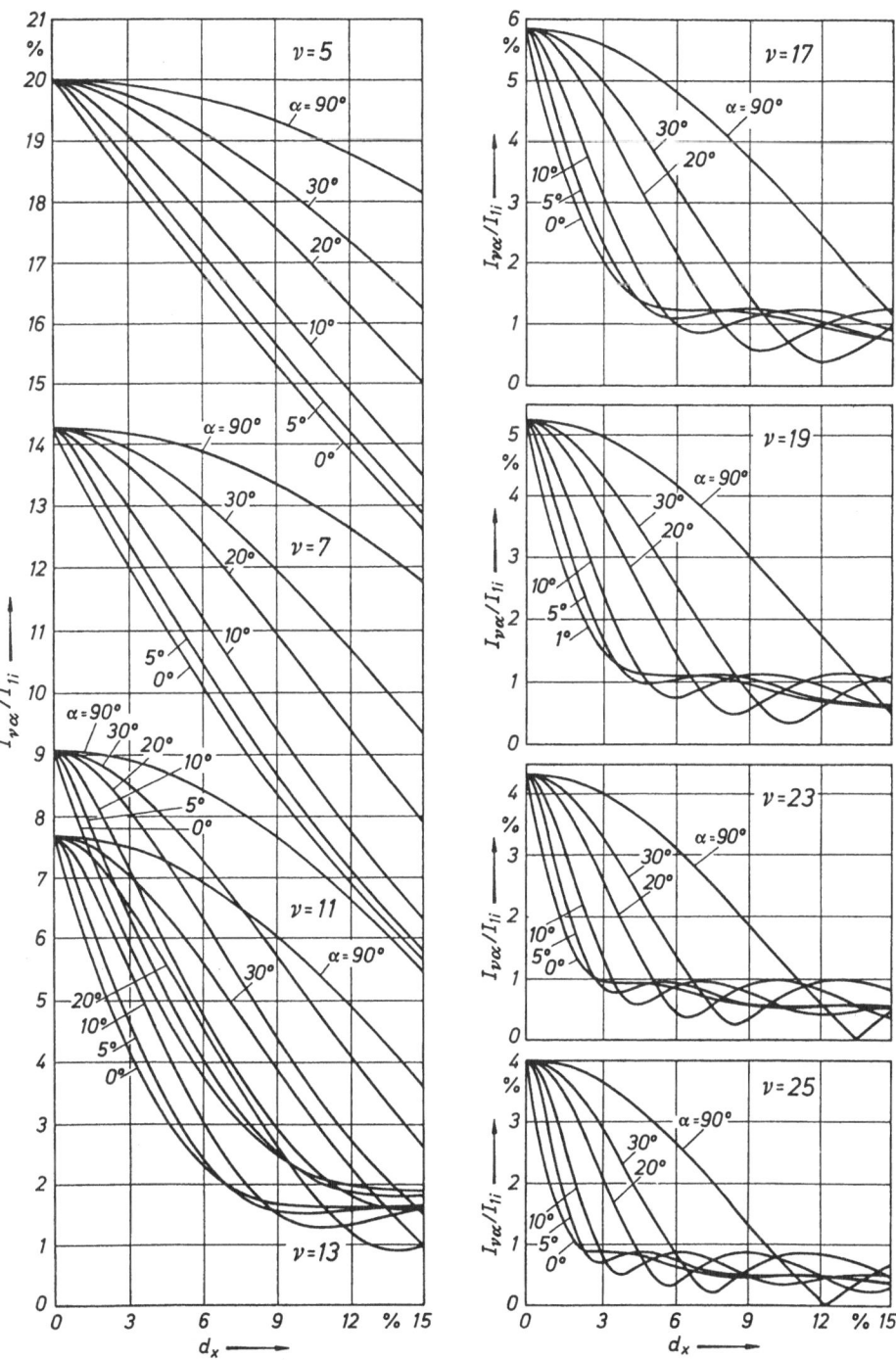

Fig. 7.27. Harmonics of the line-side line current as a function of the relative inductive dc voltage regulation d_x, with the control angle α as parameter

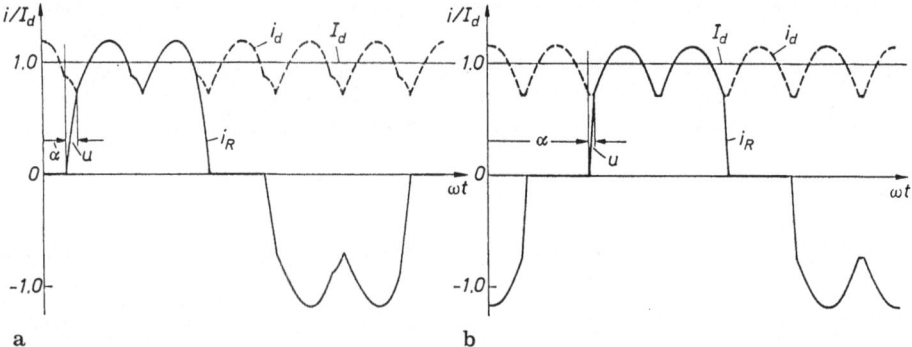

a   b

Fig. 7.28a,b. Line current and dc current of a six-pulse converter with a 15% dc ripple factor (for $d_x = 5\%$). **a** Control angle $\alpha = 20°$; **b** control angle $\alpha = 90°$

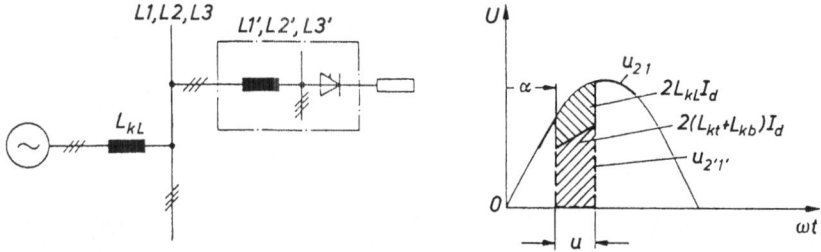

Fig. 7.29. Line voltage waveform during a commutation

In practice, however, the dc current is not usually smooth. Wavy dc current also affects the harmonics of the line-side currents. Figure 7.28 gives an example of the current waveform in the case of a six-pulse converter (three-phase ac bridge connection) assuming a dc ripple factor w_{Id} of 15%. The dc ripple factor is calculated from the equation

$$w_{Id} = \frac{\sqrt{\sum_{v=2}^{\infty} I_v^2}}{I_d}. \tag{7.99}$$

Line Voltage. During commutation phase short-circuits occur in line-commutated converters. Figure 7.29 shows the waveform of the interlinked line voltage u_{12} during the overlap angle u (commutation from phase 1 to phase 2 for symmetrical operation). The larger the leakage inductances $(L_t + L_{kb})$ of the transformer and in the converter in relation to the line inductances (L_{kL}) the smaller are the line voltage notches produced by the commutation. With a finite short-circuit power S_m of the ac power supply the converter must have a minimum reactance if the commutation notches are to be restricted to a definite maximum value.

The ratio of the power $U_{di}I_d$ of the converter to the short-circuit power S_m of the power supply therefore has considerable effect on the power supply reaction. A power ratio of up to 1% is considered normal for a line-commutated converter.

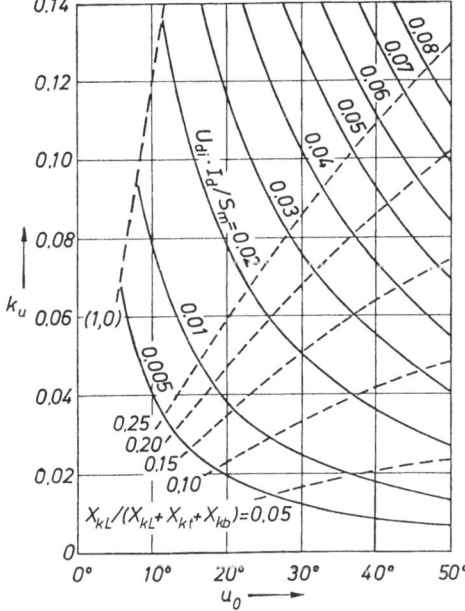

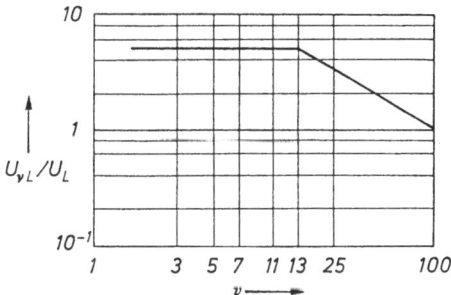

Fig. 7.31. Maximum permissible value for the line voltage harmonics in German power systems

◄ **Fig. 7.30.** Harmonic content k_u of the line voltage upon connection of an uncontrollable six-pulse converter as a function of the initial overlapping u_0, where the power ratio $U_{di} \cdot I_d / S_m$ and the reactance ratio $X_{kL} / (X_{kL} + X_{kt} + X_{kb})$ are parameters

The harmonic content

$$k_u = \frac{\sqrt{\sum U_{vL}^2}}{U_L} \tag{7.100}$$

of the line voltage can be calculated with a known power and reactance ratio. Figure 7.30 shows the harmonic content when an uncontrollable six-pulse converter is connected dependent on the initial overlapping u_0.

To give an example for limit values of harmonics in power systems the German standards are given in Fig. 7.31 (according to VDE 0160). The voltage harmonics of low ordinal number (up to $v = 13$) must not be more than 5% of the ac line voltage. A falling limit curve is valid for harmonics of higher ordinal number. The fundamental frequency content of the ac line voltage,

$$g_u = U_{1L} / U_L, \tag{7.101}$$

must be at least 99.5%. This corresponds to a harmonic content k_u of at most 10%.
DC Voltage. Table 7.8 gives numerical values for the superimposed ac voltages on the dc side for various modulation ratios, ignoring the overlapping. Table 7.9 shows the dependence of the superimposed ac voltages on the initial overlapping in the case of full modulation.

From Fig. 7.32 one can calculate the superimposed ac voltages for ordinal numbers from $v = 3$ to $v = 24$ from the control angle α and the relative inductive direct voltage regulation d_x. The ordinal numbers are obtained from Eq. (7.92) and the pulse number of the converter. The values are correct if one assumes a completely smooth dc current. The dc ripple factor affects the superimposed dc-side ac voltage.

Table 7.8. Superimposed ac voltages on the dc side with different modulation (overlap angle $u=0$)

	U_{vi}/U_{di} in % at $U_{di\alpha}/U_{di}$					
	100%	80%	60%	40%	20%	0%
2	47.1	67.8	80.6	88.5	92.8	94.2
3	17.7	34.9	43.9	49.2	52.2	53.1
4	9.43	23.8	30.7	34.8	37.0	37.7
6	4.04	14.9	19.5	22.3	23.8	24.2
8	2.74	11.0	14.5	16.5	17.7	18.0
9	1.77	9.5	12.8	14.6	15.6	15.9
10	1.43	8.67	11.5	13.1	14.0	14.3
12	0.99	7.17	9.54	10.9	11.7	11.9
14	0.73	6.16	8.17	9.42	10.0	10.2
15	0.63	5.71	7.58	8.70	9.27	9.46
16	0.55	5.31	7.04	8.09	8.64	8.80
18	0.44	4.75	6.34	7.26	7.78	7.92
20	0.35	4.22	5.60	6.44	6.86	7.00
21	0.32	4.03	5.38	6.18	6.59	6.72
22	0.29	3.83	5.10	5.86	6.26	6.38
24	0.25	3.60	4.80	5.50	5.90	6.00

Table 7.9. Variation of the superimposed ac voltages on the dc side by the initial overlapping u_0 (with delay angle $\alpha=0$)

v	U_{vi}/U_{di} in % at u_0				
	0°	10°	20°	30°	40°
3	18	19	21	24	26
6	4.0	4.9	6.0	6.1	6.0
12	1.0	1.5	1.6	3.2	4.2
18	0.4	0.7	1.4	2.0	2.5

DC Current. AC currents are likewise superimposed on the dc side current. The magnitudes of these depend on the superimposed ac voltage and the smoothing inductance L_d. With a resistive load the dc current behaves like the ac voltage.

If the momentary value of the superimposed ac currents exceeds the mean dc current, then gaps appear in the dc current. That is, with discontinuous flow, the dc-side current within each period of the line voltage is zero over certain time intervals. With discontinuous flow there are no longer commutations between principal arms. If the load current diminishes, the superimposed ac currents remain almost at their maximum. The smaller the pulse number p and the larger the control angle α the earlier gaps appear. The factor f_1 at which the direct current becomes intermittent is defined here by

$$f_1 = I_{d1} \frac{L_d}{U_{d0}} \quad (L_d \text{ in mH}). \tag{7.102}$$

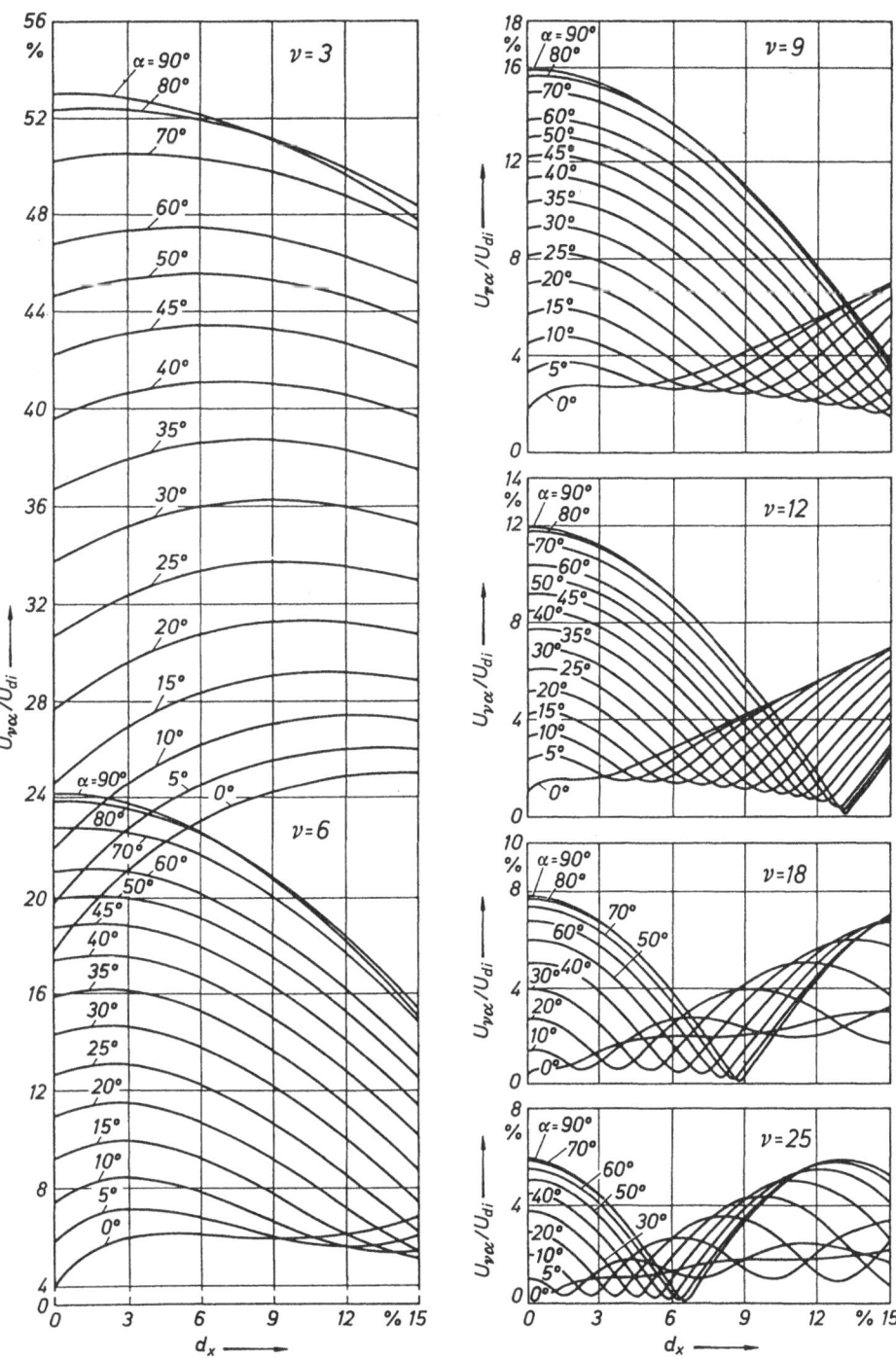

Fig. 7.32. Superimposed ac voltage on the dc side as a function of the relative inductive dc voltage regulation d_x with the control angle α being a parameter

Table 7.10. Intermittent factor f_1 of the dc current for different pulse numbers p and modulation

$U_{d0\alpha}/U_{d0}$	Intermittent factor f_1 for pulse number p			
%	2	3	6	12
100	1.72	0.43	0.054	0.006
90	2.05	0.65	0.137	0.031
80	2.34	0.82	0.180	0.044
70	2.56	0.95	0.213	0.052
60	2.74	1.05	0.238	0.059
50	2.88	1.12	0.257	0.064
40	2.99	1.17	0.272	0.068
30	3.07	1.21	0.283	0.070
20	3.13	1.24	0.291	0.071
10	3.16	1.25	0.295	0.072
0	3.17	1.26	0.296	0.072

I_{dl} is the dc current at the point when the total current becomes discontinuous Table 7.10 gives the this factor for various pulse numbers and modulation ratios. The smoothing inductance L_d (in mH) or the current I_{dl} (in A) can thereby be ascertained.

7.2 Line-commutated Cycloconverters

The line-commutated static converters dealt with up to now for rectification and inversion in fact render possible energy flow in both directions by reversing the polarity of the direct voltage. However, with these circuits the direction of current on the dc side remains the same. Such static converters are called single converters.

When the direction of current on the dc side is also to be reversible, a double or dual converter must be used. These can for example be formed by connecting two single converters in antiparallel, each of which is intended for one direction of current. Another possibility is that a second single converter valve be connected in antiparallel with each converter valve of a single converter.

7.2.1 Double Converters

Multi-quadrant Operation. To identify the possible working ranges of static converters which on the input or output side work on a dc system the current/voltage plane of the dc system is divided into four quadrants which are numbered with the roman numerals I to IV (Fig. 7.33) according to the actual polarity of the dc voltage U_d and dc current I_d. Similar polarities of dc current and dc voltage mean power output to the dc system (in quadrants I and III). If the dc current and dc voltage have opposite polarities, power is drawn from the dc system (on quadrants II and IV). The corresponding torque/speed quadrants for an electrical machine are defined in Fig. 10.6.

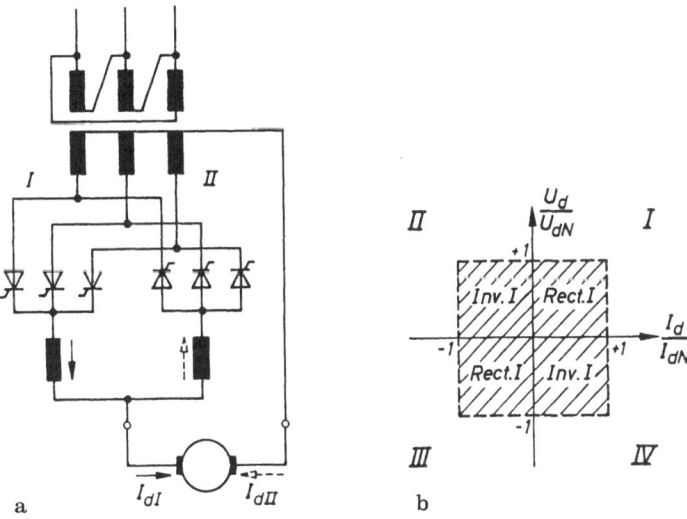

Fig. 7.33a,b. Double converter. **a** Antiparallel connection of two converters in three-pulse center tap connection; **b** dc current/dc voltage quadrants

According to their ability to operte in only one, two or four quadrants, static converters are also classified as single-quadrant converters, two-quadrant converters or four-quadrant converters.

Antiparallel Connection. The double converter illustrated in Fig. 7.33a is created by connecting two partial converters in three-pulse center tap connection in antiparallel. This produces a double converter that can operate in four quadrants of the dc current/dc voltage plane. In quadrant I partial converter I operates in the rectifier mode, and in quadrant III partial converter II operates in the rectifier mode. In both cases energy flows from the three-phase ac system into the dc system. In Quadrant IV partial converter I operates in the inverter mode, and in quadrant II partial converter II operates in the inverter mode. In both cases energy flows from the dc system into the three-phase ac system.

Therefore two-way converters convert ac current into dc current or dc current into ac current by operating alternately as rectifiers or inverters. They permit an exchange of energy in both directions. Various connection are employed to realize two-way converters [7.1, 7.21].

Crossover Connection. Besides the antiparallel connection shown in Fig. 7.33, a connection known as the crossover connection is also employed (Fig. 7.34). In this connection the two partial converters working in antiparallel are connected to separate secondary windings of the converter transformer. Due to division of the secondary side into two three-phase windings the size of the converter transformer is increased. The six-pulse crossover connection has, however, a lower circulating current than the six-pulse antiparallel connection, because the leakage inductances of the transformer act as additional circulating current chokes. The origin of a circulating current with double converters is explained in greater detail in the following section.

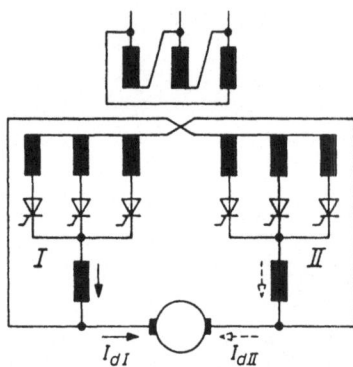

Fig. 7.34. Crossover connection

Circulating Current. The two partial converters I and II of a double converter work in parallel on the same dc busbar. They must therefore be controlled such that the output dc voltages U_{dI} and U_{dII} are as similar as possible. While one partial converter operates in the rectifier mode the other is modulated to operate in the inverter mode. Which of the two partial converters carries the current depends upon the direction of energy.

The voltage U_d on the dc side can be varied by control of the two control angles α_I and α_{II}. In order to assure that the mean dc output voltages U_{dI} and U_{dII} of the two partial converters are the same, the condition

$$\alpha_{II} = 180° - \alpha_I \tag{7.103}$$

must be satisfied. However, there is a non-zero difference between the instantaneous voltages u_{dI} and u_{dII} caused by the different curves of the dc voltage when operating in the rectifier and inverter modes. This differential voltage drives a circulating current i_{KR} between the two partial converters which must be limited by series inductances in particular the circulating current chokes L_{circ}.

In Fig. 7.35 the transient response of the direct voltages u_{dI} and u_{dII} is plotted against time. With a phase angle control constructed in accordance with Eq. (7.103) the hatched voltages are obtained as the direct voltage differences. These are designated the circulating voltage u_{circ}. The circulating voltage u_{circ} drives the circulating current i_{circ} which does not flow through the dc load but flows instead from one partial converter through the other and two phases of the converter transformer. It is driven by the differential voltage of the phases carrying current at any time and is limited only by the circulating current chokes L_{circ}. In Fig. 7.29 the circulating current i_{circ} is drawn in dotted for an instant chosen at random, namely when valves 1 and 6 are carrying current.

So long as the condition established in Eq. (7.103) is maintained, the circulating voltage u_{circ} is a pure alternating voltage. If α_{II} becomes greater than $180° - \alpha_I$, an additional direct voltage component arises which causes a direct current component limited only by the resistances in the circulating current path and is therefore only permissible in so far as no undue circulating current is generated. On operating with circulating current the direct voltage component can be used to come out of the intermittent current range. Operation with

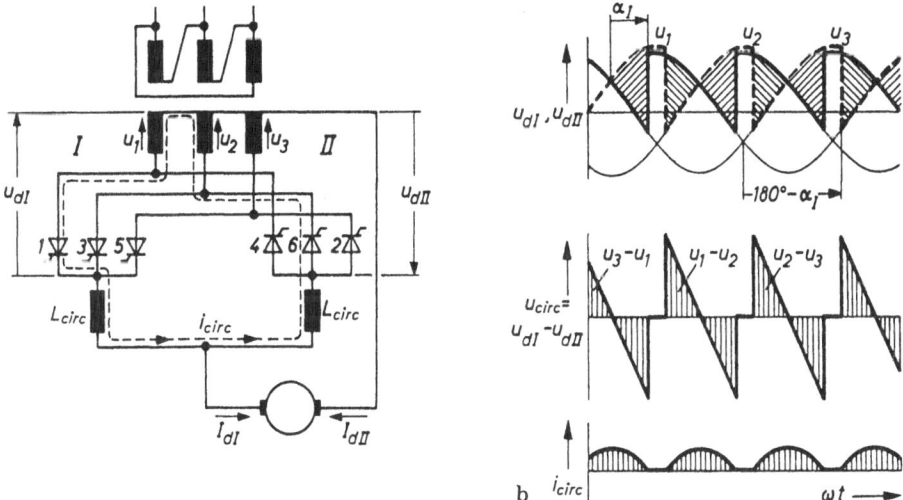

Fig. 7.35a,b. Circulating current with double converters. **a** Antiparallel connection of two M3 converters; **b** waveforms of circulating voltage u_{circ} and circulating current i_{circ}

$\alpha_{II} > 180° - \alpha_I$ would result in a higher mean dc voltage of the partial converter operating in the inverter mode.

Since due to the valve direction, a dc component of this polarity cannot appear in the circulating current, this mode of operation is basically permissible.

Three-pulse connections have considerably larger circulating currents than six-pulse connections [7.4]. In Figs.7.33, 7.34, and 7.35 three-pulse connections were shown because they are more readily comprehended. In practice, double converters are generally made up of two six-pulse single converters.

Connection without Circulating Current. At powers in the range from 10 kW to 5000 kW double static converters are often operated without circulating current. This is achieved by controlling only the partial converter needed for the direction of current required by the load. The trigger pulses for the other converter are suppressed. To do this it is necessary to detect the direction of current or passage of the current through zero in the dc load so that the control system can decide which partial converter must be triggered and which must be blocked. Passage of the current through zero is measured on either the dc or the three-phase ac side.

On reversal there is a certain dead time. In order to keep this as small as possible low-circulating-current connections with controlled circulating current are employed (see Sect. 12.4). With double converters for outputs of more than 1000 kW circulating current chokes are employed which also limit the short-circuit current in the event of faults.

Connections for Double Converters. Characteristics of double converter connections (four quadrant converter) are listed in Table 7.11. To denote connections for double converters, the brackets denoting partial connections are joined by the following letters: X for cross connections, A for anti-parallel connections, H for H-connections.

Table 7.11. Double converter

Converter connection / Nomenclature DIN 41761	Code letter	Phasor diagram of the transformer voltages (line-side, valve-side)	Principal circuit of the converter assembly	Pulse number and commutation $p\ q\ s\ g\ \delta$	D.C. voltage U_{di} — $\dfrac{U_{di}}{U_{i0}}$	$\dfrac{d_{r1}}{u_{x1}}$	w_{Ui} in %	Voltage at the converter arm $\dfrac{U_{im}}{U_{di}}$	$\dfrac{U_{i0m}}{U_{di}}$	Arm current I_p — $\dfrac{I_{peff}}{I_d}$	$\dfrac{I_{pmax}}{I_d}$	$\dfrac{I_{pav}}{I_d}$	Line current valve-side $\dfrac{I_v}{I_d}$	line-side $\dfrac{I_{Li}}{I_d}$	Line-side apparent power $\dfrac{S_{Li}}{U_{i0} I_d}$
Cross-over connection of two double three-pulse center tap connections (with interphase transformers and circulating current reactors)	(M 3.2) X (M 3.2)	△ or Y		6 3 1 2 1	0.675 $\dfrac{3}{\pi\sqrt{2}}$	0.500 $\dfrac{1}{2}$	4.2	2.094 $\dfrac{2\pi}{3}$	2.420 $\dfrac{4\pi}{3\sqrt{3}}$	0.289 $\dfrac{1}{2\sqrt{3}}$	0.500 $\dfrac{1}{2}$	0.167 $\dfrac{1}{6}$	0.289 $\dfrac{1}{2\sqrt{3}}$	0.408 $\dfrac{1}{\sqrt{6}}$	1.05 $\dfrac{\pi}{3}$
Anti-parallel connection of two double three-pulse center tap connections (with interphase transformers and circulating current reactors)	(M 3.2) A (M 3.2)	△ or Y		6 3 1 2 1	0.675 $\dfrac{3}{\pi\sqrt{2}}$	0.500 $\dfrac{1}{2}$	4.2	2.094 $\dfrac{2\pi}{3}$	2.420 $\dfrac{4\pi}{3\sqrt{3}}$	0.289 $\dfrac{1}{2\sqrt{3}}$	0.500 $\dfrac{1}{2}$	0.167 $\dfrac{1}{6}$	0.289 $\dfrac{1}{2\sqrt{3}}$	0.408 $\dfrac{1}{\sqrt{6}}$	1.05 $\dfrac{\pi}{3}$
Anti-parallel connection of two six-pulse bridge connections (with circulating current reactors)	(B 6) A (B 6)	△ or Y		6 3 2 1 1	1.350 $\dfrac{3\sqrt{2}}{\pi}$	0.500 $\dfrac{1}{2}$	4.2	1.047 $\dfrac{\pi}{3}$	1.047 $\dfrac{\pi}{3}$	0.577 $\dfrac{1}{\sqrt{3}}$	1.000 1	0.333 $\dfrac{1}{3}$	0.816 $\sqrt{\dfrac{2}{3}}$	0.816 $\sqrt{\dfrac{2}{3}}$	1.05 $\dfrac{\pi}{3}$
Anti-parallel connection of two six-pulse bridge connections	(B 6) A (B 6)	△ or Y		6 3 2 1 1	1.350 $\dfrac{3\sqrt{2}}{\pi}$	0.500 $\dfrac{1}{2}$	4.2	1.047 $\dfrac{\pi}{3}$	1.047 $\dfrac{\pi}{3}$	0.577 $\dfrac{1}{\sqrt{3}}$	1.000 1	0.333 $\dfrac{1}{3}$	0.816 $\sqrt{\dfrac{2}{3}}$	0.816 $\sqrt{\dfrac{2}{3}}$	1.05 $\dfrac{\pi}{3}$

Double converters are rarely used in center tap connections. The anti-parallel connections of two double three-pulse center tap connections are used in electrolysis with short-term energy reversal (depolarisation). Two six-pulse bridge connections connected in anti-parallel with and without circulating current chokes are used for reversible operation of dc drives up to the power limit of dc machines.

7.2.2 Cycloconverters

Double converters can be used to convert single-phase or three-phase ac current with a frequency f_1 into another frequency f_2. To achieve this their output voltage must be cyclically reversed in time with the desired output frequency f_2. In this mode of operation they perform the basic function of ac conversion. They are therefore called ac converter or, for short, converter.

Frequency conversion is performed by directly reversing the phase voltages of the primary supply system without using a dc link circuit (see Chap. 8); hence, they are called direct converters or cycloconverters, whereby all input conductors are generally connected to all output conductors via converter valves connected in antiparallel. An upper limit is set for the attainable output frequency.

Cycloconverters were already being employed experimentally in the thirties using mercury-arc rectifiers to supply railway system with 16 2/3 Hz from the 50 Hz three-phase ac supply. According to their mode of operation, cycloconverters can be classified as envelope cycloconverters or control cycloconverters.

Trapezoid Cycloconverters. With the so-called trapezoid cycloconverter, an envelope cycloconverter, the voltage curve of an output phase follows the cusps of the phase voltages of the three-phase ac supply (Fig. 7.36). The name trapezoid cycloconverter originates from the approximately trapezoidal waveform of its output voltage. A double converter is required to form the output voltage u of one phase of the trapezoid cycloconverter, as not only the output voltage but also the output current cyclically changes its polarity. The connection would, for example, correspond to that in Fig. 7.33.

The periodic time T_2 of the output voltage is determined by the number n of the voltage cusps per half cycle. Each voltage cusp has a width of T_1/p_1 dependent upon the number of pulses p_1 of the cycloconverter connection. T_1 is the supply

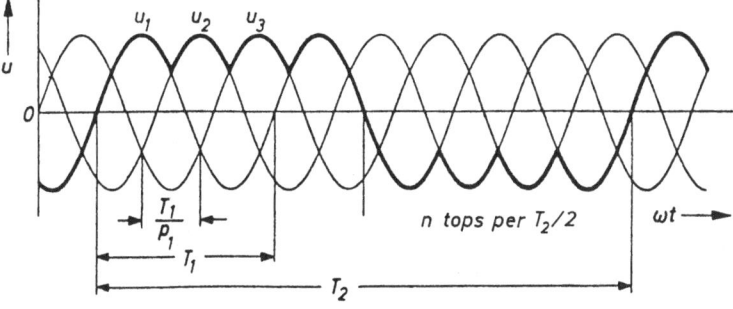

Fig. 7.36. Voltage curve with trapezoid frequency converter

system cycle. For the periodic time T_2 of the output voltage the relationship is

$$T_2 = T_1 + 2(n-1)\frac{T_1}{p_1}. \tag{7.104}$$

The ratio of output frequency f_2 to input frequency f_1 is then calculated as

$$\frac{f_2}{f_1} = \frac{1}{1 + \dfrac{2(n-1)}{p_1}}. \tag{7.105}$$

The possible output frequencies that can be set are obtained by substituting whole numbers 1, 2, 3... for the number of cusps n in this equation. So long as the output voltage curve follows the phase voltages of the input three-phase ac system in accordance with the method shown in Fig. 7.36 and is only reversed at the intersection of the phase voltages only discrete output frequencies f_2 calculated according to Eq. (7.105), can be attained. A trapezoid cylcoconverter is therefore called a fixed-frequency cycloconverter.

A multi-phase output system is created when several antiparallel connections are operated with appropriate phase displacement. With fixed-frequency trapezoid cycloconverters a symmetrical multi-phase system is produced only when the number z of cusps per cycle T_2 of the output voltage

$$z = p_1 + 2(n-1) \tag{7.106}$$

is divisible by the number of phases of the output system.

Freedom to set the output frequency is achieved with the variable-frequency trapezoid cycloconverter. Here, reversal of the polarity of the voltage is not limited to the curve of one phase voltage as with fixed-frequency trapezoid cycloconverters, but intermediate commutation onto the next phase is carried out by temporarily altering the control angles α_I or α_{II} during reversal of the polarity of the output voltage (Fig. 7.37).

Control Cycloconverters. With the control cycloconverter the output voltage of the two partial converters operating in antiparallel is controlled sinusoidally by continuously altering the control angles α_I and α_{II} during each half cycle of the output voltage [7.11, 7.20]. Figure 7.38 shows the connection and voltage of an output phase of a six-pulse control cycloconverter. Each output phase is formed by the antiparallel connection of six-pulse converters. Altogether, therefore, at least $3 \times 2 \times 6 = 36$ converter valves are required.

The output voltage is controlled to approximate as close as possible a preset sinusoidal reference. Both partial converters operate alternately in the rectifier and inverter modes. The displacement factor of the load side determines the actual direction of current in each case. With a phase displacement between the voltage and current on the output side which occurs with resistive-inductive loads or when three-phase ac machines are supplied the partial converter modulated to operate in the rectifier mode carries the current when the energy flow is from the input to the output side. The converse is the case with the partial converter modulated to operate in the inverter mode.

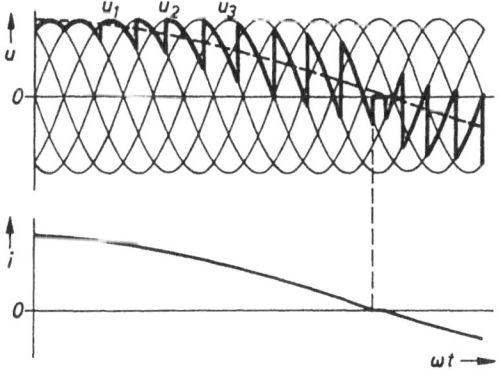

Fig. 7.37. Intermediate commutation with variable-frequency trapezoid cycloconverter

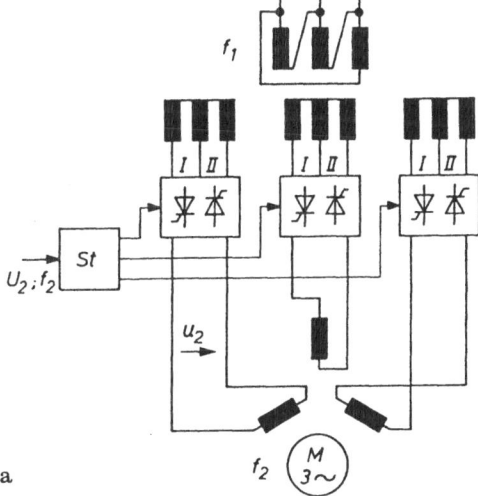

a

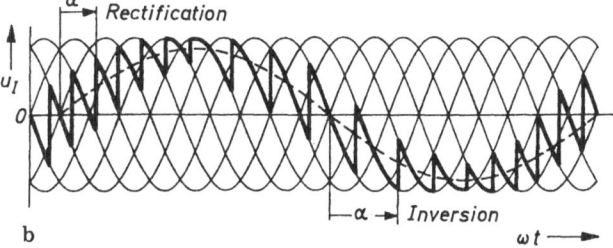

b

Fig. 7.38a,b. Connection and voltage of one phase with control cycloconverter

As is the case with double converters the difference in the output voltages u_I and u_{II} of the two partial converters creates circulating voltages u_{circ} which result in circulating currents i_{circ}. Connections without circulating current can also be used with cycloconverters to eliminate the circulating current. A dead time again occurs with these on reversal of the direction of current.

The output frequency f_2 can be continuously altered with the control cycloconverter with the essential limitation, however, that the output frequency f_2 can only be raised to about half the value of the input frequency f_1 (at 50 Hz therefore only up to a maximum of 25 Hz). Moreover, a control cycloconverter has a high reactive power consumption from the three-phase ac supply because it is predominantly operated with phase-angle control. With special circuit and control procedures also higher outut frequencies can be obtained [7.19].

7.3 Load-commutated Inverters

Externally commutated converters draw their commutating reactive power either from the ac supply or from the load.

Up to now we have only dealt with line-commutated converters.

With load-commutated inverters the load provides the commutating voltage during the period of commutation. Since the converter constantly needs inductive reactive power for natural commutation, it is a prerequisite for the operation of load-commutated converters that the load can provide this. The load current must, for this reason, have a capacitive component. This condition can be satisfied by parallel and series resonant circuits or by an overexcited synchronus machine. Loads with inductive current components e.g. induction machines, can therefore not take over the control of converters with natural commutation.

Load commutated converters behave similarly to single-phase ac or three-phase line commutated converters. Basically, therefore, the characteristics and equations derived for these converters also apply for load-commutated converters. If the load absorbs energy, the load-commutated converter operates in the inverter mode. This is the normal case. The most important load-commutated inverters are dealt with in the following. These are resonant circuit inverters [7.5, 7.6, 7.8, 7.10, 7.15, 7.22] and motor-commutated inverters.

7.3.1 Parallel Resonant Circuit Inverters

A resistive-inductive load can be augmented by means of a capacitor into a parallel or series resonant circuit. The resonant frequency f_0 of the no-loss load circuit

$$f_0 = \frac{1}{2\pi\sqrt{LC}} \tag{7.107}$$

is called the rated frequency. The resonant frequency f_R of the freely oscillating load circuit subject to losses with the damping element γ is

$$f_R = \frac{\omega_0}{2\pi}\sqrt{1-\delta^2}, \tag{7.108}$$

whereby

$$\delta = \frac{R}{2\omega_0 L} \tag{7.109}$$

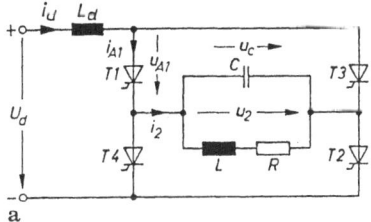

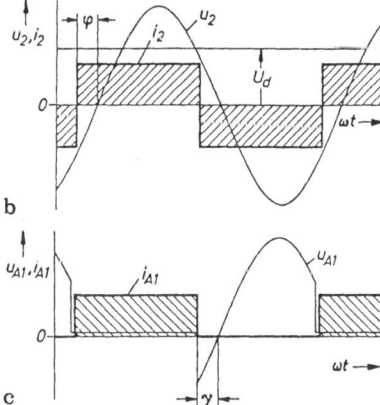

Fig. 7.39a,b. Parallel resonant circuit inverter. **a** Single-phase bridge connection (single converter); **b** voltage and current waveforms

These equations apply not only for a parallel but also for a series resonant circuit. The operating frequency f_B with which the resonant circuit inverter is operated is preset by the control system. In order that the resonant circuit has a capacitive current component the operating frequency with the parallel resonant circuit must lie above the resonant frequency and with the series resonant circuit below it.

With the parallel resonant circuit inverter the resistive-inductive load L and R is augmented by a parallel capacitor C into a parallel resonant circuit (Fig. 7.39). In each arm of the inverter there is a controlled converter valve whose current has an approximately rectangular waveform. Since the parallel resonant circuit does not permit sudden voltage changes, the inverter needs a smoothing inductance L_d on the dc side. The voltage u_2 of the load side is approximately sinusoidal. The current thereby transfers in direct commutation from one converter valve into the following one. The load current has a rectangular waveform and leads the load voltage by the phase angle φ. This is necessary to ensure the margin angle γ.

The sinusoidal curve of load voltage u_2 which is identical to the capacitor voltage u_C can be calculated from the energy balance between the dc and load sides. For a half cycle

$$U_d I_d \frac{T}{2} = I_d \int_{-\gamma/\omega}^{\frac{\pi-\gamma}{\omega}} \hat{u}_2 \sin\omega t \, dt. \tag{7.110}$$

From this the peak value $\hat{u}_2$ of the load voltage can be calculated as a function of the dc voltage U_d and the margin angle γ

$$\hat{u}_2 = \sqrt{2}U_2 = \frac{\pi}{2\cos\gamma} U_d. \tag{7.111}$$

The same result is obtained from Eq. (7.5) which was derived for line-commutated converters. For the two-pulse bridge connection considered here

$$\hat{u}_2 = \sqrt{2}U_2 = \frac{\pi}{2} U_{di} = \frac{\pi}{2\cos\alpha} U_{di\alpha}. \tag{7.112}$$

This result is identical to the Eq. (7.111) derived from the energy balance when the control angle α is replaced by the margin angle $\gamma = 180° - \alpha$.

As Eqs. (7.111) and (7.112) show, an increase in the margin angle at constant output dc voltage U_d results in a voltage increase. For this reason the power output to the load circuit can be altered only to a limite degree by controlling the angle of advance $\beta = u + \gamma$. The output power can be controlled to a great extent by adjusting the dc voltage U_d.

7.3.2 Series Resonant Circuit Inverters

With the series resonant circuit inverter the resistive-inductive load L and R is augmented by a series capacitor C into a series resonant circuit (Fig. 7.40). In each arm of the inverter lies a controllable converter valve to which an uncontrollable diode is connected in antiparallel so that current can be carried in both directions.

The series resonant circuit forces an approximately sinusoidal load current i_2, which is controlled alternately by the thyristors and the antiparallel diodes. The load voltage u_2, and hence also the valve voltage u_A, has an approximately rectangular waveform. The current commutates from the uncontrollable valve to the respective controllable valve connected in antiparallel. The load current leads the load voltage by the phase angle φ. This is necessary to maintain the required margin angle γ.

As the series resonant circuit does not permit sudden current changes, as stable an output voltage as possible is needed. While the thyristors are carrying the current (hatched zone) energy flows into the load circuit and while the diodes are carrying the current energy flows back into the dc circuit. The output power can be controlled by altering either the margin angle γ or the input direct voltage U_d.

The load current i_2 can be calculated in the same way as with the parallel resonant circuit inverter from the energy balance between the dc and ac sides during a half cycle

$$U_d I_d \frac{T}{2} = U_d \int_{\gamma/\omega}^{\frac{\pi+\gamma}{\omega}} \hat{i}_2 \sin\omega t \, dt. \tag{7.113}$$

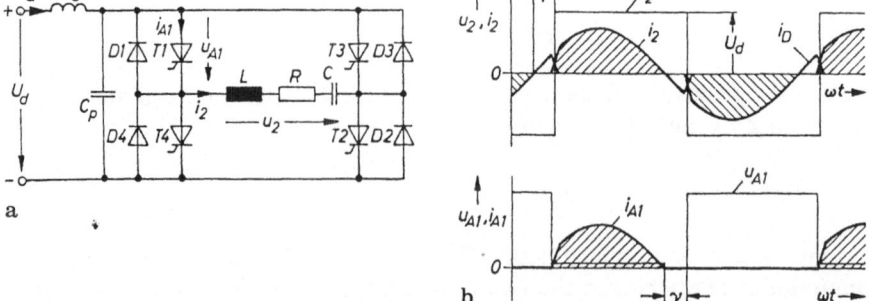

Fig. 7.40a,b. Series resonant circuit inverter. **a** Single-phase bridge connection (double converter); **b** voltage and current waveforms

The peak value $\hat{i}_2$ of the load current is obtained from

$$\hat{i}_2 = \frac{\Pi}{2\cos\gamma} I_d. \tag{7.114}$$

The attainable upper frequency limit of resonant circuit inverters is determined chiefly by the circuit-commutated recovery time of the thyristors. Operating frequencies of more than 10 kHz can be attained (see Sects. 13.1.4 and 13.3). For the oscillations of the load to build up particularly in the case of parallel resonant circuit inverters a starting device is required to provide the commutating voltage

Table 7.12. Formulae for resonant circuit inverters

Frequencies:

Resonant frequency $f_0 = \dfrac{1}{2v\sqrt{LC}}$ $\qquad \omega_0 = 2\pi f_0$

Rated frequency $f_R = \dfrac{\omega_0}{2\pi}\sqrt{1-\vartheta^2}$ mit $\vartheta = \dfrac{R}{2\omega_0 L}$

Operating frequency f_B

Clock frequency f_T

	Parallel resonant circuit		Series resonant circuit	
	Single-phase	Three-phase	Single-phase	Three-phase
Calculation of fundamental frequency component:				
Fundamental frequency component of the ac voltage U_1	$\dfrac{\pi}{2\sqrt{2}\cos\gamma} U_d$	$\dfrac{\pi}{3\sqrt{2}\cos\gamma} U_d$	$\dfrac{2\sqrt{2}}{\pi} U_d$	$\dfrac{\sqrt{6}}{\pi} U_d$
Fundamental frequency component of the ac current I_1	$\dfrac{2\sqrt{2}}{\pi} I_d$	$\dfrac{\sqrt{6}}{\pi} I_d$	$\dfrac{\pi}{2\sqrt{2}\cos\gamma} I_d$	$\dfrac{\pi}{3\sqrt{2}\cos\gamma} I_d$
Mean value of the current in the controllable valve arm I_{pav}	$\dfrac{1}{2} I_d$	$\dfrac{1}{3} I_d$	$\dfrac{I_1\sqrt{2}}{2\pi}(1+\cos\gamma)$	$\dfrac{I_1\sqrt{2}}{2\pi}(1+\cos\gamma)$
Mean value of the current in the un-controllable valve arm I_{pav}	–	–	$\dfrac{I_1\sqrt{2}}{2\pi}(1-\cos\gamma)$	$\dfrac{I_1\sqrt{2}}{2\pi}(1-\cos\gamma)$
Peak off-state voltage at the converter arm U_{im}	$U_1\sqrt{2}$	$U_1\sqrt{2}$	U_d	U_d
Fundamental frequency active power P_1	$I_1 U_1 \cos\gamma$ $\sim \dfrac{U_d^2}{R\cos^2\gamma}$	$\sqrt{3} I_1 U_1 \cos\gamma$	$I_1 U_1 \cos\gamma$ $\sim \dfrac{U_d^2\cos^2\gamma}{R}$	$\sqrt{3} I_1 U_1 \cos\gamma$

needed for the first commutations after switching on the load-commutated inverter. For this, capacitive energy stores on the load and dc side are precharged.

Resonant circuit inverters with series-connected rectifiers are called resonant circuit converters (see Figs. 13.15 and 13.16). Formulae for single-phase and three-phase series resonant circuit inverters are listed in Table 7.12 which also contains the frequencies as defined above. The operating frequency f_B not only depends on the resonant frequency but also on the angle of advance β or the margin angle γ. The clock frequency f_T is the frequency with which the principal arms are periodically activated to conduct. f_T is usually identical with the operating frequency.

7.3.3 Motor-commutated Inverters

Motor-commutated inverters are load-commutated inverters that draw their commutating reactive power from an appropriately magnetized synchronous machine serving as a load [7.3, 7.9, 7.12, 7.13, 7.14, 7.18].

Their connection generally also renders reversal of the energy flow possible. The connection of motor-commutated inverters corresponds to that of line-commutated converters for rectification and inversion. By connecting a line-commutated rectifier and a motor-commutated inverter in series, an ac freqeuency converter is created with a synchronous machine as load (Fig. 7.41). Generally an energy store (a smoothing inductance L_d) is provided in the dc link circuit which decouples the energy of the converter I on the line side from the energy of the converter II on the load side.

The converter I on the line side operates as a line-commutated rectifier when the connected synchronous machine is operating as a motor. It generates the direct

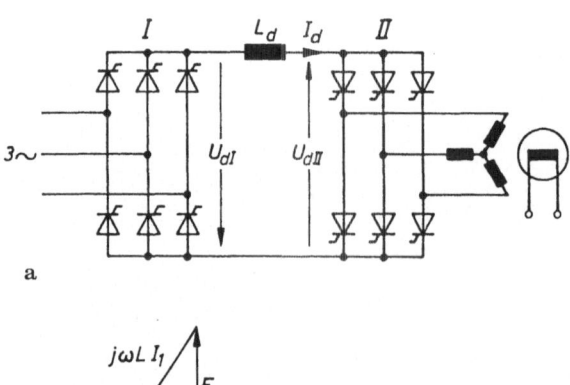

a

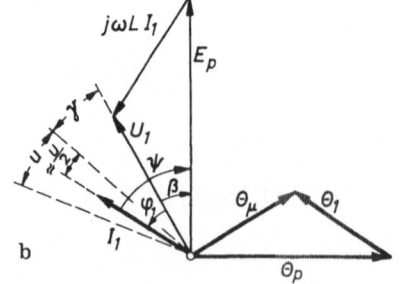

b

Fig. 7.41a,b. Motor-commutated inverter (converter motor). **a** Cycloconverter with dc link circuit; **b** phasor diagram of machine parameters

voltage U_{dI} which is adjustable by phase angle control via the control angle α. The current I_d in the dc link circuit is smoothed by the inductance L_d.

The converter on the load side operates as a load-commutated inverter. It generates the direct voltage U_{dII}. As operation is in the inverter mode the mean value of this dc voltage is negative. In steady-state operation $U_{dII} = -U_{dI}$.

If the direction of energy reverses, converter II must be modulated into operation in the rectifier mode and converter I into operation in the inverter mode. The dc voltages U_{dII} and U_{dI} thereby change their polarity. The direction of the current I_d is maintained. The synchronous machine then operates as a generator.

The synchronous machine load can only control converter II, when the machine current has a capacitive component i.e. when it leads the voltage. In Fig. 7.41b the phasor diagram of the synchronous machine is illustrated for this operating condition U_1 is the voltage of the machine, I_1 the stator current. I_1 leads with respect to U_1 by the phase angle φ_1. Thus an overlap angle u produces the indicated margin angle γ. This must not be less than a certain minimum, namely the hold-off intervall necessary for the valves of converter II, as otherwise inverter failure would occur.

Margin angle and overlap determine the power factor (for fundamental) $\cos \varphi_1$ of the synchronous machine. In order to keep the power factor high (but < 1) also with changing speed and torque the hold-off interval (e.g. the margin angle γ) can be controlled in a closed loop to a preset minimum value [12.12].

The other parameters illustrated in the phasor diagram are E_p, the voltage induced by the field of the magnetic wheel, and L_1, the effective machine inductance. Θ_p are the ampere-turns of the magnetic wheel Θ_1 the ampere-turns of the stator, and Θ_μ the resultant magnetizing ampere-turns. The angle β between U_1 and E_p is the load, or magnetic wheel, angle. ψ is the so-called internal phase angle between the stator current I_1 and the voltage E_p induced by the magnetic wheel. The torque developed by the synchronous machine is proportional to the cosine of this angle

$$M \sim \Theta_1 \cdot \Theta_p \cdot \cos \psi. \tag{7.115}$$

With a given stator and field current, a synchronous machine therefore develops maximum torque when the internal phase angle ψ becomes zero.

Since when the synchronous machine is standing still, no commutating voltage is at first present on the secondary side, and special steps must be taken for starting [7.17]. One possible method is to run up by adjusting the converter on the input side in synchronism with the low starting frequency. Applications of the converter-fed synchronous machine are described in Sects. 13.1 and 13.2.3.

8 Self-Commutated Converters

Self-commutated converters need no separate ac voltage source for commutation. More often than not commutating voltage is provided by an energy store (usually a quenching capacitor) belonging to the converter or is formed by increasing the resistance of the converter valve to be turned off (e.g. a power transistor or a thyristor capable of being turned off).

Self-commutated converters are designed for all types of conversion of electrical energy as well as for energy flow in one or in both directions.

The forced commutation existing with self-commutated converters has already been dealt with briefly in Chap. 5 (see Fig. 5.5). In self-commutated converters with converter valves which cannot be turned off via a gate electrode, commutation is generally indirect. When only one auxiliary arm takes part this is called single-stage commutation. Generally, however, commutation is in two or more stages when, besides the principal arm, two or more auxiliary arms take part.

8.1 Semiconductor Switches for DC

Compared to switching a single-phase ac circuit where the alternating current periodically changing its direction passes through zero after each half-cycle switching a dc circuit is different in that the current can only be interrupted by a counter-voltage applied by the switch. The semiconductor switch needs a turn-off arm. A semiconductor switch for dc therefore belongs to those converters which have forced commutation.

8.1.1 Closing a DC Circuit

A dc circuit is closed using a semiconductor switch by triggering the thyristor. Figure 8.1 shows the current and voltage waveforms on closing a dc circuit with a resistive-inductive load (see Fig. 5.1).

After triggering of the thyristor at instant t_0 the following differential equation applies:

$$(L+L_\sigma)\frac{di}{dt}+Ri=U_d. \tag{8.1}$$

Its solution produces an exponential function rising with the time constant $\tau=(L_\sigma+L)/R$ to the final value U_d/R (see Eq. (5.1)). The maximum rate of rise of current occurs at the instant of switching, t_0. It is $U_d/(L+L_\sigma)$ (see Eq. (5.3)).

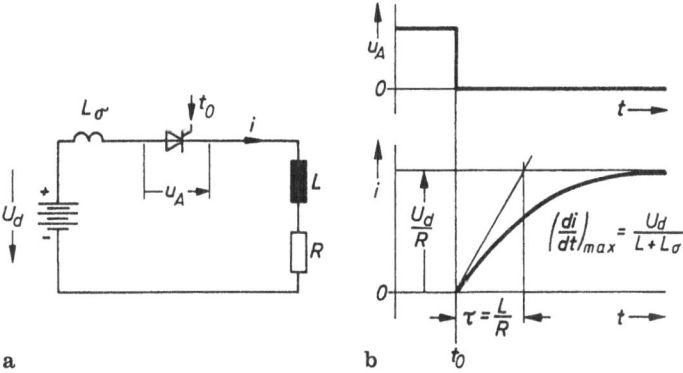

Fig. 8.1a,b. Closing a dc circuit. **a** Circuit; **b** voltage and current waveforms

Generally, the inductance L in the load circuit will be large enough to limit the rate of rise of current in the semiconductor switch to safe values. In special cases, however, e.g. on switching resistances with a low-inductance supply cable the rate of rise of current occuring at the instant of switching on must be checked. This also applies to the connection of an RC circuit across the thyristor where only the series resistance R_B limits the capacitor discharge current and the rate of rise of the current is determined by the leakage inductances.

8.1.2 Opening a DC Circuit

Interruption of a dc circuit using a thyristor switch is not necessarily possible, because to turn off a thyristor the anode current must fall below its holding current. Moreover, after successful interruption of the current, there must be a hold-off interval t_c during which the anode voltage is negative before the thyristor can again block positive off-state voltage. This hold-off interval must be greater than the turn-off time of the thyristor.

With power transistors or gate turn-off thyristors, a dc circuit can be interrupted via the gate terminal. For this, however, the magnetic energy $Li^2/2$ stored in the inductances must be dissipated in the power semiconductor switch.

Figure 8.2 shows the process of opening a dc circuit. It should first be assumed that the semiconductor switch capable of being turned off generates a constant countervoltage U_S from the instant t_0 onwards. The current i can be calculated by the following equation:

$$u_S(t) = U_S = U_d - L_\sigma \frac{di}{dt}. \tag{8.2}$$

It falls linearly to zero because the freewheeling diode D short-circuits the load circuit as soon as the switch voltage U_S is higher than the dc voltage U_d. During the opening process the energy dissipated in the semiconductor switch is

$$W = \frac{1}{2}U_S I \Delta t = \frac{1}{2}U_d I \Delta t + \frac{1}{2}L_\sigma I^2. \tag{8.3}$$

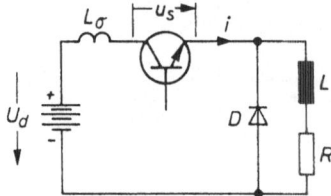

Fig. 8.2. Opening a dc circuit

The energy dissipated in the switch is thus made up of two components:

1. A component $U_d I \, \Delta t/2$ varying with the opening time Δt, and
2. A further component $L_\sigma I^2/2$ which originates from the magnetic energy of the leakage inductance L_σ and varies with the opening time Δt.

With a switching transistor this energy is dissipated as heat in the transistor. The same applies for a gate turn-off thyristor.

Capacitive Quenching. The current I in a normal thyristor cannot be interrupted via the gate terminal. Instead a turn-off arm is required. This turn-off arm comprises an energy store, the quenching capacitor C [8.19, 8.21].

Interruption of the current in a thyristor with a quenching capacitor is illustrated in Fig. 8.3. It is a prerequisite that the quenching capacitor be charged with the polarity shown in the drawing. If the auxiliary switch shown (which is generally realized by an auxiliary thyristor) is closed at instant t_1, the current i_A rapidly commutates from the thyristor T to the quenching capacitor. The rate of rist of the current in the quenching capacitor or the rate of fall of the current in the main thyristor T is determined by the leakage reactance L'_σ. The maximum rate of rise of current is

$$\frac{di}{dt_{max}} = \frac{U_C}{L'_\sigma}. \tag{8.4}$$

When the leakage reactance L'_σ is very small the rate of rise of current is high, which can be critical for the stressing of the auxiliary thyristor. The first commutation stage is completed at instant t_2. Because of the high rate of fall of current and the storage charge in the main transistor the current i_A passes through zero and continues to flow in the reverse direction. After 1 to 2 µs during which the anode current is negative the current falls to zero at a high rate (see Fig. 3.9).

After the thyristor current has been quenched the load current I which is first maintained by the inductance L continues to flow via the turn-off arm and charges the quenching capacitor C in the opposite polarity. During the hold-off interval t_c the main thyristor is reverse biased. t_c must be larger than the recovery time t_q for otherwise the thyristor would switch on again even without a trigger pulse as the anode voltage u_A becomes positive.

At instant t_3 the capacitor voltage exceeds the dc voltage U_d. Thus the second stage of commutation is initiated since the voltage across the freewheeling diode becomes positive. The equivalent switch in the freewheeling arm therefore closes at instant t_3. From t_3 to t_4 the load current I is transferred from the quenching capacitor to the freewheeling arm. Thereafter the load current I decays according

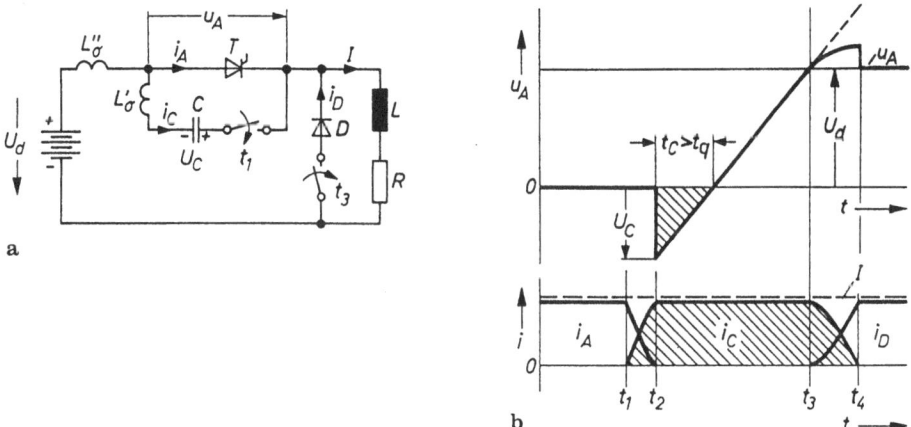

a

b

Fig. 8.3a,b. Capacitive quenching. **a** Circuit; **b** voltage and current waveforms

to an exponential function

$$i = \frac{U_d}{R} e^{-\frac{t-t_4}{L/R}}. \tag{8.5}$$

At this point the dc load has been isolated from the direct-voltage U_d due to opening of the semiconductor switch. However, the current in the load is not zero, but only gradually decays via the freewheeling circuit.

Calculation of the Quenching Capacitor. After interruption of the current at instant t_2 the thyristor T is reverse biased for the period t_c. This period of time t_c is called the hold-off interval. It can be calculated from

$$\frac{du_C}{dt} = \frac{i_C}{C} \tag{8.6}$$

assuming constant capacitor current I

$$t_c = \frac{CU_C}{I}. \tag{8.7}$$

From this one obtains the equation for the quenching capacitor C

$$C = \frac{It_c}{U_C}, \tag{8.8}$$

from which the necessary value of the quenching capacitor can be calculated using the condition $t_c > t_q$. The largest load current occuring should be substituted for I and the capacitor voltage at the instant of turn-off for U_C.

With many turn-off circuits the capacitor voltage at the instant of quenching equals the dc voltage U_d. The hold-off interval t_c must be greater than the circuit-commutated recovery time t_q of the thyristor to be turned off by a safety factor of between 1.3 and 1.5. Equation (8.8) shows that the capacitance of the necessary quenching capacitor grows in direct proportion to the hold-off interval or recovery time. For self-commutated converters therefore preferably F-type thyristors (also called high-speed thyristors or inverter thyristors) are employed whose recovery time t_q is particularly short (less than 60 μs) [8.6].

With thyratrons and mercury-arc rectifiers the practical realization of connectios of self-commutated converters is difficult because of their longer recovery time compared to thyristors and the consequent greater expense of quenching.

8.2 Semiconductor Power Controllers for DC

A semiconductor switch for dc circuits can be employed for purposes other than for switching the circuit on and off at any instant. If the switch is cyclically triggered and quenched at a particular switching frequency, the power drawn by a load from a direct voltage source can be controlled. Such a static converter is called a dc *power controller* or *chopper* [8.9, 8.10].

The switching frequency with which the main or auxiliary thyristor is cyclically triggered is designated the pulse frequency f_p.

DC power controllers perform the basic function of dc-to-dc conversion.

8.2.1 Current and Voltage Waveforms

The basic connection and the current and voltage waveforms of a dc power controller are illustrated in Fig. 8.4. The load side is connected to the constant direct voltage source U_1 via a thyristors switch S capable of being turned off.

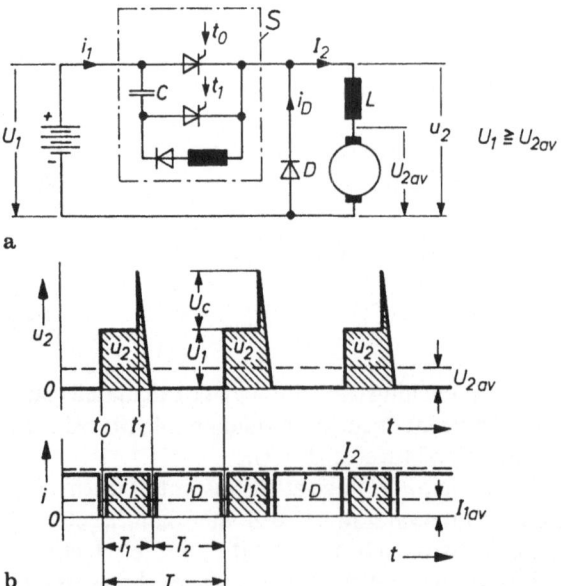

Fig. 8.4a,b. DC power controller. **a** Driving circuit; **b** voltage and current waveforms

The thyristor switch has a turn-off arm consisting of a quenching capacitor C and an auxiliary thyristor. Via the oscillatory circuit consisting of an inductance and a blocking diode the quenching capacitor C is charged to the opposite polarity required for quenching when the main thyristor is switched on. It is assumed that there is a large smoothing inductance L on the load side as well as a freewheeling arm with the freewheeling diode D.

Pulse Control Factor. If the semiconductor switch S is cyclically switched by triggering it at instant t_0 and the turn-off thyristor at moment t_1, pulse-shaped voltage blocks u_2 are produced on the load side. Their height equals the direct voltage U_1 and their width equals the 'on' time T_1. At the instant of turn-off t_1 an additional voltage peak occurs on the load side. This is generated by the quenching capacitor because it lies in series with the dc voltage U_1 during quenching. The pulse control factors of the cyclically actuated semiconductor switch is defined by

$$\lambda = \frac{T_1}{T_1 + T_2} = \frac{T_1}{T}. \tag{8.9}$$

The mean direct voltage U_{2av} on the load side can be calculated from the phase control factor and the dc voltage U_1

$$U_{2av} = \frac{1}{T} \int_0^T u_2 dt = \frac{T_1}{T_1 + T_2} U_1. \tag{8.10}$$

Assuming complete smoothing of the current I_2 the rectangular current blocks illustrated in Fig. 8.4a are produced and during the 'on' time T_1 current is drawn from the direct-voltage source U_1. During the 'off' time T_2 the load current I_2 flows via the freewheeling arm. The mean direct current I_{1av} can be calculated from the phase control factor and the load current I_2

$$I_{1av} = \frac{1}{T} \int_0^T i_1 dt = \frac{T_1}{T_1 + T_2} I_2. \tag{8.11}$$

8.2.2 Transformation Equations

Ignoring the switching losses the following energy balance between the input and output side of a dc power controller is produced

$$\frac{1}{T} \int_0^T u_1 i_1 dt = \frac{1}{T} \int_0^T u_2 i_2 dt. \tag{8.12}$$

If a large smoothing inductance L is assumed, a constant current I_2 flows on the load side. From Eq. (8.12) then

$$\frac{U_1}{T} \int_0^T i_1 dt = \frac{I_2}{T} \int_0^T u_2 dt \tag{8.13}$$

or with Eqs. (8.10) and (8.11)

$$U_1 I_{1av} = U_{2av} I_2. \tag{8.14}$$

Using the determining Eq. (8.9) for the phase control factor the transformation equation of a dc power controller are obtained from

$$U_{2av} = \frac{T_1}{T} U_1 = \lambda U_1 \qquad (8.15)$$

and

$$I_{1av} = \frac{T_1}{T} I_2 = \lambda I_2. \qquad (8.16)$$

These equations correspond to the transformation equations with a single-phase transformer. However, with a single-phase transformer the transformation ratio w_1/w_2 (the relationship between the numbers of primary and secondary turns) is constant and can only be altered by changing taps. On the other hand with the dc power controller the pulse control factor λ can be steplessly adjusted between 0 and 1 by altering the instants of triggering t_0 and t_1.

Therefore with a dc power controller a transformation of voltage and current mean values occurs. On the side with the higher direct voltage U_1 pulse-shaped current blocks i_1 flow. On the other side with a continuous current I_2, there are pulse-shaped voltage blocks u_2.

Owing to the pulse-shaped current blocks the direct-voltage source U_1 must have only a small internal inductance. If this condition is not satisfied, smoothing capacitors must be provided.

8.2.3 Energy Recovery and Multi-quadrant Operation

With the connection shown in Fig. 8.4 energy flows from the direct voltage source U_1 into the load. When the direction of energy flow is reversed i.e. when energy should be returned from the load into the direct voltage source, this connection must be modified [8.22, 8.30].

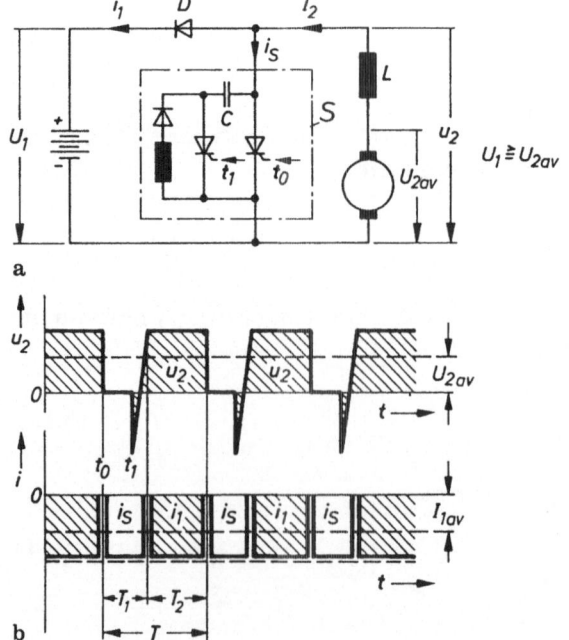

a

b

Fig. 8.5a,b. Energy recovery with dc chopper. **a** Braking circuit; **b** voltage and current waveforms

Current must then flow from a direct voltage source with a mean value U_{2av} into one with a lower mean value U_1. This task is performed by the connection illustrated in Fig. 8.5 which needs the same semiconductor switches as the connection in Fig. 8.4. In this case the quenchable thyristor switch S lies in parallel with the dc side u_2. If the main thyristor is triggered at instant t_0, the load current I_2 rises storing magnetic energy in the smoothing inductance L. After the semiconductor switch S is turned off the load current I_2 flows back into the direct voltage source via the blocking diode even though the dc voltage U_1 is higher than U_{2av}. The required differential voltage is provided by the smoothing inductance L. On renewed triggering of the semiconductor switch S it carries again the current and with the blocking diode D prevents any short-circuiting of the direct voltage source U_1.

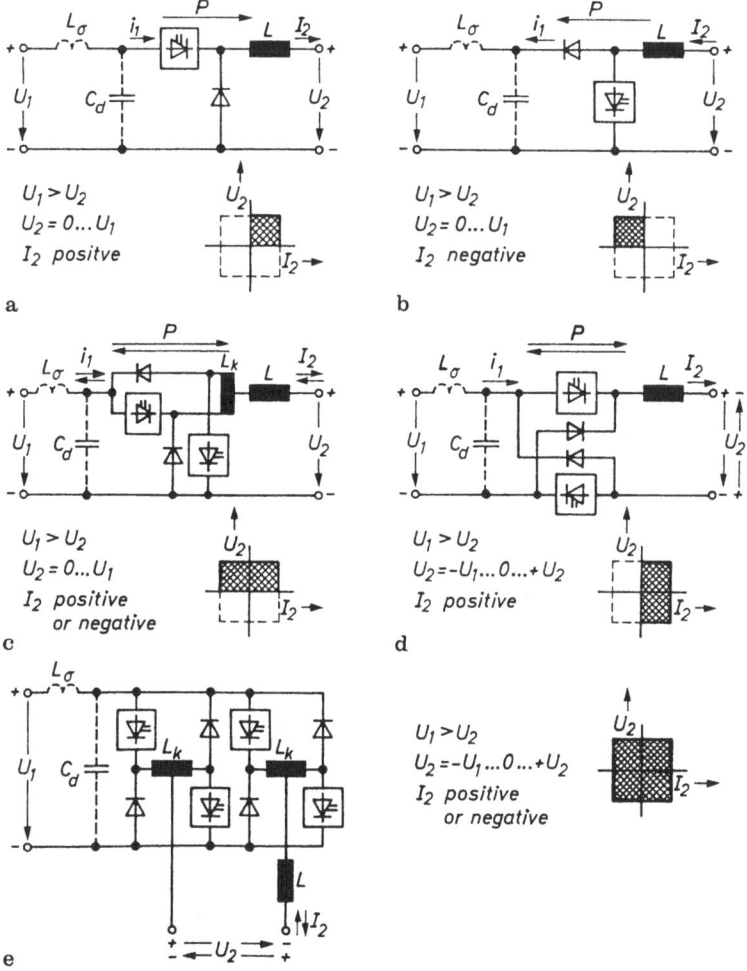

Fig. 8.6a–e. Extension of the dc chopper into four-quadrant operation. a,b Single-quadrant operation; c two-quadrant operation with current reversal; d two-quadrant operation with voltage reversal; e four-quadrant operation (self-commutated inverter)

The voltage and current waveforms for energy recovery are shown in Fig. 8.5b. Here again, pulse-shaped voltage blocks u_2 occur on the load side with the lower mean value U_{2av} while pulse-shaped currents i_1 flow on the other side. The transformation Eqs. (8.15) and (8.16) also apply with this connection. They can be used for the regenerative braking of dc motors down to very low speeds.

The two connections dealt with up to now and shown in Figs. 8.4 and 8.5 each permit only single-quadrant operation, because the polarity of the voltage U_2 and the direction of the current I_2 across the load are preset and cannot be altered.

In Fig. 8.6a and b the two connections are reproduced once again. P indicates the direction of energy flow. The possible operating range in the dc voltage/dc current plane is hatched in.

The connections can, however, be combined so that multi-quadrant operation can be realized. In Fig. 8.6c and d two connections are given for two-quadrant operation, c for current reversal and d for voltage reversal. In both cases the direction of energy flow P can be altered. With the connection for current reversal a center-tapped commutating choke L_k is necessary to decouple the quenchable thyristor switch from the antiparallel diodes. These commutating chokes prevent the quenching current from flowing away unhindered via the antiparallel diode.

The connection shown in Fig. 8.6e renders four-quadrant operation of the dc power controller possible. Not only the voltage U_2 but also the current I_2 on the load side can assume both directions. This connection is already a pure inverter connection. To generate a voltage system of variable frequency f_2 on the load side it is only necessary to reverse the quenchable thyristor switches in rhythm with the desired frequency f_2. This connection is that of a self-commutated inverter in single-phase bridge connection. This will be dealt with in greater detail later.

8.2.4 Capacitive Quenching Circuits

Besides the capacitive quenching circuit using a quenching thyristor and an oscillatory circuit with inductance and blocking diode (the only capacitive quenching circuit considered so far) a series of other quenching circuits are also used (Fig. 8.7).

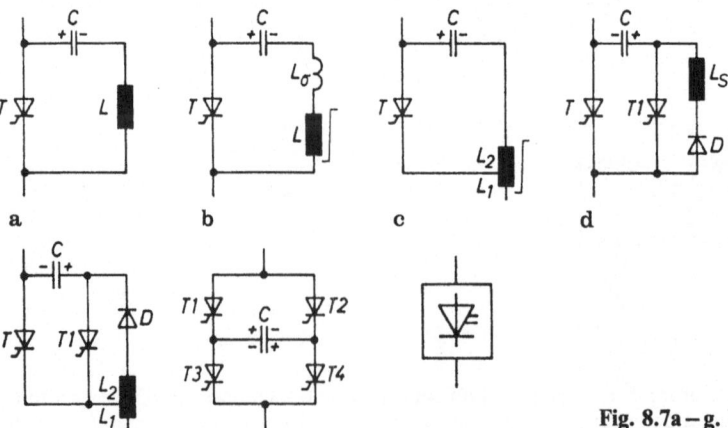

Fig. 8.7a–g. Various capacitive quenching circuits

On cyclic operation in dc power controllers the quenching capacitor in each capacitive quenching circuit is automatically charged or recharged in the opposite direction to the polarity needed for quenching. If in the simplest case this occurs via a resistance, losses arise (at least $Cu^2/2$ on each charge). This method is therefore uneconomic for cyclic operation and is not practical at high frequencies. The connections shown in Fig. 8.7 therefore all work without resistance, and ideally also without charging losses.

With circuit a an LC resonant circuit lies in parallel with the thyristor T. When the capacitor C is initially charged to the polarity shown it reacharges in the opposite polarity via the inductance L when the thyristor is turned on. When the current in the LC resonant circuit swings back the thyristor current is interrupted so long as the amplitude of the resonant circuit current is greater than the load current. With this circuit the load current is adjusted by altering the pulse frequency f_p.

Employing an oscillatory circuit choke L with saturable iron core (rectangular hysteresis loop) produces circuit b. This circuit is called the Morgan circuit after the inventor [8.2]. The unsaturated inductance L_σ first delays recharging of the quenching capacitor C in the opposite direction after triggering of the thyristor T thereby increasing the 'on' time. After successful remagnetization of the saturable inductance L the capacitor recharges in the opposite direction with the leakage inductance L_σ of the quenching circuit. After renewed remagnetization of the inductance L with the opposite polarity the thyristor current is then interrupted.

Current-sensitive charging of the quenching capacitor is achieved by tapping the saturable inductance (circuit c). The 'on' time of the thyristor T can be additionally influenced by premagnetisation of the saturation inductance.

Circuit d shows the quenching circuit already dealt with employing an auxiliary thyristor and oscillatory circuit. The quenching process begins with the indicated polarity of the voltage on capacitor C by triggering the auxiliary thyristor T1. The quenching capacitor then recharges to the opposite polarity. On renewed triggering of the main thyristor T the quenching capacitor C charges up again via the inductance L_s to the polarity required for the next quenching process. The blocking diode D prevents the capacitor from swinging back again.

Curcuit e represents a variant of this oscillatory circuit. The quenching capacitor C is charged to a voltage dependent upon the load current via a tapped inductance. At high currents the voltage induced by the load current rises, and hence the quenching voltage rises producing a higher voltage stress across the thyristor.

With circuit f the pair of thyristors T1 and T3 and the pair T2 and T4 carry the load current alternately. The current in thyristor T1 is interrupted by triggering thyristor T2. Conversely, on the next turn-off process thyristor T1 interrupts the current in thyristor T2 via the quenching capacitor C. With such a push-pull quenching circuit each recharging process of the quenching capacitor C in the opposite direction is utilized to interrupt the current. The quenching capacitor is therefore recharged in the opposite direction at only half the pulse frequency f_p. With all other circuits the quenching capacitor is recharged at the pulse frequency itself.

The symbol shown in circuit g is generally used for a valve arm with any associated quenching turn-off circuit i.e. for a quenchable converter valve.

8.2.5 Control Techniques

A dc chopper can be controlled in various ways. The following control techniques are used: pulse duration control, pulse frequency control, and two-step control (Fig. 8.8).

With *pulse duration control* the periodic time T is constant. The 'on' period T_1 between triggering of the main thyristor and triggering of the quenching thyristor is varied (Fig. 8.8a).

With *pulse frequency control* the 'on' period T_1 between triggering of the main thyristor and triggering of the quenching thyristor is kept constant while the periodic time T is varied (Fig. 8.8b).

With *two-step control* the instants of triggering and quenching are made to vary as the momentary value of the current or voltage across the load (Fig. 8.8c). The quenching thyristor is triggered when the current or voltage across the load exceeds a preset reference value.

The main thyristor is triggered when these drop below another preset reference value.

Two-step control can only be used when there is an energy store in the load circuit. It operates neither with constant pulse frequency f_p nor with constant 'on'

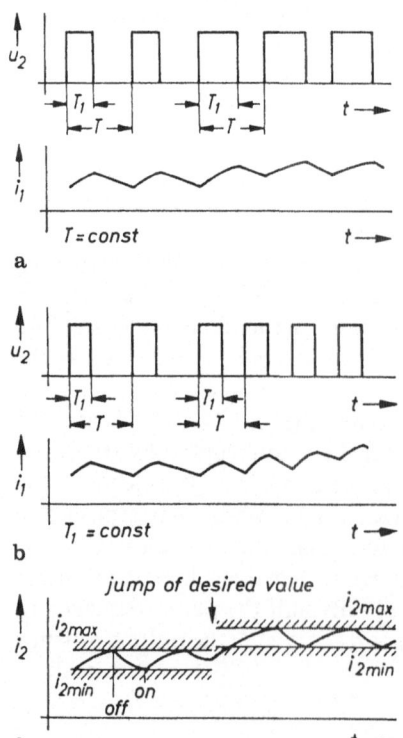

Fig. 8.8a – c. Control techniques with dc chopper. **a** Pulse duration control; **b** pulse frequency control; **c** two-step control of the current

time T_1. With this control technique the desired mean value of load current or voltage is preset as reference value. The actual value of current or voltage must be detected on the load side.

8.2.6 Calculation of Smoothing Inductance and Smoothing Capacitor Values

To convert voltage or current mean values a dc power controller needs at least one energy store namely a smoothing inductance on the side with the smaller direct voltage mean value. Up to now this smoothing inductance has been assumed to be vary large, and hence the load current I_2 has been assumed to be constant.

The size of smoothing inductance needed is now calculated (Fig. 8.9). With the main thyristor triggered the load current is given by the differential equation

$$L\frac{di_2}{dt} = U_1 - U_2, \tag{8.17}$$

if resistive voltage drops are ignored.

When the semiconductor switch opens the load current flows through the freewheeling diode. In this case the following differential equation applies

$$L\frac{di_2}{dt} = -U_2. \tag{8.18}$$

From these two equations the smoothing inductance required in the load circuit can be calculated for a specified pulse frequency $f_p = 1/T$ and a permissible current ripple amplitude Δi_2. From Eqs. (8.17) and (8.18) one obtains the determining equations

$$L = \frac{(U_1 - U_2)T_1}{\Delta i_2} \tag{8.19}$$

and

$$L = \frac{U_2 T_2}{\Delta i_2}. \tag{8.20}$$

The maximum current ripple amplitude occurs when $T_1 - T_2 = T/2$. Then $U_2 = U_1/2$. This yields

$$L = \frac{U_1 T}{4\Delta i_2} = \frac{U_1}{4f_p \Delta i_2} \tag{8.21}$$

from which the required smoothing inductance can be calculated.

A smoothing capacitor C_d is required when the direct voltage source U_1 has too high an internal inductance L_σ. The buffer capacitor then supplies the pulse-

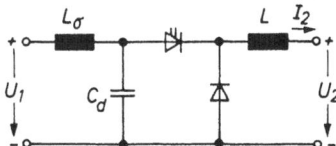

Fig. 8.9. DC chopper with smoothing capacitor C_d and smoothing inductance L

shaped current blocks needed by the dc chopper. Practically all direct voltage sources with the exception of storage batteries have such high internal inductances that a smoothing capacitor is required for the connection of a dc chopper.

Assuming for simplification that the smoothing capacitor supplies the entire alternating portion of the current needed by the dc power controller on the primary side while only the direct portion flows from the dc voltage source U_1 a voltage ripple Δu_c takes place across the smoothing capacitor C_d. This can be calculated for a given 'on' and 'off' period and a given load current I_2:

$$\Delta u_C = \frac{1}{C_d} \frac{T_1 T_2}{T_1 + T_2} I_2. \tag{8.22}$$

The maximum voltage ripple again occurs when $T_1 = T_2 = T/2$. From Eqs. (8.22) the equation for the smoothing capacitor can then be obtained

$$C_d = \frac{I_2 T}{4 \Delta u_C} = \frac{I_2}{4 f_p \Delta u_C} \tag{8.23}$$

from which the value of the smoothing capacitor can be calculated for a specified permissible voltage fluctuation Δu_{Cperm}.

If electrolytic capacitors are used as smoothing capacitors, their size depends upon the voltage fluctuations permitted for electrolytic capacitors in the data sheets.

8.2.7 Pulse-controlled Resistance

Gate turn-off thyristor switches can also be arranged in parallel or in series with resistances R. This creates the possibility of varying the effective resistance R^* as a function of the control factor $\lambda = T_1/T$. Such a special form of dc chopper is also called a pulse-controlled resistance.

Pulse-controlled Resistance in Parallel Connection. If a quanchable thyristor switch is arranged in parallel with a resistance R (Fig. 8.10), the effective resistance R^* can be infinitely varied between zero (with continuously turned on-thyristor) and R (with the thyristor turned-off). An energy store in the form of an inductance L is necessary to smooth the load current I.

With the current I assumed to be constant the current i_C in the quenching capacitor becomes

$$i_C = I + \left(\frac{U_{C1}}{R} \right) e^{-\frac{t - t_1}{\tau}} \tag{8.24}$$

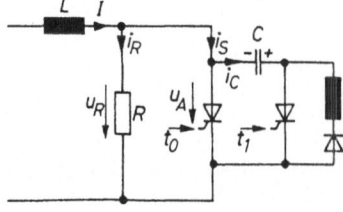

Fig. 8.10. Pulse-controlled resistance in parallel connection

where $\tau = RC$ when the quenching capacitor C is charged at the instant of turning on the auxiliary thyristor t_1 to the voltage U_{C1} with the polarity indicated in Fig. 8.10.

The voltage u_R across the resistor R after the instant of quenching t_1 is given by

$$u_R = Ri_R = RI - (RI + U_{C1})e^{-\frac{t-t_1}{\tau}}. \tag{8.25}$$

Due to the parallel connection of thyristor and resistance this voltage equals the voltage u_A across the main thyristor.

The end of the hold-off internal Δt is obtained from the condition $u_A = u_R = 0$. From Eq. (8.25) one obtains the hold-off interval

$$t_c = RC \ln\left(1 + \frac{U_{C1}}{RI}\right) \tag{8.26}$$

and the required quenching capacitor

$$C = \frac{t_c}{R \ln\left(1 + \dfrac{U_{C1}}{RI}\right)}. \tag{8.27}$$

The effective resistance R^* can be infinitely adjusted between zero and R under the assumption that $\tau = RC \ll T$

$$R^* = \frac{T_1}{T} R = \lambda R. \tag{8.28}$$

Pulse-controlled Resistance in Series Connection. As shown in Fig. 8.11 a quanchable thyristor switch can also be connected in series with a resistance R.

For this connection the hold-off interval can be obtained from

$$t_c = RC \ln\left(1 + \frac{U_{C1}}{U}\right) \tag{8.29}$$

when the quenching capacitor C is charged up at the instant of quenching t_1 to the indicated polarity. The capacitance C of the required quenching capacitor is obtained from

$$C = \frac{t_c}{R \ln\left(1 + \dfrac{U_{C1}}{U}\right)}. \tag{8.30}$$

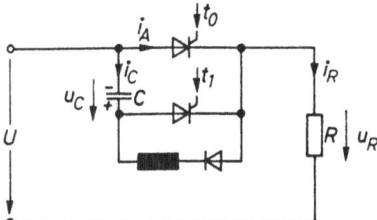

Fig. 8.11. Pulse-controlled resistance in series connections

Assuming $\tau = RC \ll T$ the effective resistance is

$$R^* = \frac{R}{T_1/T} = \frac{R}{\lambda}. \tag{8.31}$$

Therefore, the effective resistance R^* can be adjusted between R and infinity when the control factor λ varies from 1 to 0.

8.2.8 Analysis of a Capacitive Quenching Process

A capacitive quenching process in two commutation steps is analyzed, taking the oscillatory circuit as an example. Figure 8.12a shows the circuit considered. The load consists of an inductance L in series with a resistance R. Let a leakage inductance L'_σ be present in the turn-off arm, and let the leakage inductance of the direct voltage source U_d be L''_σ.

The following values are assumed: $U_d = 500$ V, $R = 5\ \Omega$. $L + L_\sigma = 1$ mH, $L_\sigma = L'_\sigma + L''_\sigma = 15\ \mu H$ with $L'_\sigma = 5\ \mu H$ and $L''_\sigma = 10\ \mu H$. Resistive losses are ignored (apart from the load resistance).

The waveforms of current and voltage during a quenching process are now simple to calculate. To this end, several permissible simplifications are made which do not impair the accuracy of the result.

If a hold-off interval $t_c = 50\ \mu s$ is required for the main thyristor, the quenching capacitor value can be calculated according to Eq. (8.8).

$$C = \frac{I_2 t_c}{U_{C1}} = \frac{100\ \text{A} \cdot 50\ \mu s}{500\ \text{V}} = 10\ \mu F. \tag{8.32}$$

If at instant t_1 (when the capacitor voltage $U_{C1} = -U_d$) the auxiliary thyristor T2 is triggered, the load current I_2 commutates from the main thyristor T1 to the auxiliary thyristor T2. The rate of rise of current then occurring is obtained from

$$\frac{di_C}{dt} = \frac{U_{C1}}{L'_\sigma} = \frac{500\ \text{V}}{5\ \mu H} = 100\ \text{A}/\mu s. \tag{8.33}$$

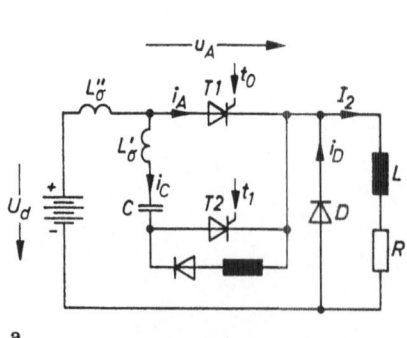

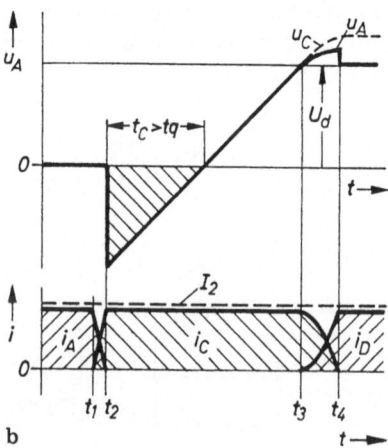

a   b

Fig. 8.12a,b. Voltage and current waveforms during capacitive quenching

At instant t_2 this commutation process is completed. The quenching capacitor C then carries the load current I_2. The commutating time $t_2 - t_1$ is approximately

$$t_2 - t_1 \approx \frac{L'_\sigma I_2}{U_d} = \frac{5 \text{ μH } 100 \text{ A}}{500 \text{ V}} = 1 \text{ μs.} \tag{8.34}$$

During this time a voltage decay Δu_C has occurred across the capacitor C

$$\Delta u_C \approx \frac{\frac{1}{2} I_2 (t_2 - t_1)}{C} = \frac{\frac{1}{2} 100 \text{ A } 1 \text{ μs}}{10 \text{ μF}} = 5 \text{ V.} \tag{8.35}$$

From t_2 to t_3 the quenching capacitor C recharges almost linearly with the opposite polarity. The capacitor voltage u_C is

$$u_C \approx -U_{C2} + \frac{I_2 (t - t_2)}{C} \tag{8.36}$$

where $U_{C2} = U_d - \Delta u_C = 495$ V.

By setting u_C in Eq. (8.36) equal to zero the actual hold-off interval t_c is obtained as

$$t_c \approx \frac{C U_{C2}}{I_2} = \frac{10 \text{ μF } 495 \text{ V}}{100 \text{ A}} = 49.5 \text{ μs.} \tag{8.37}$$

The current i_D in the freewheeling diode D is initiated at instant t_3 when the capacitor voltage u_C begins to be higher than the dc voltage U_d. Then

$$t_3 - t_2 \approx \frac{C(U_{C2} + U_d)}{I_2} = \frac{10 \text{ μF }(495 \text{ V} + 500 \text{ V})}{100 \text{ A}} = 99.5 \text{ μs.} \tag{8.38}$$

Afterwards the current commutates in a sinusoidal quarter oscillation from the quenching capacitor C into the freewheeling diode D. The capacitor voltage is given by equation

$$u_C = U_d + \sqrt{\frac{L_\sigma}{C}} I_2 \sin v_0 (t - t_3) \tag{8.39}$$

and the capacitor current by equation

$$i_C = I_2 [1 - \cos v_0 (t - t_3)] \tag{8.40}$$

where

$$v_0 = \frac{1}{\sqrt{L_\sigma C}} \tag{8.41}$$

is the angular frequency of the series resonant circuit consisting of L_σ and C. The capacitor voltage and current perform a quarter oscillation from t_3 to t_4. So

$$t_4 - t_3 = \frac{T_0}{4} = \frac{\pi}{2 v_0} = \frac{\pi}{2} \sqrt{L_\sigma C} = \frac{\pi}{2} \sqrt{15 \text{ μH } 10 \text{ μF}} = 19,2 \text{ μs.} \tag{8.42}$$

The overvoltage across capacitor C at instant t_4 still has to be calculated. From Eq. (8.39)

$$\hat{u}_C = U_d + \sqrt{\frac{L_\sigma}{C}} I_2 = 500 \text{ V} + \sqrt{\frac{15 \text{ μH}}{10 \text{ μF}}} 100 \text{ A} = 500 \text{ V} + 122 \text{ V} = 622 \text{ V.} \tag{8.43}$$

After completion of the quenching process it follows that as a result of the leakage reactance L_σ, a voltage increase of approximately 25% of the dc voltage U_d arises in the quenching capacitor C. This stresses the thyristors as reverse voltage. L_σ must therefore be kept as low as possible and this can be achieved, for example by an additional smoothing capacitor C_d which would to a great extent eliminate the leakage reactance L''_σ of the dc voltage source.

8.2.9 Construction of an Energy Balance-sheet

The overvoltage across the quenching capacitor C at instant t_4 can also be obtained by constructing an energy balance-sheet. At instant t_3 the electrical energy stored in the quenching capacitor is $CU_d^2/2$, the magnetic energy stored in the leakage inductance L_σ is $L_\sigma I_2^2/2$. At instant t_4 the capacitor energy is $C\hat{u}_C^2/2$. The magnetic energy is zero because no more current flows. This is, however, not sufficient for the preparation of the energy balance-sheet, because between t_3 and t_4 the direct voltage source U_d supplies additional energy

$$U_d \int_{t_3}^{t_4} i \; dt = U_d \Delta Q = U_d C (\hat{u}_C - U_d). \tag{8.44}$$

The complete energy balance-sheet is therefore

$$\frac{1}{2}C\hat{u}_C^2 = \frac{1}{2}CU_d^2 + \frac{1}{2}L_\sigma I_2^2 + U_d C (\hat{u}_C - U_d). \tag{8.45}$$

From this the peak value $\hat{u}_C$ of the voltage across the quenching capacitor is

$$\hat{u}_C = U_d + \sqrt{\frac{L_\sigma}{C}} I_2 \tag{8.46}$$

which is the same result as in Eq. (8.43).

In general it is sufficient to calculate commutation processes in self-commutated converters approximately i.e. with permissible simplifications. Besides the energy balance other considerations can be of use in the approximate calculation of processes in circuits.

1. The continuity conditions for currents in inductances and voltages across capacitors. Neither electrical quantity makes jumps even at instants of switching (see Sect. 5.1)
2. The current rise in a circuit is obtained from the sum of the effective voltage divided by the inductance. Thus

$$\frac{di}{dt} = \frac{\Sigma U}{L}. \tag{8.47}$$

3. The voltage rise across a capacitor is obtained from the ratio of current to capacitance

$$\frac{du_C}{dt} = \frac{I}{C}. \tag{8.48}$$

Table 8.1. Formulae for dc chopper

Quenching capacitor:

(communication capacitor) $C_k = \dfrac{I \cdot t_e}{\Delta U_C}$

t_e = hold-off interval

Mean voltage value: $U_{2av} = \dfrac{T_1}{T_1 + T_2}$ $U_1 = \dfrac{T_e}{T} U_1$

U_1 = dc voltage on side 1, U_{2av} = (mean value of the) dc voltage on side 2

Mean current value: $I_{1av} = \dfrac{T_1}{T_1 + T_2}$ $I_2 = \dfrac{T_1}{T} I_2$

I_{av} = (mean value of the) dc current on side 1, I_2 = dc current on side 2

Pulse control factor: $\dfrac{T_1}{T_1 + T_2} = \dfrac{T_1}{T} = T_1 \cdot f_p$

T_e = "on" time, T_a = "off" time, T = (pulse) cycle duration, f_p = pulse frequency

Power: $U_1 I_{1av} = U_{2av} I_2$

Smoothing inductance:
(on side 2) $L_2 = \dfrac{T U_1}{4 \Delta i_{2\,perm}} = \dfrac{U_1}{4 f_p \Delta i_{2\,perm}}$

Smoothing capacitor:
(on side 1) $C_1 = \dfrac{T I_2}{4 \Delta u_{1\,perm}} = \dfrac{I_2}{4 f_p \Delta u_{1\,perm}}$

Pulse-controllable resistance:

Parallel connection: $C_k = \dfrac{\Delta t}{R \ln\left(1 + \dfrac{U_{C1}}{RI}\right)}$ $R^* \approx \dfrac{T_1}{T_1 + T_2} R$ (von 0...R)

Series connection: $C_k = \dfrac{\Delta t}{R \ln\left(1 + \dfrac{U_{C1}}{U}\right)}$ $R^* \approx \dfrac{T_1 + T_2}{T_1} R$ (von R...∞)

R = resistance, R* = effective resistance, U_{C1} = voltage across the quenching capacitor at quenching time t_1

Construction of an energy balance-sheet for the start and end of a commutation process often saves the construction and solution of differential equations. Besides the magnetic and electrical stores the energy portions supplied by or drawn from the voltage sources during the commutation period must also be taken into consideration.

Table 8.1 lists the formulae for dc power controller (dc chopper) including pulse controlled resistances.

8.3 Self-commutated Inverters

Self-commutated inverters fulfill the basic function of inversion i.e. the conversion of direct current into alternating current. Since with these inverters commutation is carried out by energy stores belonging to the converter (quenching capacitors)

or by increasing the resistance of the converter valve to be quenched, they are not dependent upon a separate source of alternating voltage e.g. an ac power supply system or the load.

The ac voltage generated can therefore generally be varied over a wide range of frequency. With multi-phase self-commutated inverters a multi-phase ac voltage system of variable frequency can also be generated. Under certain conditions besides frequency variation control of the single-phase or multi-phase ac output voltage is also possible.

8.3.1 Single-phase Self-commutated Inverters

Parallel Inverter. The circuit known as a parallel inverter is illustrated in Fig.8.13. This is a self-commutated inverter in center tap connection with which the quenching capacitor C_k is arranged in parallel between the controllable valves, which relieve each other in carrying the current.

The basic circuit was invented by Alexanderson in 1923 [8.1]. Similar circuits with mechanical circuit-interrupting contacts were used even earlier in calling systems in telephone engineering to generate ac current from a dc battery (the so-called polarity reversers or vibrating converters). Due to the reverse-current diodes D1 and D2 first used by Peterson, the parallel inverter is also in a position to supply reactive current i.e. ac current of any phase angle. The commutating choke L_k decouples the quenchable converter valves from the uncontrollable reverse-current diodes [8.3].

With a resistive-inductive load the commutation process takes place in several stages.

Under the assumption that thyristor T1 is carrying the current the quenching capacitor (with the indicated voltage polarity) due to triggering of thyristor T2 discharges through T1 and interrupts its current. Afterwards the load current temporarily flows through the capacitor C_k and thyristor T2 recharging C_k with the opposite polarity. As soon as the capacitor voltage u_C is higher than $2U_d$ the current commutates into the reverse-current diode D2. The current on the dc side then reverses its direction. Finally the current on the load side also changes its direction, so the current in D2 is extinguished and T2 eventually assumes the load current. The next commutation occurs correspondingly with reversed indices.

In order that no circulating currents flowing through the thyristors and diodes (via T1 and D1 or T2 and D2) can form in the commutating choke L_k the reverse-current diodes are connected to tappings on the converter transformer.

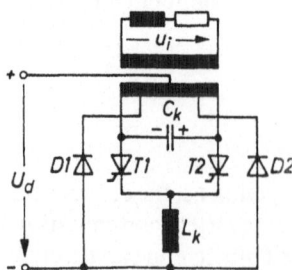

Fig. 8.13. Parallel inverters in single-phase center tap connection

The ideal output ac voltage is square-shaped. Its amplitude $U_i = 2U_d$ (with a turns ratio $w_1/w_2 = 1$). The rms value U_{1i} of the fundamental oscillation is obtained from

$$U_{1i} = \frac{4\sqrt{2}}{\pi} U_d. \qquad (8.49)$$

With a negative counter-voltage on the load side the quenching capacitor C_k can lose part of its voltage after successful recharging to $2U_d$. In order to prevent this, blocking diodes are connected in series with the thyristors behind which the quenching capacitor is connected.

Two-pulse Bridge Connection. A single-phase self-commutated inverter can also be constructed in bridge connection [8.14]. With the extension of the dc power controller to four-quadrant operation such a connection has already been produced (see Fig. 8.6e). This connection corresponds to a two-pulse bridge connection. It is indicated with the symbol for any turn-off arms.

Figure 8.14a shows a further connection for a self-commutated inverter. With this arrangement of the quenching capacitors it is a parallel inverter in two-pulse bridge connection. The reverse-current diodes (also called the regenerative arms) again render alternating current of any phase position (double converter) possible. The blocking diodes prevent the quenching capacitors discharging.

The voltage generated on the ac side is square-shaped (Fig. 8.14b). With constant direct voltage U_d the ideal output ac voltage is also constant. With $U_i = U_d$ the fundamental oscillation for the rms value U_{1i} is

$$U_{1i} = \frac{2\sqrt{2}}{\pi} U_d. \qquad (8.50)$$

Its frequency can be preset by the control system. It can be infinitely varied from zero to an upper limiting value. This is determined amongst other things by the hold-off interval necessary for the thyristors.

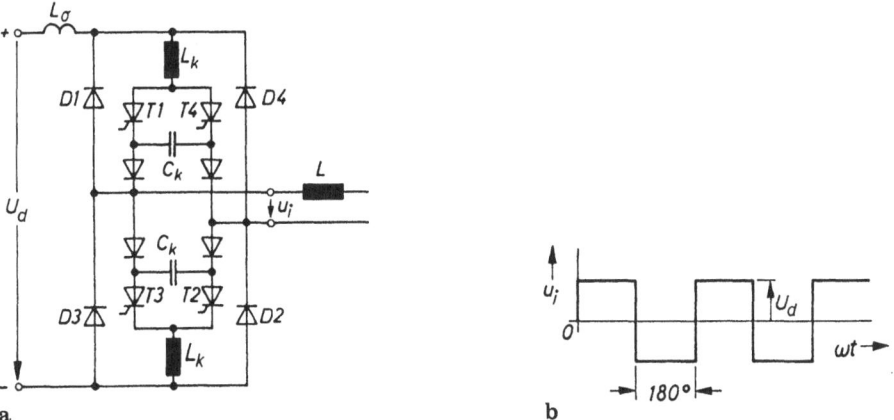

Fig. 8.14a,b. Parallel inverters in single-phase bridge connection

8.3.2 Multi-phase Self-commutated Inverters

Three-phase AC Bridge Connection. Three-phase ac voltage systems can be created by self-commutated inverters in multi-phase connections. Figure 8.15a shows the basic circuit of a three-phase self-commutated inverter. It is a six-pulse bridge connection with turn-off and regenerative arms. Such an inverter supplies the ac voltage indicated in Fig. 8.15b. With a resistive load only the thyristor arms carry current. If reactive power occurs on the load side, the diodes (regenerative arms) also periodically participate in carrying the current. On reversal of the energy direction the diodes take over carrying the current.

At constant dc voltage U_d the output ac voltage is also constant with multi-phase inverters provided no special steps (such as phase angle control or pulse control) are taken (see Sect. 8.3.3).

For the three-phase ac bridge connection with 120° square-wave voltage the rms value U_i of the line voltage on the valve side

$$U_i = \sqrt{\frac{2}{3}} U_d \qquad\qquad (8.51)$$

and for the fundamental oscillation

$$U_{1i} = \frac{\sqrt{6}}{\pi} U_d. \qquad\qquad (8.52)$$

The frequency of the three-phase voltage system on the output side can be freely preset via the control system.

Arrangement of the Quenching Capacitors. The connection indicated in Fig. 8.15 by the general symbol for a thyristor with associated turn-off arm can be realized in various ways.

As with the parallel inverter the quenching capacitors can be arranged between the mutually commutating phases. This is called sequential phase quenching. With this circuit each thyristor to carry current quenches the previous one.

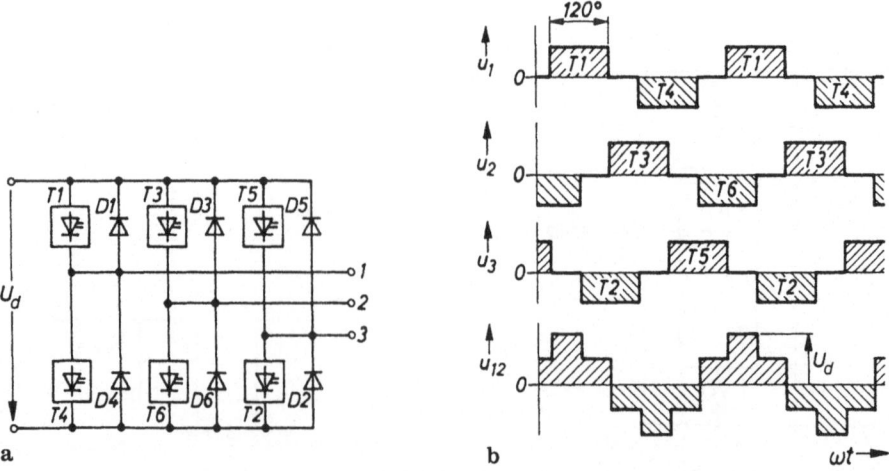

Fig. 8.15a,b. Voltage waveform of an inverter in three-phase bridge connection

However, circuit with auxiliary thyristors can also be constructed with the help of which the arm of the bridge actually carrying current is quenched via a commutating capacitor C_k [8.37, 8.39, 8.40]. Figure 8.16 shows three different quenching circuits. The commutating circuit for thyristor Tl is, in each case, drawn in dotted.

Quenching the six arms of the bridge from a common commutating capacitor C_k via auxiliary thyristors is called total quenching (Fig. 8.16a).

With phase quenching the upper and lower arms of the bridge of one phase of the inverter are quenched alternately via a commutating capacitor C_k and two auxiliary thyristors (Fig. 8.16b). Altogether three commutating capacitors and six quenching thyristors are required.

Each arm of the bridge can also be allocated its own commutating capacitor C_k (Fig. 8.16c). In this case the process is called individual quenching.

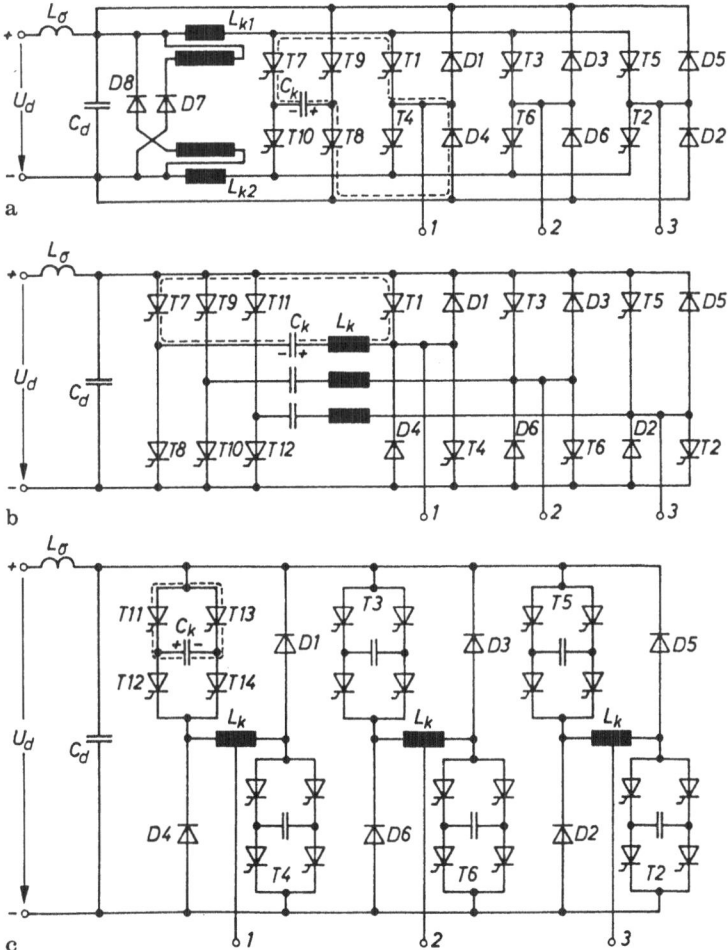

Fig. 8.16a–c. Various quenching circuits with self-commutated inverters in three-phase bridge connection. **a** Total quenching; **b** phase quenching; **c** individual quenching

8.3.3 Voltage Control

Self-commutated inverters generate a single-phase or three-phase voltage system of adjustable frequency. According to Eqs. (8.49), (8.50), and (8.52) the ac output voltage is proportional to the dc voltage U_d. If the ac output voltage is also variable, voltage control can be carried out either on the dc side in the inverter itself or on the ac side.

In Fig. 8.17 an ac converter with a dc voltage link circuit is illustrated [8.6]. If the line-commutated rectifier on the input side can be controlled, then the voltage U_d in the dc voltage link circuit can be set as can the ac output voltage U_2. A prerequisite is that the commutating devices operate even at the lowest dc voltage U_d.

On the ac side the voltage can, of course, be altered by means of a regulating transformer. Its size grows, however, as the frequency becomes smaller.

The ac converter shown in Fig. 8.17 has a dc voltage link circuit i.e. the dc voltage in the link circuit is impressed by a smoothing capacitor C_d. Instead of a capacitor a smoothing inductance L_d can also lie in the link circuit and impress the current. Then an ac converter with a dc current link circuit is created [8.11, 8.31]. The different properties of ac converters with impressed voltage (dc voltage link circuit) or with impressed current (dc current link circuit) will be discussed in more detail in Sect. 11.3.

Phase Control and Phase-shifting Technique. With constant dc voltage U_d in the link circuit voltage control must be carried out in the inverter itself [8.16, 8.17, 8.34, 8.36]. With phase control the conduction interval in the converter arms is shortened to the controlled conduction interval as a function of the control angle α.

In control using the phase-shifting technique ac voltage of two uncontrollable inverters are phase shifted by an angle α and added (Fig. 8.18).

Shortening the voltage blocks causes the amplitude of the fundamental of the ac output voltage to be reduced. However, at the same time the harmonics in the output voltage grow relative to the fundamental. For this reason these voltage control techniques can only be employed within a limited range.

Pulse Technique. In control according to the pulse technique the converter arms are triggered and quenched several times in each period of the fundamental frequency. This is analogous to the method of operation of the dc power controller dealt with earlier. The pulse technique produces a sequence of individual conduction and idle intervals in the converter arm, the ratio of which determines the output voltage.

In Fig. 8.19 various pulse techniques for voltage control with self-commutated inverters are reproduced. Depending upon the connection either two voltage levels $+U_d$ and $-U_d$ are possible or three voltage levels $+U_d$, 0, and $-U_d$ are possible. The mean value of the voltage of a half-cycle can be controlled by altering the control factor $\lambda = T_1/(T_1 + T_2)$. Pulse techniques with three voltage levels have the advantage that the energy does not pulse unnecessarily between the load and the dc voltage link circuit [8.26, 8.27, 8.29, 8.33].

If operation is not with a constant control factor λ, but the duration of the applied voltage blocks is matched to the waveform of the sinusoidal voltage

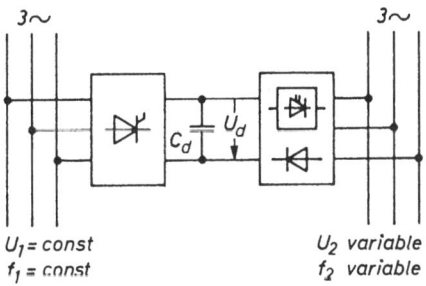

$U_1 = const$
$f_1 = const$

U_2 variable
f_2 variable

Fig. 8.17. AC converter with dc voltage link

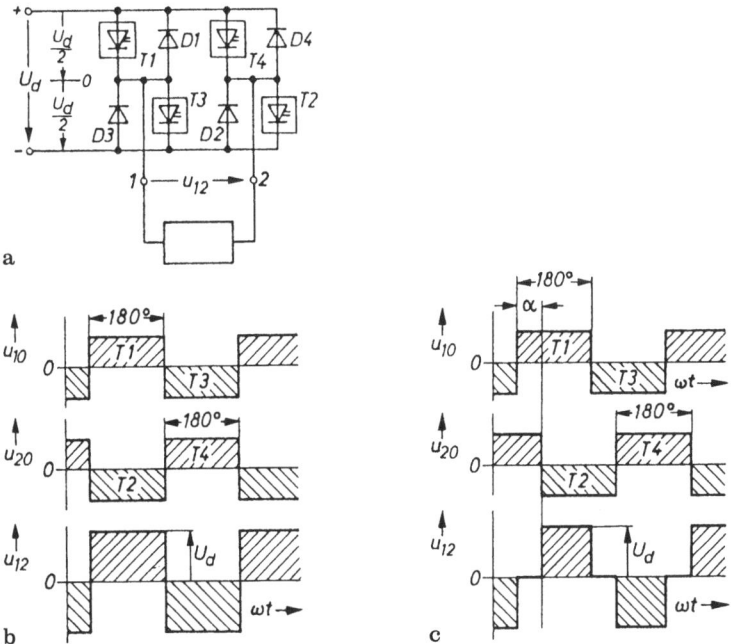

Fig. 8.18a − c. Voltage adjustment by phase control or by the phase-shifting technique. **a** Two-pulse bridge connection; **b** fully controlled; **c** partially controlled

reference, a good approximation to the sinusodial fundamental is obtained. The fundamental frequency of the output voltage thus generated is said to be sub-harmonic [8.12]. Besides the fundamental frequency only harmonics of the chosen pulse frequency f_p and still higher harmonics occur at the load. Due to the inductance on the load side a good approximation to a sine wave is obtained for the current waveform.

The pulse technique can also be extended to a direct two-step control of the load current (see Fig. 8.8c). The load current then fluctuates within a preset current interval Δi by the (generally sinusoidal) preset current reference.

8.3.4 Pulse Width Modulated (PWM) Inverter

A self-commutated inverter whose output voltage or current is controlled in open or closed loop according to the pulse techniques is called a pulse width modulated (PWM) inverter. With this inverter the total number of non-simultaneous commutations during a cycle is increased (without increasing the number of semiconductor switches) by repeated turning on and off at the pulse frequency f_p. This increase can be used to reduce current and voltage harmonics, because it corresponds to an increase in the pulse number (see Sect. 11.6). With line-commutated converters an increase in the pulse number is only possible by means of a corresponding increase in the number of converter arms.

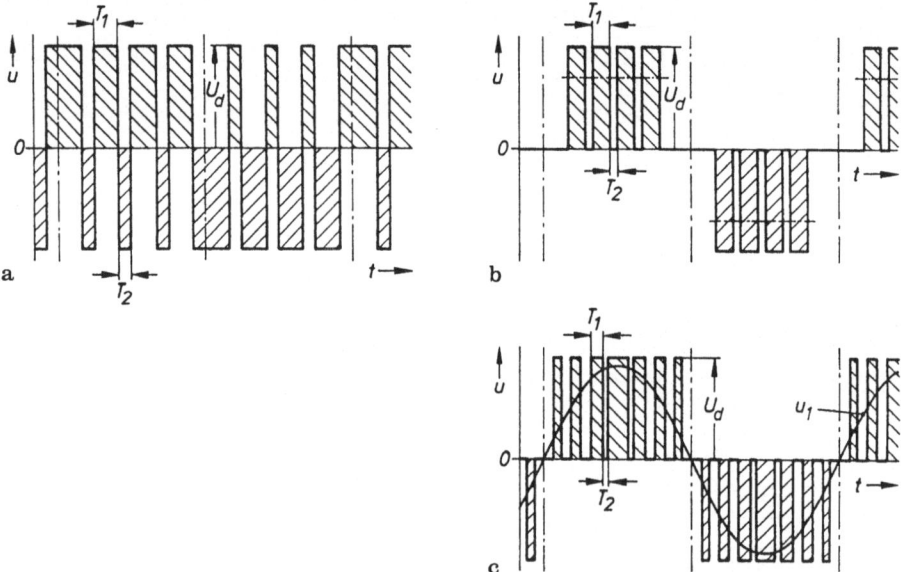

Fig. 8.19a–c. Voltage control by the pulse technique. **a** Two voltage levels $+U_d$ and $-U_d$; **b** three voltage levels $+U_d$, 0, and $-U_d$; **c** control factor $T_1/(T_1+T_2)$ variable according to sine function

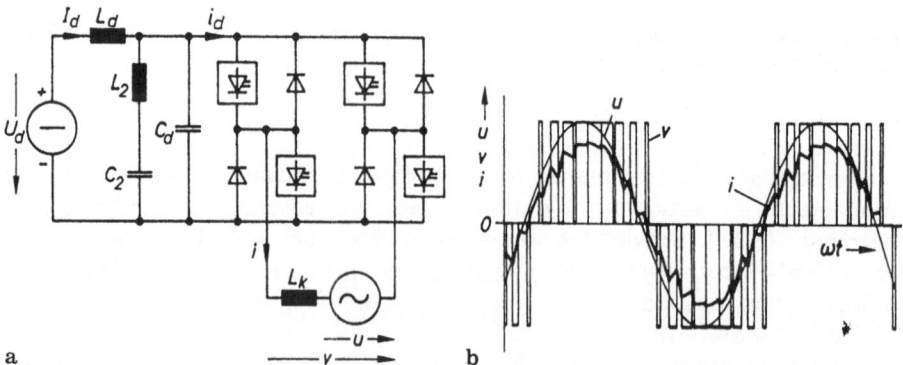

Fig. 8.20a,b. Single-phase PWM inverter. **a** Single-phase bridge connection (double converter); **b** voltage and current waveforms

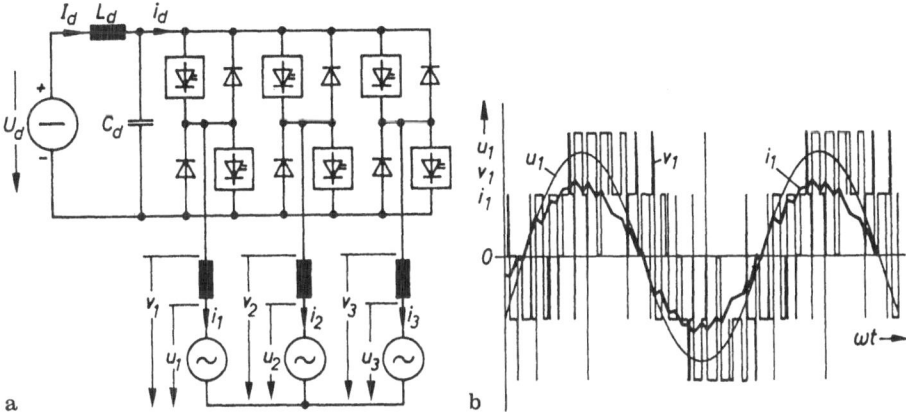

Fig. 8.21a,b. Three-phase PWM inverter. **a** Three-phase ac bridge connection (double converter); **b** voltage and current waveforms

Single-phase Bridge Connection. Figure 8.20 shows the coupling of a dc voltage source U_d with an ac voltage source u via a pulse-controlled inverter in single-phase bridge connection [8.18]. Basically this connection corresponds to that of a dc power controller extended to four-quadrant operation (see Fig. 8.6e). Each arm of the bridge consists of a quenchable converter valve and an antiparallel diode. The current i in the ac voltage source is assumed to be continuously adjustable so long as the condition $U_d \geq \sqrt{2}U$ is satisfied. The current harmonics are determined by the inductance L_k on the ac side. Therefore, the inductance must not drop below a minimum value. Since with sinusoidal current, the power in the single-phase ac voltage source u pulses at double the frequency, a sinusoidal current component of double frequency is superimposed upon the dc current I_d.

This can be supplied by a resonant circuit tuned to this frequency.

Three-phase Bridge Connection. Figure 8.21 shows the connection and voltage and current waveforms of a pulse-controlled inverter in three-phase bridge connection. Each of the six arms of the bridge again consists of the antiparallel connection of a quenchable converter valve and a diode. Here again, the currents i_1, i_2 and i_3 are controlled to be sinusoidal. With a sinusoidal current waveform the sum of the phase powers drawn on the ac side is constant i.e. equal to the power supplied by the dc voltage source U_d.

In Sect. 11.6 the energy conditions with single and multi-phase pulse-controlled converters are discussed again in more detail.

PWM inverters can be employed to control the speed of rotating-field machines [8.4, 8.5, 8.13, 8.25, 8,32, 8.35] which need a multi-phase ac system of variable frequency and a voltage proportional to the speed (see Sect. 13.1.1).

8.3.5 Converter with Sector Control

In Sect. 7.1.7 the reactive power of line-commutated converters was discussed. With phase control and a control angle α with natural commutation there is a

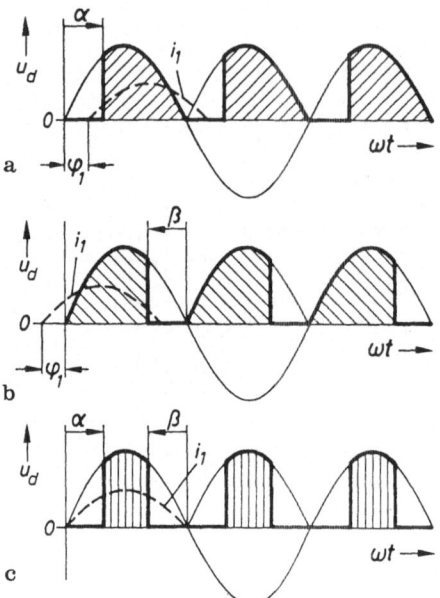

a

b

c

Fig. 8.22a – c. Sector control. **a** Delay phase control; **b** advance phase control; **c** delay and advance phase control

phase displacement φ_1 of the fundamental oscillation of the supply system current with respect to the voltage. With fully-controlled connections and smoothed dc current without vacancies φ_1 is equal to the control angle α (see Eq. (7.80)). For the single-phase or three-phase ac supply system such a load with inductive reactive power is usually undesirable.

Besides the reactive power-saving half-controllable connections described in Sect. 7.1.8 a new technique called sector control can be employed to avoid this (Fig. 8.22). The dc voltage u_d of a converter in single-phase bridge connection and the fundamental frequency component i_1 of the ac current on the line side are plotted assuming resistive load on the dc side (no smoothing inductance L_d).

With normal phase control and a control angle α the fundamental component of the line current is retarded by the phase angle φ_1 (Fig. 8.22a). If, however, the phase control is carried out from the end of the voltage half-cycle beginning with the termination angle β, there is a phase advancement of the fundamental component of current (Fig. 8.22b). In the supply system capacitive instead of inductive reactive power flows. This technique, however, cannot be used with natural commutation with line-commutated converters (see Fig. 7.18). Instead forced commutation must be employed. Finally, formation of the dc voltage can be carried out by simultaneously employing α and β control. When α and β are the same size the fundamental component i_1 of the line current remains in phase with the line voltage. Here again phase control can only be by means of forced commutation at the end of the half-cycle [8.15, 8.23, 8.24, 8.28, 8.38].

Quenchable Asymmetric Bridge Connection. Figure 8.23 shows realization of the sector control technique with the asymmetrically half-controllable bridge connection employed in converter locomotives and motor coaches of the Federal German Railways [8.20]. By incorporating a quenching capacitor C_k and two auxiliary thyristors T1' and T2', the main thyristors T1 and T2 can be turned off

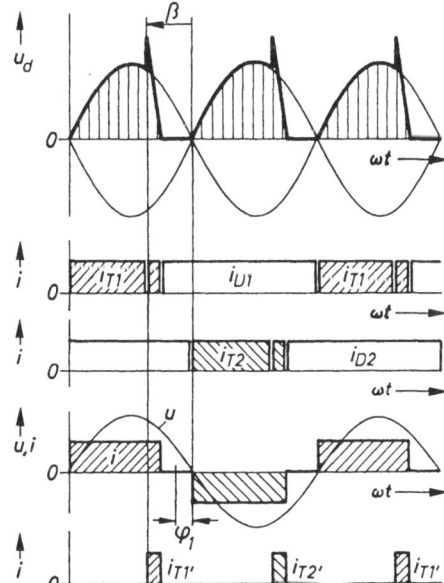

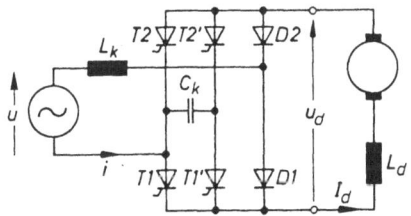

Fig. 8.23. Quenchable asymmetric bridge connection

Fig. 8.24. Voltage and current waveforms for the quenchable asymmetric bridge

without having to wait for the commutating voltage from the ac supply system. The connection is called the quenchable asymmetric bridge connection. It considerably reduces the reactive power consumption of the converter.

The fundamental power factor cos φ_1 can be made capacitive. The power factor λ is thereby brought nearer to the desired value, namely 1.

Figure 8.24 shows the voltage and current waveforms of quenchable asymmetric bridge connection. An operating condition is depicted with the termination angle β beginning at the end of the half-cycle. The current in the controllable valves T1 and T2 is thereby shortened. It is lengthened in the uncontrollable diodes D1 and D2 (see Fig. 7.24). During the quenching processes the current temporarily flows via the alternate of the quenching thyristors T1' and T2'. The ac current i on the line side is advanced by the phase angle φ_1 with respect to the ac voltage u. The ac voltage source is loaded with a capacitive component.

With sector control it should be remembered that the single-phase or three-phase ac system generally contains non-negligeable inductances L_k. The quenching capacitor C_k must therefore interrupt the current in the line inductance L_k and the magnetic energy stored there, $L_k i^2/2$, must be absorbed. In order that no excessive transient voltages are thereby generated and, moreover, no unnecessary power oscillations occur the circuit shown in Fig. 8.23 has been modified to include a capacitive intermediate energy store coupled via diodes (see Fig. 13.21).

Quenchable Half-Controllable Bridge Connections. Figure 8.25 shows several variations of half-controllable bridge connections in which the control branches are made quenchable by means of quenching capacitors C_k and auxiliary connections.

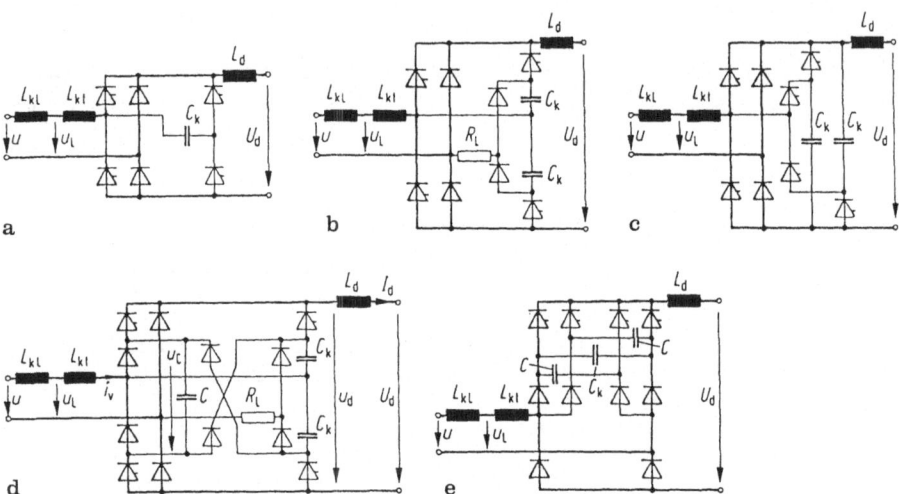

Fig. 8.25a–e. Quenchable half-controllable bridge connections. **a** Non-uniform double-way connection with one quench capacitor; **b** non-uniform double-way connection with separate quench arms; **c** non-uniform double-way connection with separate quench arms and slow quenching; **d** non-uniform double-way connection with separate quench arms and a unipolar storage capacitor; **e** non-uniform double-way connection (one ac pole controlled) with one quench capacitor and separate unipolar storage capacitors

Circuit a employs a common quenching capacitor C_k and two auxiliary thyristors. In circuit b each of the controllable main branches contains a quenching capacitor C_k and an auxiliary thyristor in addition to diodes and a resistor R_L to charge the quenching capacitor. In circuit c the ac voltage source u is in series with each of the control arms. The rate of the current rise in the control arm is therefore slowed due to the inductive component of the mains and transformer. In all of these circuits the quenching capacitors C_k must absorb the magnetic energy stored in the mains and transformer inductances which is released when the mains current is interrupted. This leads to relatively large values of capacitance. Both circuits d and e employ storage capacitors C. These capacitors (which need only withstand voltage in one direction) store the magnetic energy built up in the mains and transformer inductances. The values of the quenching capacitors C_k can therefore be reduced. However the number of auxiliary diode branches must be increased. In addition, blocking diodes must be added to the quenchable main branches.

Figure 8.26 illustrates the voltage and current curves of circuit d. The current i_v begins to flow when one of the main thyristors is triggered at a phase angle $\alpha = 0$. It commutates with an initial overlap u_0 from the non-controlled main branch to the triggered main branch. The quenching process is initiated by the triggering of the auxiliary thyristor at a phase angle of $\pi - \beta$. When the voltage u_c exceeds the momentary value of the mains voltage the mains current begins to decrease becoming zero at $\pi - \gamma$. Before the end of the half-cycle the storage capacitor C is discharged through the two primary thyristors onto the dc load. The dc voltage u_d

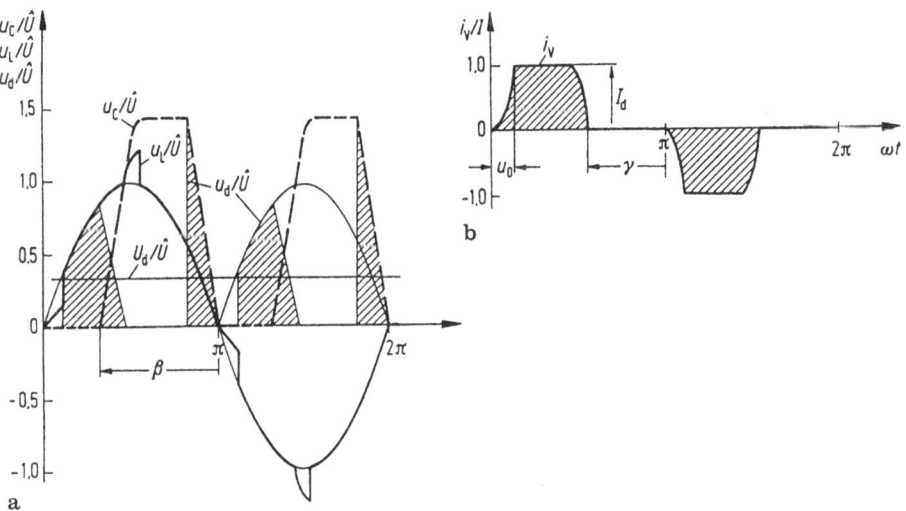

Fig. 8.26a,b. Quenchable half-controllable bridge connections with separate quench arms and a unipolar storage capacitor (see Fig. 8.25a). **a** Voltage u_L on line side, voltage u_c on capacitor and ac voltage u_d; **b** current i_V on valve side of transformer; parameters: $I_d = I_{dN}$, $U_d = 0.5\, U_{di}$, $u_{kt} = 5\%$, $u_{kL} = 1\%$, and idealized quenching process are assumed

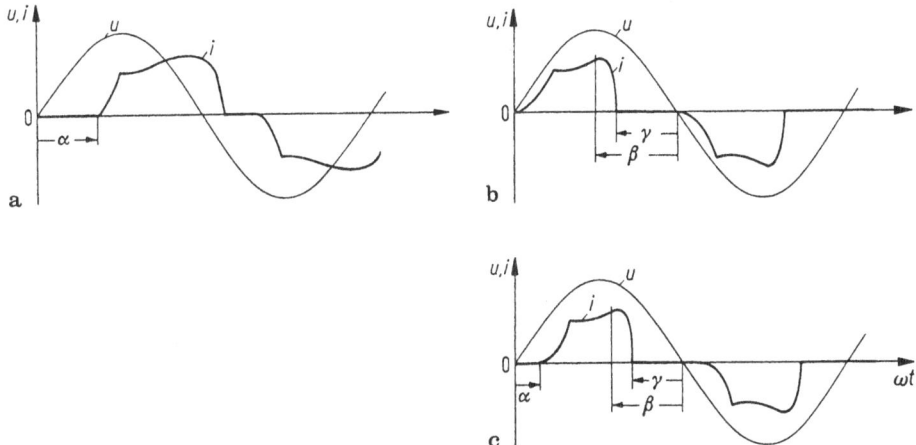

Fig. 8.27a–c. Waveform of the current in single phase mains with half-controllable bridge connections. **a** Delay phase control; **b** advance phase control; **c** sector control

is composed of the shaded areas in the voltage diagram. The line side current leads the line phase voltage, e.g., its fundamental wave has a capacitive component.

Phase and Sector Control. Figure 8.27 illustrates the line current for the various control methods. Using phase control with delay angle α (a) the line current i lags the line voltage u (inductive). Using phase control with a lead angle β (b) the line current has a positive phase shift (capacitive). This mode of operation assumes quenchable main branches. With sector control (c) a combination of delay phase

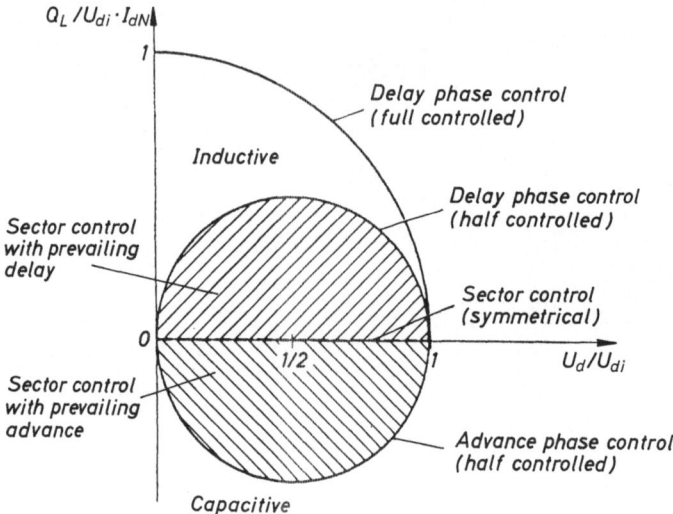

Fig. 8.28. Reactive power of the fundamental wave of current for different control methods
(angle of overlap $u = 0$ and smoothing inductance $L_d \to \infty$)

control (with a delay angle α) and advance phase control (with a lead angle β) is
used. The fundamental oscillation of the current i can be held in phase with the line
voltage u. This method requires quenchable main branches as well.

Reactive Power and Power Factor. Figure 8.28 shows the range of reactive power
(related to the fundamental wave of the current) possible for each of the different
control methods. Using a delay phase control with the half-controllable circuit
(ignoring any overlap and assuming sufficient smoothing) the reactive power
(for fundamental wave of current) describes a semi-circle in the region of
inductive reactive power. For the advance phase control this semi-circle lies in the
region of capacitive reactive power. Using sector control the operating point may
be set anywhere within the shaded area of the diagram. The reactive power factor
$\cos \varphi$ of the fundamental wave of current can be set to be inductive, unity or
capacitive. Due to the waveshape (non-sinusoid) of the current distortion power
loss is present. The (total) power factor λ is less than unity.

Applications. Circuits employing advance phase control and sector control offer
many advantages especially in the area of traction. Compared to half-controllable
bridge circuits they achieve a considerable improvement in the power factor for
single-phase railway systems.

8.4. Reactive Power Converters

Besides the four fundamental functions (rectification, inversion, dc conversion,
ac conversion, see Fig. 1.4) converters can be used for other purposes e.g. to
generate continuously controllable reactive power ([15], Sect. 6.3). Reactive
power converters generate inductive or capacitive reactive power. They can be

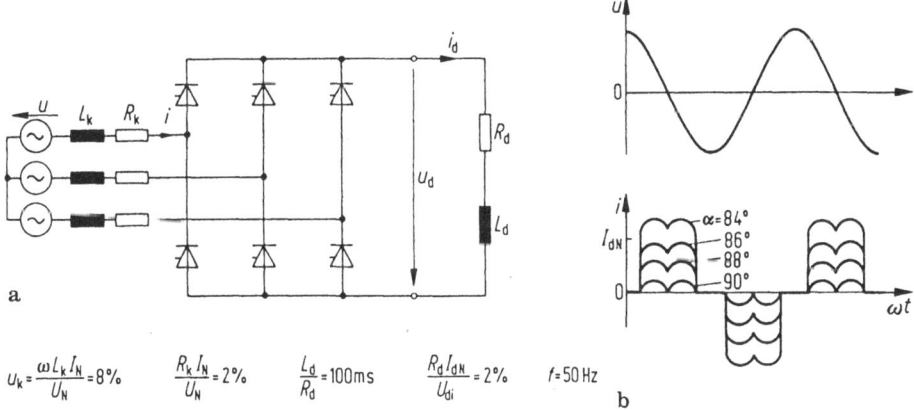

$$U_k = \frac{\omega L_k I_N}{U_N} = 8\% \qquad \frac{R_k I_N}{U_N} = 2\% \qquad \frac{L_d}{R_d} = 100\,\text{ms} \qquad \frac{R_d I_{dN}}{U_{di}} = 2\% \qquad f = 50\,\text{Hz}$$

Fig. 8.29a,b. Reactive power converter with inductive storage. **a** Circuit with assumed parameters; **b** voltage and current waveforms

externally or internally (self-)-commutated. The reactive power absorbed or supplied can be continuously regulated by the control. If one ignores their losses, reactive power converters absorb no active power on average. For the average dc power

$$U_d I_d = 0. \tag{8.53}$$

To generate reactive power, however, inductive or capacitive storage is required.
Reactive Power Converters with Inductive Storage. Figure 8.29 shows a reactive power converter in three-phase bridge connection with inductive storage L_d on the dc current side. The figure shows a line-commutated converter short-circuited on the dc side via the smoothing inductance.

The dc current I_d can be adjusted by the control angle α. In the three-phase system the ac current i has a phase shift of approximately 90° (inductive) with respect to the ac voltage.

The losses in the converter and the smoothing inductance must be made up by the three-phase system (with control angle α less than 90°). Such a line-commutated reactive power converter loads the three-phase system with inductive reactive power. With natural commutation capacitive reactive power is not possible (see Fig. 7.18).
Reactive Power Converters with Capacitive Storage. Figure 8.30 shows a reactive power converter in a three-phase bridge connection with capacitive storage on the dc voltage side. The circuit shows a self-commutated inverter (see Fig. 8.15) with turn-off and regenerative arms. There is only a smoothing capacitor C_d on the dc voltage side where there is a pure ac current i_d i.e. i_d has an average value of zero. Inductive or capacitive reactive currents i indicated in Fig. 8.30b occur in the three-phase system. The magnitude of the reactive current is determined by slightly adjusting the phasing of the inverter voltage v relative to the line ac voltage u (by triggering the quenchable principal arms with a control angle δ). This alters the magnitude of the dc voltage U_d at the smoothing capacitor C_d. U_d decreases with inductive reactive power in the three-phase system and increases with capacitive

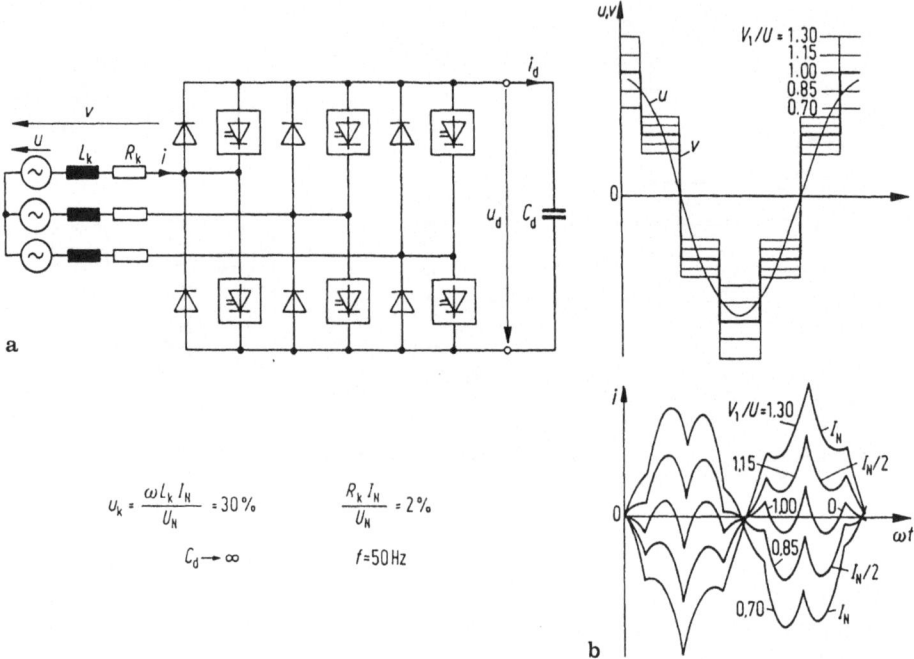

$$u_k = \frac{\omega L_k\, I_N}{U_N} = 30\%$$ $$\frac{R_k\, I_N}{U_N} = 2\%$$

$$C_d \to \infty$$ $$f = 50\,\text{Hz}$$

Fig. 8.30a,b. Reactive power converter with capacitive storage. **a** Circuit with assumed parameters; **b** voltage and current waveforms

reactive power. The following approximation holds:

$$U_d \approx 2\sqrt{2}\,U\,(1 \mp u_k I/I_N). \tag{8.54}$$

Applications. Reactive power converters can be used where it is necessary to continuously adjust the reactive current. This can be the case in three-phase systems with loads creating fluctuating reactive power demands such as arc ovens, welding machines and drives. Reactive current impulses and hence the resulting line voltage fluctuations can be eliminated within a few periods.

9 Power Systems for Converters

Converters either control the exchange of energy between power systems and loads or they couple power systems to one another. Several power system characteristics of significance for the operation of converters are described in the following. The different types of loadings for converters are covered in Chap. 10. The energy conditions are discussed in Chap. 11.

Power systems for converters are the ac or dc sources of electrical energy. A general treatment of power systems is difficult as the parameter factors can often be very complex.

9.1 Characteristics of Electrical Power Systems

Besides the type of impressed voltage (dc voltage, single-phase or multi-phase ac voltage) the characteristic variables are the generally complex internal impedance, the waveform, and tolerances of the voltage, any asymmetry in multi-phase, frequency changes, overvoltages of varying energy content as well as short circuit behaviour or other disturbances.

Various types of power systems must be differentiated here: public supply systems, industrial power systems, traction systems, ship's supply systems or isolated power systems. In single-phase and three-phase ac power systems the energy is generated by alternators whose voltages can as a reasonable approximation be taken to be sinusoidal. The frequency of large supply systems and industrial power systems is practically constant. Their short-circuit power is determined by the internal impedances. In high-power systems these are essentially the system reactances including the reactances of the alternators and transformers. In cable systems the resistive voltage drops must also be taken into consideration. Ship's supply systems and other isolated systems generally have greater voltage fluctuations and sometimes also larger frequency fluctuations as well as mains frequencies deviating from 50 Hz or 60 Hz e.g. 400 Hz. Their short-circuit power referred to the connected rating of the connecting loads is low (weak systems).

The energy for dc power systems is either generated via rectifiers from single-phase or three-phase ac (substations), in some cases also by dc generators, or made available by electrochemical generators like batteries or fuel cells. In special cases the energy can also be generated by MHD generators or solar cells. Both the steady state and fault behavior of the systems have a considerable effect upon the operation and design of any converters to which they are connected. For example

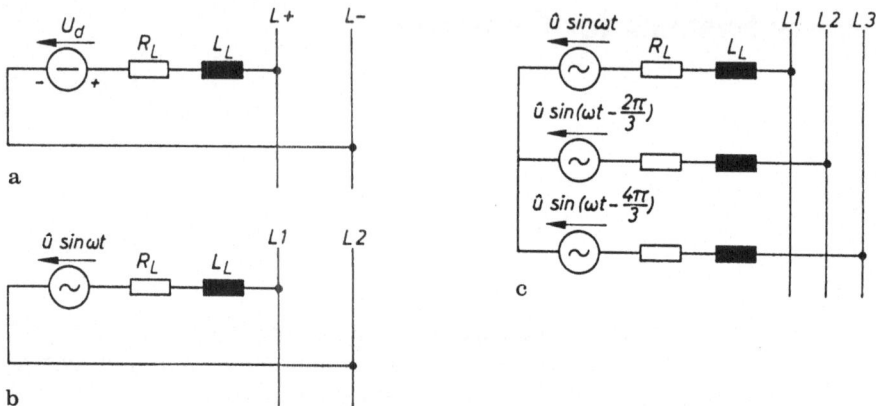

Fig. 9.1a–c. Equivalent circuit for dc and single-phase and three-phase ac systems. **a** DC mains; **b** ac mains (single phase); **c** ac mains (three phase)

overvoltages can destroy the semiconductor components. On the other hand the converters with their valves performing cyclic switching operations affect the waveforms of the system current and voltage. They generate harmonics which, because of the requirements of other loads, must not exceed certain values. Moreover, they can load single-phase and three-phase systems with considerable reactive power depending upon the degree of phase control. Usually tolerances are given for public power systems; e.g. in Germany the electrical operating conditions for heavy-current installations using electronic apparatus are stated in the VDE Regulations (VDE 0160 and VDE 0558).

In Fig. 9.1 the simplest equivalent circuits are illustrated for dc and single-phase and three-phase ac systems. Besides the impressed voltage these take into account only the system resistances R_L and system inductances L_L which are assumed to be constant. In many cases such simple equivalent circuits are adequate when considering the operating behavior of any converters connected. When loaded voltage waveforms which deviate from the impressed voltages arise at the connection points $P-N$, $1-2$ or $1-2-3$. In addition distortion of the system voltage occurs when harmonic currents exist. DC and three-phase ac systems are balanced systems. The power flow can be considered constant if the distortion power is neglected. The single-phase ac system supplies an instantaneous power pulsating at twice the system frequency (see Chap. 11).

The simplest equivalent circuits of system drawn in Fig. 9.1 give no information about how system voltage fluctuations or overvoltages can occur in actual systems. Moreover they do not take into consideration the effect of line capacitances on overvoltages. The processes in supply systems are accessible to accurate mathematical treatment only in special cases, when all the parameters needed are known. Supply systems are usually interconnected. They are subject only to statistically known loads of various types. Switching processes or atmospheric disturbances cause overvoltages which spread across the system from the point of origin as travelling waves. A general equivalent circuit must take this into consideration.

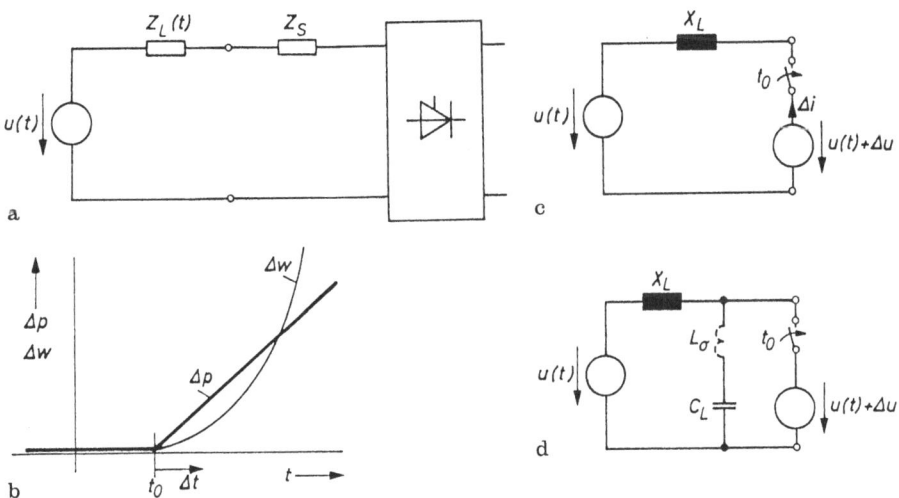

Fig. 9.2a–d. Generation of overvoltage in an inductive system

Figure 9.2a shows such a general equivalent circuit with an arbitrary impressed voltage u(t) and an impedance $Z_L(t)$ varying with time to which a converter with the input impedance Z_S on the ac side is connected. Such a general equivalent circuit, however, provides little information since the time functions of the impressed voltage and the effective impedances are not known. For high-frequency currents and overvoltages of low power the line capacitances and, of course, the capacitors of any RC circuits present would also have to be taken into account.

A simple calculation can show that in inductive systems considerable overvoltages can be generated temporarily without consuming large amounts of energy. In addition it can be assumed that a temporary overvoltage Δu occurs at a point in the system as a result for instance of switching processes (Fig. 9.2b). If a system reactance X_L is assumed between the point considered on the system and the impressed voltage u(t), a transient overvoltage of constant amplitude Δu produces a transient current

$$\Delta i = \frac{\Delta u}{L_L} \Delta t. \tag{9.1}$$

The power needed to generate the overvoltage is

$$\Delta p = \Delta u \cdot \Delta i = \frac{\Delta u^2}{L_L} \Delta t. \tag{9.2}$$

It increases linearly from zero with the duration of the overvoltage. The energy provided at the point of the fault is

$$\Delta w = \int \Delta p dt = \frac{1}{2} \frac{\Delta u^2}{L} \Delta t^2 = \frac{1}{2} L \Delta i^2. \tag{9.3}$$

The two curves are illustrated in Fig. 9.2c. Without taking line capacitances C_L into consideration, therefore, low-energy faults can also lead to considerable temporary overvoltages. When considering line capacitances or capacitors of RC circuits energy is drawn from the overvoltage generators by charging of the capacitances. A voltage change Δu must bring about a change in the electrical energy from $Cu^2(t)/2$ to $C[u(t)+\Delta u]^2/2$. Resistors effect a conversion into heat and lead to frequency-dependent damping.

9.2 DC System

DC systems contain batteries or other sources of dc such as mains rectifiers or dc machines as voltage sources. They generally have resistive-inductive internal impedances; batteries have additionally also non-linear current/voltage characteristics.

Permissible tolerances are generally stated for the range of fluctuation of the input voltage when converters are connected. When batteries are the voltage source this tolerance is $\pm 15\%$ of the rated input voltage. For all other sources of dc this tolerance is, e.g. in Germany, $+5\%$ and $-7,5\%$ of the rated input voltage (according to VDE 0558). Moreover the permissible ripple of the input voltage is limited. In systems supplied from rechargeable batteries, the range of fluctuation may be up to 15% of the rated input voltage.

Overvoltages are dangerous for the converter valves, limiting values are specified for the permissible peak input voltages as a function of the duration of overvoltage. Figure 9.3 indicates the permissible overvoltage ratio

$$\frac{U_{dN}+\Delta U}{U_{dN}}$$

as a function of the overvoltage duration (according to VDE 0558). ΔU is a non-periodic deviation from the rated input voltage U_{dN}. In the range of short-time overvoltages the continuous limiting curve applies for rated dc voltages up to 50 V. For higher input voltages the permissible short-time overvoltage is limited to a value that can be calculated from the equation

$$\left(\frac{U_{dN}+\Delta U}{U_{dN}}\right)_{Limit} = \frac{1400\ V}{U_{dN}} + 2.3. \tag{9.4}$$

The dashed line in Fig. 9.3 applies for $U_{dN}=110$ V.

The inductance of a dc supply system can also be limited for the connection of converters. For example in the German VDE regulations 0558 it is stipulated to have a maximum value of 0.75 mH for dc chopper converters. At input voltages up to 260 V the input overvoltage energy may be up to 4 Joules (4 Ws). This input overvoltage energy would charge an equivalent capacitor C calculated in accordance with equation

$$\frac{1}{2}CU^2_{Limit} - \frac{1}{2}CU^2_{dN} = 4\ Ws \tag{9.5}$$

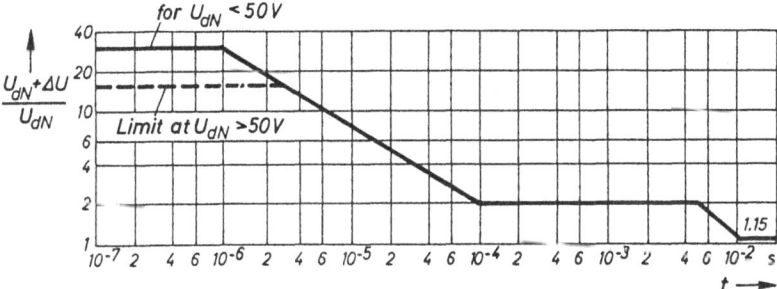

Fig. 9.3. Maximum permissible system overvoltage ratio as a function of the duration of overvoltage

and connected to the dc supply system in place of the converter from the rated voltage U_{dN} to the voltage U_{Limit} indicated in Fig. 9.3. At input voltages above 260 V the input overvoltage energy can exceed 4 Joules. Therefore the capacity to withstand stresses must be determined for each application.

9.3 Single-phase and Three-phase AC Systems

For single-phase and three-phase ac systems voltage tolerance are likewise specified in the long and short-time ranges for the connection of converters. Moreover, the permissible deviation from the sinusoidal waveform in the ideal case is given. Besides this the effect on the waveform of the system ac voltage should be considered particularly during the commutation processes of converters [9.1, 9.7, 9.8, 9.9, 9.10, 9.11, 9.12, 9.13, 9.14].

In the German VDE regulations 0160 e.g. fluctuation of the rms value of the ac voltage between 90% and 110% of the rated voltage of the system is specified as the tolerance of the system ac voltage in the long-time range. Rated voltages for the system ac voltages of power supply units and power sections are 220 V, 380 V, 500 V, and 660 V. For the power sections there are also standardized values for high voltage.

Besides the voltage changes in the long-time range short-time non-periodic overvoltages occur whose magnitude varies with time. These can range from 10 to 100 times the normal value due to atmospheric faults and from 1.5 to 2 times the normal value due to the load being taken off alternators. In high and medium-high-voltage systems the higher overvoltages must be limited by surge diverters to at most 2.5 times the value of the rated voltage. Switching by switches or fuses in low-voltage systems generate overvoltages that are normally less than 2.5 times the value of the rated voltage. Figure 9.4 shows the permissible non-periodic overvoltage

$$\frac{\hat{U}_N + \Delta U}{\hat{U}_N}$$

as a function of time (according to VDE 0160).

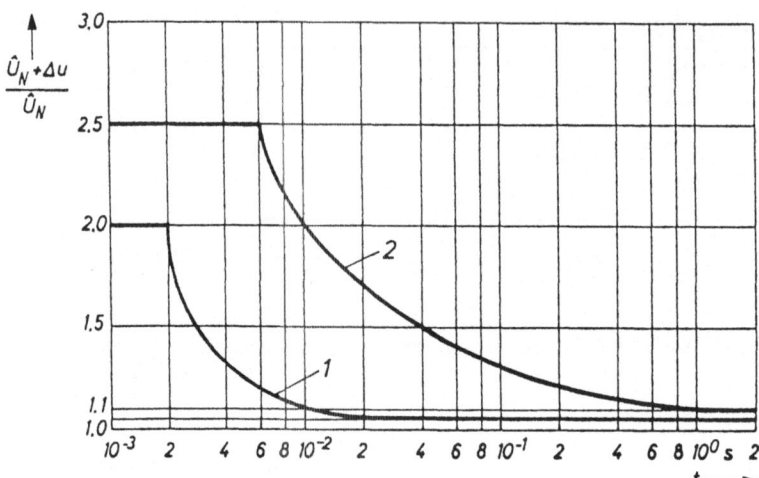

Fig. 9.4. Permissible non-periodic overvoltage as a function of the duration of overvoltage. *1* Operability curve; *2* limiting curve

When power electronics apparatus is operated in single-phase and three-phase systems it is henceforth presupposed that short-time non-periodic overvoltages remain below the limiting curve 2 illustrated in Fig. 9.4. On overvoltages in the zone between curves 2 and 1 operation may be interrupted by the protective devices operating but no damage must occur to the static converters. Below curve 1 they must remain able to operate on overvoltages. The permissible overvoltage with any transient response is determined by converting the voltage waveform lying above the peak value of the rated voltage into a rectangle of height and area equal to the peak amplitude and area of the transient waveform.

Operation of protective devices in the zone between curves 1 and 2 is difficult to realize in practice. The operability curve 1 and limiting curve 2 have not yet been given a place amongst the international standards. Efforts are being made to standardize the maximum permissible system overvoltage ratio as a function of the duration of overvoltage (as in Fig. 9.3). The same holds for single-phase and three-phase systems.

According to curve 1 the apparatus must withstand twice rated voltage for 2 ms. As it is not always economical to design the semiconductor components for twice the rated voltage particularly with low-power converters, series filters connected on the line side can be used to limit the short-time overvoltage in the system to values which are not dangerous for the static converter.

Power electronics apparatus must also be designed to withstand non-periodic overvoltages up to the peak value of the test voltage for as long as 10 μs without operational failure. Temporary dips in the system ac voltage which occur for example on system short circuits and temporary interruption must also be withstood in operation so long as the voltage does not drop by more than 15% of the rated voltage for a maximum of 0.5 s. On larger or longer voltage dips the converters may switch off via the protective device without damage occurring.

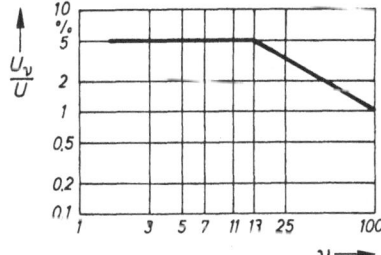

Fig. **9.5.** Maximum permissible value for the harmonics in the system voltage

Certain minimum requirements are likewise set on the waveform of the system ac voltage as harmonics in the input voltage can have an effect on for example the control device, the regulator, and the control characteristic of the converter. Moreover they affect the converter transformer [9.3, 9.5].

In general limiting values are given for every individual harmonic in the system ac voltage. For example Fig. 9.5 reflecting VDE 0160 shows that the lower-order voltage harmonics (up to $\nu = 13$) must not exceed 5% of the system ac voltage.

The falling limiting curve shown in Fig. 9.5 applies for the permitted higher-order harmonics.

The fundamental frequency component of the system ac voltage

$$g_u = \frac{U_1}{U} \tag{9.6}$$

must be at least 99.5%, which corresponds to a harmonic content (distortion factor)

$$k_u = \frac{\sqrt{U_2^2 + U_3^2 + \cdots}}{U} = \frac{\sqrt{U^2 - U_1^2}}{U} = \sqrt{1 - g_u^2} \tag{9.7}$$

of up to 10%.

For other loads, such as synchronous machines, induction motors or condensers, higher harmonics than stated in Fig. 9.5 would be permissible in the system ac voltage without their operating characteristic being impaired.

The voltage waveform drawn in Fig. 9.6 is given for the additional deviation in the system ac voltage from the instantaneous value of the fundamental component. According to german regulations the deviation may be up to 20% of the peak value of the fundamental. The permissible width b varies with the permissible harmonic content (see Fig. 9.5). Such dips in the system ac voltage occur for example during the commutating voltage is temporarily short-circuited via the valves relieving each other in carrying the current (see Sect. 7.1.3).

On one hand, therefore, certain minimum rules are set for the waveform of the system ac voltage when power electronics apparatus is to be connected. On the other hand the converters have an effect upon the waveform of the system ac voltage due to their periodic switching operations [9.2]. So generally converters generate harmonics not only in the system current but also in the system voltage (see Sect. 7.1.9).

In Fig. 9.7a the equivalent circuit is given for a converter in three-phase bridge connection on a three-phase power system with inductive internal impedance. As a

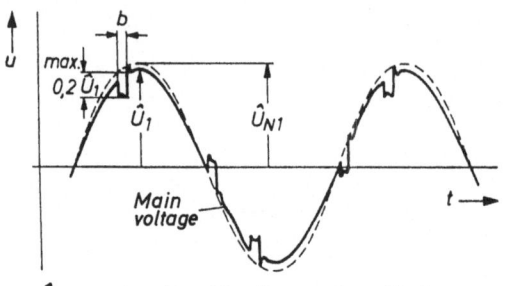

$\hat{U}_1$ peak value of fundamental oscillation
$\hat{U}_{N1}$ peak value of fundamental oscillation
of rated voltage

Fig. 9.6. Permissible momentary dips in
the system ac voltage (according to
VDE 0160)

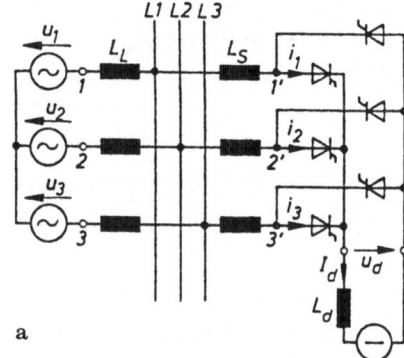

a

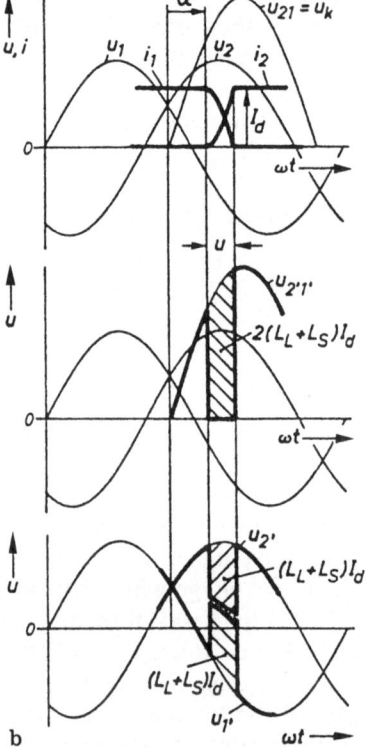

b

Fig. 9.7a,b. Voltage waveform during commutation. **a** Three-phase bridge connection; **c** voltage and current waveforms

first approximation it is enough to neglect resistance and take only the inductive
reactances into account when considering the effects upon the system. The system
then contains inductances L_L and the converter equivalent inductances L_S which
comprise the control reactances referred to, the supply side of a converter
transformer if there is one present line reactances in the converter and series-
connected reactances. During the commutating time u the valves relieving each
other short-circuit the voltage between connection points 1', 2' and 3'. As an
example the waveform of voltage $u_{2'1'}$ during commutation of valve 1 to valve 2 is

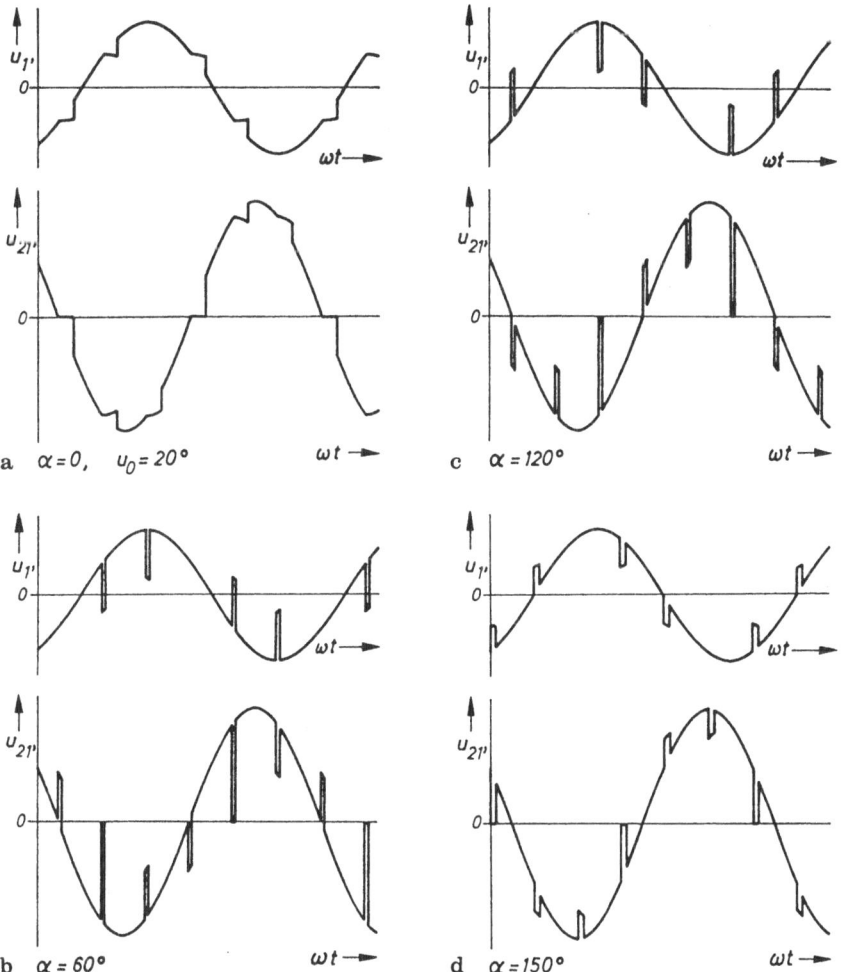

Fig. 9.8a – d. Commutation dips in the voltage across a converter in three-phase bridge connection at various delay angles

indicated. The shaded area under the curve during commutation is $2(L_L + L_S)I_d$. When the reactances in the individual phases are equally large, the voltage potential of connection points 1' and 2' lies midway between the impressed voltages u_1 and u_2.

Figure 9.8 shows commutation dips in the voltage caused by a converter in three-phase bridge connection operating at various control angle α. An initial overlap $u_0 = 20°$ is assumed. The largest voltage dips result from the line voltage being short-circuited (i.e. brought to zero) twice per cycle. The other four voltage dips per cycle originate from the commutation of other valves.

The voltage curve shown in Figs.9.7 and 9.8 refers to the potential of connection points 1' and 2'. The system ac voltage, however, lies between 1, 2 and

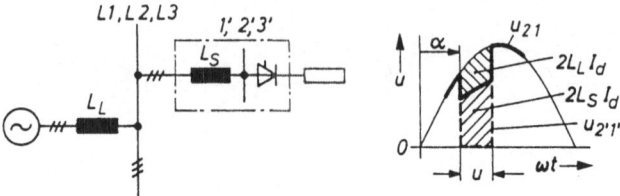

Fig. 9.9. System voltage waveform during commutation

3. The commutation dips occurring there vary with the ratio of the inductances L_L and L_S. The inductances lying in series act, therefore, as voltage dividers. The curve u_{21} of the line system voltage during the commutating time u is shown in Fig. 9.9. The larger the converter reactance X_S is in relation to the system reactance X_L the smaller the voltage dips occurring in the system due to commutation. In the limiting case with a robust system with infinitely large short-circuit power X_L would become zero. In that case there would be no effects on the waveform of the system voltage due to operation of the converter. With a system of finite short-circuit power

$$Q = 3U_s I_k = \frac{3U_s^2}{X_L} \tag{9.8}$$

the converter must have a minimum reactance X_S if the commutation dips are to be limited to a certain maximum value.

U$_s$ is the star voltage (phase voltage) and I_k is the short-circuit current in the system. The relationship

$$\frac{u'}{u} = a = \frac{X_S}{X_L + X_S} \tag{9.9}$$

applies where u is the momentary value of the system voltage on no load and u' is the momentary u-value of the system voltage during commutation. According to equation

$$X_{Smin} = \frac{a}{1-a} \frac{3U_s^2}{Q} \tag{9.10}$$

with a known system short-circuit power Q and desired voltage ratio a the required minimum reactance X_{Smin} of the converter can be worked out. When several converters are connected to a system it is generally considered that for statistical reasons not all converters commutate simultaneously at full power.

The power ratio of the converter as the ratio of ideal dc power of the converter $U_{di}I_{dN}$ to system short-circuit power Q is therefore of considerable influence upon the system reactions occurring. Normally a power ratio up to 1% applies when line commutated static converters are connected. The commutating inductances or converter transformers of line-commutated converters ought to produce a relative short-circuit voltage u_{ks} of at least 4%. The system voltage dip during commutation is then a maximum of 20% of the peak value. In the case of converters for

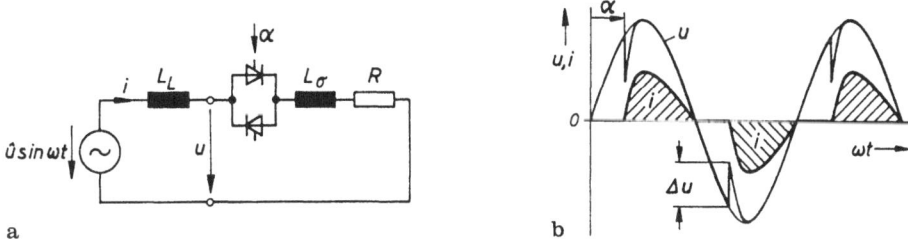

Fig. 9.10a,b. System voltage dip with a single-phase ac power controller with resistive load

traction motors on single-phase systems the dc power can be up to 10% of the system short-circuit power [9.4, 9.6].

Other generators of harmonics are single-phase and three-phase ac power controllers. Figure 9.10 shows as an example the system voltage dip with a single-phase ac power controller driving a resistive load. Here again, the magnitude of the system voltage dip varies according to the ratio of system inductance L_L to leakage inductance L_σ on the load side including the ac power controller. Then

$$\Delta u = \frac{L_L}{L_L + L_\sigma}. \tag{9.11}$$

10 Loads for Converters

Converters couple power systems to loads or power systems to one another and carry out the basic functions recification, dc-to-dc conversion, inversion or ac-to-ac conversion (see Sect. 1.2). Rectifiers or dc-to-dc converters are converters with a dc voltage output which generally have an ac ripple voltage superimposed upon it.

Inverters or ac-to-ac converters generate a single-phase or multi-phase ac voltage across the load which besides the fundamental component also contains various harmonics. In special cases subharmonics can also occur.

Not only in the case of converters with a dc output but also those with a single-phase or three-phase ac output the type of the load affects the stresses and operating characteristics of the converter. In the case of rectifiers or dc-to-dc converters the type of load characterizes the property of the dc load when connected to a dc voltage with superimposed ac ripple to absorb a current the ac content of which is smaller than equal to or larger than the ac ripple voltage content of the output voltage of the converter. In the case of inverters and ac-to-ac convertersthe most important distinguishing feature of the types of load is the displacement angle φ_1 between the fundamental component of the current and the fundamental component of the output voltage. Further distinguishing features are non-linearities in the load, countervoltages or energy recovery.

Figure 10.1 shows the general coupling between a power system and a load via a converter. The relationship between the voltage u and the current i at the output

Table 10.1. Types of load with direct current

Resistance load L/R or RC < 1,35 ms
    AC current content $\approx$ AC voltage content of the load
    Loading with electric arc
    DC current stabilisation necessary

Inductive load (L) $L/R \geqq 1,35$ ms
    AC current content $\ll$ AC voltage content of the load

Load with counter-voltage
    AC current content $\gg$ AC voltage content of the load
    Battery load (B), Motor load (M)
    Capacitive load (C) $RC \geqq 1,35$ ms

Distorted load
    relative current oscillation range > 50%

Mixed load

Table 10.2. Types of load with alternating current

Active load $-5° \leqq \varphi_1 \leqq +5°$, i.e. $\cos \varphi_1 > 0,996$

Load with lagging current (L)
  inductive load $0° < \varphi_1 \leqq +90°$, i.e. $1 > \cos \varphi_1 \geqq 0$
  Specification of displacement angle range possible (e.g. $+15° \leqq \varphi_1 \leqq +45°$, written 15L45)

Load with leading current (C)
  capacitive load $-90° \leqq \varphi_1 < 0°$, i.e. $1 > \cos \varphi_1 \geqq 0$
  Specification of displacement angle range possible (e.g. $-60° \leqq \varphi_1 \leqq -45°$, written 60C45)

Extended load range
  Specification of extended range displacement angle (e.g. $-15° \leqq \varphi_1 \leqq +15°$, written 15CL15)

Distorted load
  Harmonic content k of the load current $> 5\%$ or peak value of the load current $\hat{\imath} > 110\%$ of the peak value of the fundamental oscillation $\hat{\imath}_1$

Harmonic content k of the load current	Peak value $\hat{\imath}$ of load current	Notation
up to 5%	up to 1,1 $\hat{\imath}_1$	none
up to 10%	up to 1,2 $\hat{\imath}_1$	V10
up to 50%	up to 2 $\hat{\imath}_1$	V50
up to 100%	over 2 $\hat{\imath}_1$	V100

Load with dc current content (D)
  DC current content d of the load current $= \dfrac{\text{dc current content}}{\text{rms value of the current}} > 2\%$

DC current content d of the load current	Notation
up to 2%	none
up to 10%	D10
upt to 50%	D50
up to 100%	D100

Motor load (M)
  Ratio $m = \dfrac{\text{starting current (when stationary)}}{\text{rated current (for rated voltage and reference frequency)}}$

$\dfrac{\text{starting current}}{\text{rated current}}$	Notation
up to 3	M3
up to 7	M7
up to 10	M10

  Motor load without energy recovery
  energy recovery in no operational state (averaged over a period)
  Motor load with energy recovery
  generator operation as well as motor operation, then energy recovery (averaged over a period) and $\varphi_1 > 90°$
Periodically varying load
  By periodic on-off switching of the load sub-harmonics in the load current

---

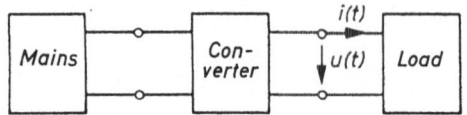

Fig. 10.1. Coupling of a loading to a power system via a converter

of the converter, i.e. at the input to the load is affected not only by the characteristics of the converter but also by those of the load. For this reason knowledge of the characteristics of the load is required in order to predict for instance the characteristics of the converter. For example the voltage u or current i can be impressed. Impressed voltages occur with voltage sources of low internal impedance or capacitances. Currents are impressed by inductances. Types of load are resistive, inductive or capacitive load, load with countervoltage such as battery or motor load with and without energy recovery as well as distorted or periodically variable load.

10.1 Resistance, Inductance, and Capacitance as Load

A resistive load corresponds to a load with an essentially resistive impedance R. In this case the waveform of the load current corresponds to that of the load voltage. With converters with a dc output the ac current content is approximately equal to the ac voltage content of the voltage across the load. With phase control and resistive load operation with intermittent flow of dc current frequently occurs.

With converters with an ac output and active load the current and voltage are in phase. The displacement angle φ_1 approaches zero.

With inductive load in the case of converters with dc output the ac content is considerably smaller than the ac voltage content at the load.

For 50 Hz systems the load is considered as inductive when the time constant L/R of the load is greater than 1.35 ms.

For converters with an ac output and inductive load the fundamental of the current lags behind the fundamental of the voltage by the phase angle φ_1; by 90° on pure inductive load and between 0° and 90° on mixed resistive-inductive load.

In the case of converters with a dc output capacitive load acts like a countervoltage. For 50 Hz systems the load is considered as capacitive when the time constant RC of the load is greater than 1.35 ms.

With converters with ac output on capacitive load the fundamental of the current leads the fundamental of the voltage. With pure capacitive load $\varphi_1 = -90°$.

For an active load R when u is the voltage, i the current, p the power, and w the energy, the known relationships are

$$u = Ri, \tag{10.1}$$

$$i = \frac{u}{R}, \tag{10.2}$$

$$p = ui = \frac{u^2}{R} = Ri^2, \tag{10.3}$$

$$w = \int p\,dt = \int ui\,dt. \tag{10.4}$$

The waveforms of current and voltage against time are the same. The electrical energy is converted into heat in the resistance R. Voltage changes Δu result in instantaneous current changes Δi. No energy is stored.

For an inductive load L the known relationships are

$$u = L\frac{di}{dt}, \tag{10.5}$$

$$i = \frac{1}{L}\int u\,dt, \tag{10.6}$$

$$p = ui = \frac{di}{dt}\int u\,dt = \frac{u}{L}\int u\,dt = Li\frac{di}{dt}, \tag{10.7}$$

$$w = \frac{1}{2}Li^2 \quad (L = const). \tag{10.8}$$

The voltage u determines the rate of change of current in an inductive load. Voltage changes Δu cause a change in the rate of rise of current but cause no sudden changes in current. The inductance stores magnetic energy which can be given up again to the circuit.

For a capacitive load C the known relationships are

$$u = \frac{1}{C}\int i\,dt, \tag{10.9}$$

$$i = C\frac{du}{dt}, \tag{10.10}$$

$$p = ui = \frac{du}{dt}\int i\,dt = \frac{i}{C}\int i\,dt = Cu\frac{du}{dt}, \tag{10.11}$$

$$w = \frac{1}{2}Cu^2 \quad (C = const). \tag{10.12}$$

The current absorbed by the capacitor is directly proportional to the rate of change of the voltage. On changes in the rate of rise of voltage, jumps in the current Δi occur. Sudden jumps in the voltage Δu result in high transient currents. Electrical energy is stored in the capacitor.

Figure 10.2 shows the behaviour of the various types of load on a voltage pulse Δu during the period Δt. With an active load R such a voltage pulse causes an identical current pulse. With an inductive load L it causes a current change and an associated change in the stored magnetic energy. With a capacitive load C a steep voltage pulse Δu is not permissible.

Figure 10.3 shows the conditions with an assumed current pulse Δi during the period Δt. With an active load R a similar type of voltage pulse is produced. With an inductive load L a steep current pulse causes high voltages. With a capacitive load C it causes a change in the capacitor voltage and hence a change in the stored electrical energy.

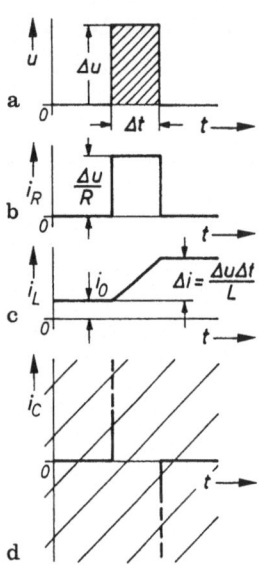

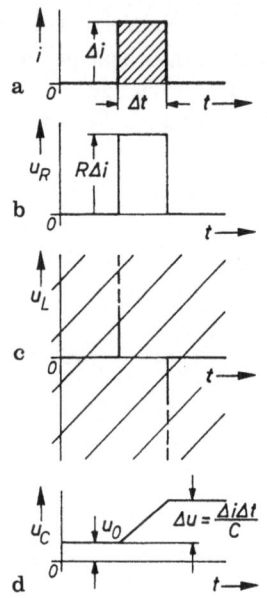

Fig. 10.2a—d. Response of R, L and C load to a voltage pulse. **a** Voltage pulse; **b** load current with R; **c** load current with L; **d** load current with C

Fig. 10.3a—d. Response of R, L and C loads to a current pulse. **a** Current Pulse; **b** load voltage with R; **c** load voltage with L; **d** load voltage with C

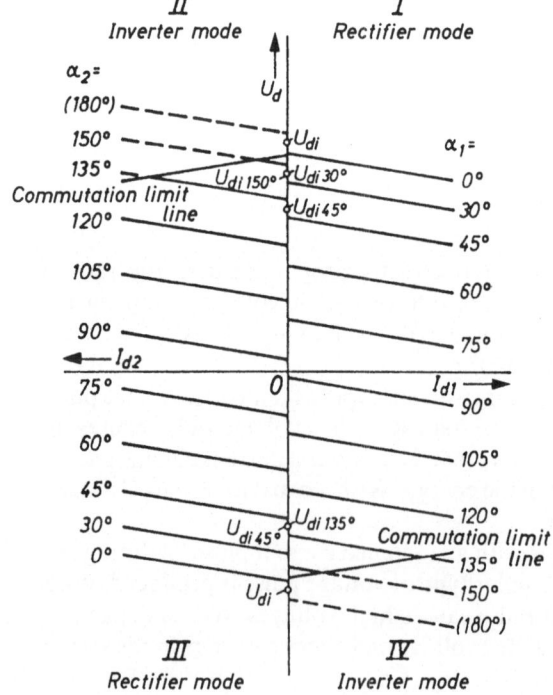

◄ **Fig. 10.4.** Linearized current/ voltage waveforms on the dc side of a line-commutated double converter

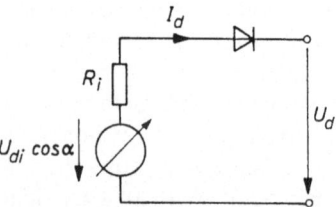

Fig. 10.5. Simplified equivalent circuit for the dc side of a line-commutated single converter

10.2 Internal Impedance of the Converter

The internal impedance of the converter affects the relationship between voltage and current at the output of the converter. The load characteristics of a line-commutated converter were considered in Sect. 7.1.4. Compared to the ideal no-load direct voltage U_{di} there is a voltage drop made up of the inductive direct voltage regulation D_x, the resistive direct voltage drop D_r, and the on-state voltage of the converter valves. Figure 10.4 shows once again linearized current and voltage characteristics on the dc side of a line-commutated double converter. This is a four-quadrant converter. Operation is in the rectifier mode in quadrants I and III and in the inverter mode in quadrants II and IV. The parameter is the control angle α. Such current/voltage characteristics of line-commutated converters can be reproduced approximately by the equivalent circuit with equivalent resistance R_i illustrated in Fig. 10.5. Internal impedances can also be calculated approximately for other types of converter.

10.3 Motor Load

Owing to their impressed voltages starting characteristic and other dynamic characteristics, electrical machines represent a special type of load. They convert electrical energy into mechanical energy or vice versa. It must be decided, therefore, whether operation is purely in the motor mode or generator mode or both [11].

Often electrical machines draw considerably higher currents when starting from standstill than during rated operation. This is particularly true for squirrel-cage induction motors. The load is a motor load without energy recovery when the electrical machine effects no energy recovery averaged over a period in any operating state. The load is a motor load with energy recovery when the electrical machine besides operating in the motor mode acts as a generator and averaged over a period of time returns energy to the source of supply.

For the supplying converter this means a reversal of the direction of energy flow and hence requires a design capable of two- or four-quadrant operation.

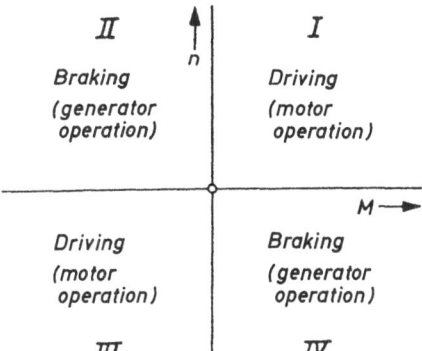

Fig. 10.6. Torque/speed quadrants for an electrical machine

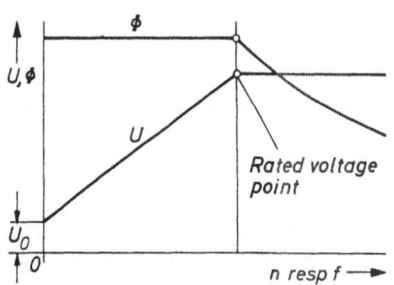

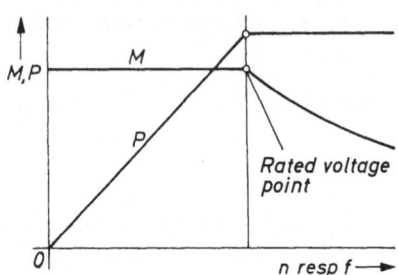

Fig. 10.7. Curve of the terminal voltage U of electrical machines at rated current and constant flux Φ below the rated voltage point

Fig. 10.8. Waveforms of the torque M and power P of electrical machines at rated current and constant flux Φ below the rated voltage point

The possible torque/speed quadrants are illustrated in Fig. 10.6. In drive technology quadrants I and III usually designate driving with the torque M and speed n always having identical signs. The electrical machine works as a motor in these quadrants.

The electrical machine works as a generator or brake in quadrants II and IV. Transition from operation as a motor to operation as a generator corresponds to reversal of the direction of energy flow. The torque/speed range attainable with converter-fed drives is given for various types of drive in Sect. 13.1.1.

The speed of dc machines is adjusted by open or closed-loop control of the terminal dc voltage, that of induction machines by altering the frequency which generally requires that the voltage be altered in direct proportion to the frequency [10.3, 10.4, 10.5].

Assuming constant machine flux Φ and rated current, the relationship between the terminal voltage U and the speed of the machine n or the supply frequency f illustrated in Fig. 10.7 applies not only for dc machines but also for three-phase machines. The voltage demand rises in direct proportion to n or f. When stationary an initial value U_0 is needed to cover the resistive voltage drops.

The magnitude of the voltage provided by the converter is, of course, limited. The **rated voltage point** is defined as the speed or frequency at which the full converter voltage is available ignoring a voltage control reserve. Above this **rated voltage point** the terminal voltage remains constant and the machine flux drops from the rated value hyperbolically as the speed increases. This is the field-weakening range of operation. This applies not only for dc machines but also for three-phase ac machines [10.1, 10.2].

In Fig. 10.8 the curve of the torque M and power P is illustrated for the conditions assumed in Fig. 10.7. In the speed range below the **rated voltage point** with constant machine flux and constant current, the torque M likewise remains constant. The power P rises linearly with speed. Above the **rated voltage point** the power remains constant. The torque drops hyperbolically as the speed increases.

When these assumptions are made the relationships shown in Figs. 10.7 and 10.8 apply not only for dc but also for three-phase ac drives. Below the **rated voltage type point** higher torques are possible, if the current or flux are increased correspondingly.

The back e.m.f. or countervoltage of electrical machine loads must be considered when determining the behavior of converters driving such loads.

The behavior of the back e.m.f. is determined by magnetic diffusion in the machine and thus exhibits characteristic time constants.

10.4 Battery Load

When converters are employed as charging rectifiers for batteries the battery characteristics during charge must be considered. Depending upon the type of batteries to be charged and the charging time available, different charging techniques are employed to prevent the batteries from beginning to gas, particularly during the final phase of the charging process. Figure 10.9 shows two battery charging characteristics.

On charging according to the W characteristics there is a fixed relationship between the charging voltage and the charging current. The charging current decays as the battery voltage increases. Another technique is charging according to an IU characteristic. With this technique, namely, constant-voltage charging with current limitation, the batteries are first charged at constant current. When gassing begins the constant-voltage control is initiated and the charging current drops. The batteries are conserved by limitation of the charging voltage to an adjustable maximum value according to the type of battery.

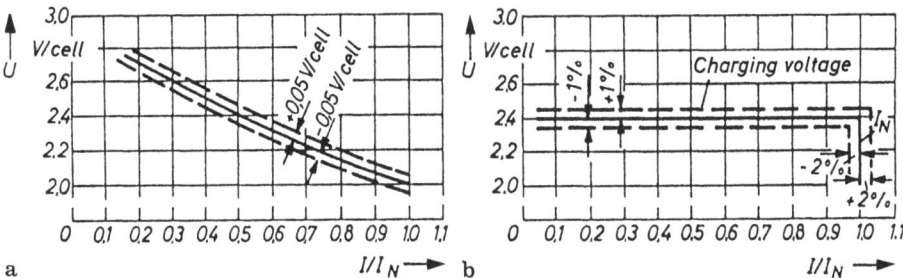

Fig. 10.9a,b. Charging characteristics of batteries. **a** Resistance characteristic; **b** IU characteristic

10.5 Distorting Load

For converters with a dc output a distorting load is defined to be a load which on connection to a constant dc voltage, absorbs a dc current with superimposed ac current the relative amplitude of which is larger than 50%. Such a distorting load can for example be produced by a converter working in the inverter mode when it is commutated on the dc side (see Sect. 11.3.2).

For converters with a single-phase ac output a distorting load produces an harmonic content of the load current greater than 5% or a peak value of the load current $\hat{\imath}$ greater than 110% of the peak value of the fundamental component $\hat{\imath}_1$.

Distorting loads can be caused by non-linear elements in the circuit e.g. by saturation phenomena in iron circuits. Converters themselves can also represent distorting loads.

10.6 Types of Duty and Classes of Load

The types of duty and classes of loads for converters are usually standardized (German norms are in DIN 41 756). The type of duty is characterized by the transient response of the load, the most important feature of the various types of duty being whether the various components in the converter attain their steady-state temperature. Generally the steady-state temperatures are different for the different components in a converter. Each component must be sized for the highest final temperature attained during the worst case load cycle. Only infrequently can the load cycle for a converter be exactly defined. Therefore, types of duty are defined with respect to an idealized description of the load cycle, e.g. continuous duty, intermittent duty or short-time duty.

For continuous duty the duration of load current I_B is long enough for all components to attain their steady-state temperature.

Short-time duty is when the duration t_B, of load current I_B is so short that not all components attain a steady-state temperature. During the interval when no current is flowing the components cool down to the temperature of the cooling medium.

For intermittent duty the load current I_B flows for a period t_B but the duration of the no-load period is so short that not all components cool down to the temperature of the cooling medium. The load period is stated as the relative duty factor (λ) with the longest load cycle. Then

$$\lambda = \frac{t_{load}}{t_{cycle}}.$$ (10.13)

In Fig. 10.10 the curve of the current I and temperature rise ϑ of a component is given for continuous duty, short-time duty, and intermittent duty, where t_{load} is the load period, t_{cycle} the cycle time, $\Delta\vartheta$ the steady-state temperature rise, and ϑ_e the final temperature rise relative to the temperature of the cooling medium.

The load cycle for a converter-fed drive for instance can be subject to periodic fluctuations about an average value. On alternating load duty the load cycle is described by the current and duration of all load periods. Parameters are the cycle time t_{cycle} and the rms value I_Q of the load current. Figure 10.11 shows an example of alternating load duty. Converters must be able to sustain a short time overload. The overcurrent I_B as a percentage of the basic load current I_G varying with the duration of overcurrent is given in duty classes I to IV. For electro-chemical plants for instance duty class II applies with 150% overcurrent with an occasional one-minute period of additional short-time load. For industrial duty class IV applies with 125% overcurrent for two hours or 200% overcurrent for 10 seconds. Duty class VI applies for arduous traction duty with 150% overcurrent for two hours or 300% overcurrent for one minute (see Table 10.3).

Table 10.3a. Standard duty classes for converters

Duty class	Most suitable application	Excess current I_{Load} as a % of the nominal current I_d	Duration of the additional short-term load (excess current duration) t_B
I	Electrochemical processes	100%	–
II	Electrochemical processes	150%	1 min occasionally (with faults)
III	Light industrial and light traction substation service	150% or 200%	2 min 10 s
IV	Industrial service	125% or 200%	2 h 10 s
V	Medium traction substation and mining	150% or 200%	2 h 1 min
VI	Heavy traction substation	150% or 300%	2 h 1 min

Table 10.3b. Examples of load cycles

Duty class	Most suitable applications	Assumed typical load conditions for the duty class. Load current in relation to rated direct current
I	Electrochemical processes, etc.	
II	Electrochemical processes, etc.	
III	Light industrial and light traction substation service	
IV	Industrial service	
V	Medium traction substation and mining	
IV	Heavy traction substation	

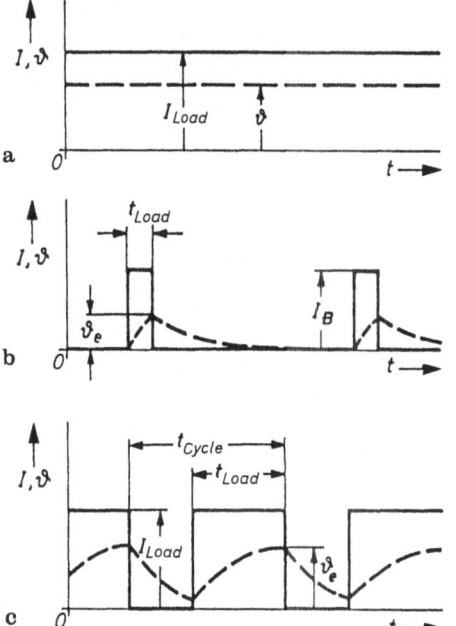

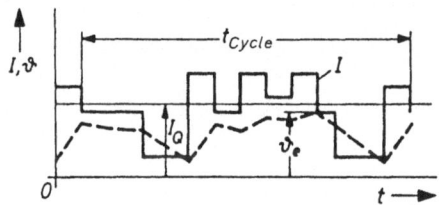

Fig. 10.10a – c. Waveforms of current and temperature rise for various types of duty. **a** Continuous duty; **b** short-time duty; **c** intermittent duty

Fig. 10.11. Waveforms of current and temperature rise for alternating load duty

Owing to the low thermal capacity of the power semiconductors, including their heat sinks, the overcurrents act on semiconductor diodes and thyristors as continuous currents even over relatively short periods (e.g. > 30 s) and must be taken into consideration when designing the converters.

10.7 Service Conditions

Service conditions for converters are external conditions which influence the load capacity, operation and the constructional performance. (The German standards VDE 0160 contain regulations for the fitting of electrical parts to heavy current installations and VDE 0558, parts 1 – 3, contains information about operating conditions of converters.)

General requirements. Converters must operate as intended. Specified type characteristics must correspond to normal operating conditions. Unusual operating conditions can be explicitly agreed to so as to be consistent with these nominal values.

Under normal cooling conditions it is assumed that there is no additional heating of the converter unit by nearby sources (e.g. heat radiation).

For converter units using air as the cooling medium the surrounding temperature should be between $-10°C$ and $+40°C$. For self-air cooling the upper limit is $+45°C$. The cooling medium temperature is the same for external air cooling as for self-air cooling, namely, between $-10°C$ and $+35°C$. It is assumed that the cooling air can circulate freely. For converter units using water cooling the surrounding temperature should be between 0°C and 40°C. The water temperature should be between 5°C and 25°C. If oil is the cooling medium and there are no other significant factors, the temperature of the incoming oil should be between -5°C and $+30$°C.

Specified type characteristics are generally valid for resistance load (see Sect. 10.1) as well as for long operation (see Sect. 10.6) and full modulation ($\alpha = 0$). Site altitudes of up to 1000 m above sea level are regarded as normal.

Electrical operating conditions. In lines supplied by synchronous machines the effective value of the ac voltage may vary between 90% and 110% of the nominal line voltage. For converters permissible short-term breaks in the line ac voltage are given in Fig. 9.6. The permissible width b depends on the permissible relative harmonics content (see Fig. 9.5). With voltage decreases of larger or longer duration the operation of converters might be interrupted by the response of protection devices.

The frequency of the line ac voltage may differ by at most ± 1% from its value. In three-phase lines the voltage of neither the negative sequence system nor the zero sequence system may be more than 2% of that of the positive sequence system.

In lines supplied by accumulator batteries the mean voltage may be between 85% and 115% of the nominal voltage. If it falls below 85%, operation might be interrupted by protection devices. For all other dc sources the rated input voltage is set between $+5$% and $-7,5$% of the actual input voltage. For dc inverters the permissible input ac voltage is the rated input voltage ± 10% (according to German standards VDE 0558). For the superpositions of the dc voltage input a variation of ≤ 15% of the rated input voltage is permissible (for accumulators at most 10% of the rated input voltage).

Unusual operating conditions. Unusual operating conditions necessitate different measures being taken and must be expressly stated by the user. They occur if the permissible tolerances of the line or voltage are exceeded.

Unusual operating conditions can also include unusual mechanical requirements, water cooling with condensation or with pipes being blocked by particles, hard particles in the cooling air, dangerous gases in the environment, risk of explosion, radioactivity, unusual cooling agent temperatures or environmental temperatures, sudden and significant temperature variations, high humidity, and site altitudes of more than 1000 m above sea level.

11 Energy Conditions

Consideration of the energy conditions in power electronic systems, that is the variation with time of the power in the various components of a system and the work (energy) gained from this by integration over the period, leads to general, and often readily accessible information [11.8].

In Sect.8.2.9 an energy balance at the beginning and end of a commutation process has already been carried out in order to determine the response of a capacitive turn-off circuit. In electrical circuits magnetic energy can be stored in inductances and electrical energy in capacitors. Resistors convert electrical energy into heat. Energy sources and loads are generally represented by equivalent circuits with impressed voltages and internal impedances, both of which are functions of time (see Chaps. 9 and 10).

11.1 Energy Sources

Electrical energy is predominantly generated by means of electrical machines (generators) in which mechanical energy is converted into electrical energy according to the principle of electro-dynamics discovered by Siemens. In addition to this there are forms of direct energy conversion in which electrical energy is generated from chemical energy or other forms of primary energy such as heat or light directly i.e. without passing through electrical machines.

primary energy	generation of electrical energy
fossil fuels	electrical machines
(mineral oil, coal, mineral gas)	(generators, dynamos)
nuclear fuels	turbo generator
water energy	water-power generator
sun radiation	diesel engine generator
wind energy	airborne supply system
geo-thermal energy	(light dynamo, charging generator ...)
	direct energy conversion
	accumulator
	galvanic cell
	isotope battery
	thermo-electricity
	photo-electricity
	MHD generator
	(magneto hydro dynamics)

Fig. 11.1. Primary energy and generation of electrical energy

In Fig. 11.1 the most important forms of primary energy are listed alongside the techniques for generating electrical energy. With electrical machines, predominantly single-phase or three-phase ac current is generated. DC current is generated only in special cases. DC energy conversion produces dc voltage or dc current.

11.2 Waveform of Power against Time

The instantaneous value of power p is the product of the instantaneous value of the voltage u across a circuit component and the instantaneous value of the current i in that component. With single-phase ac voltages and currents this product generally assumes positive and negative values during each cycle. Positive values of p indicate energy flow in one direction and negative values indicate energy flow in the opposite direction [11.23].

For a circuit with sinusoidal ac voltage and sinusoidal current displaced in phase by an angle φ, the waveform of power p against time is illustrated in Fig. 11.2. The result is a power pulsation about the mean value at a frequency of twice the system frequency (see Sect. 7.1.7). The mean power (active power) is

$$P = \frac{1}{T} \int_0^T uidt. \qquad (11.1)$$

The amplitude of the alternating component of the power oscillation is the apparent power

$$S = UI. \qquad (11.2)$$

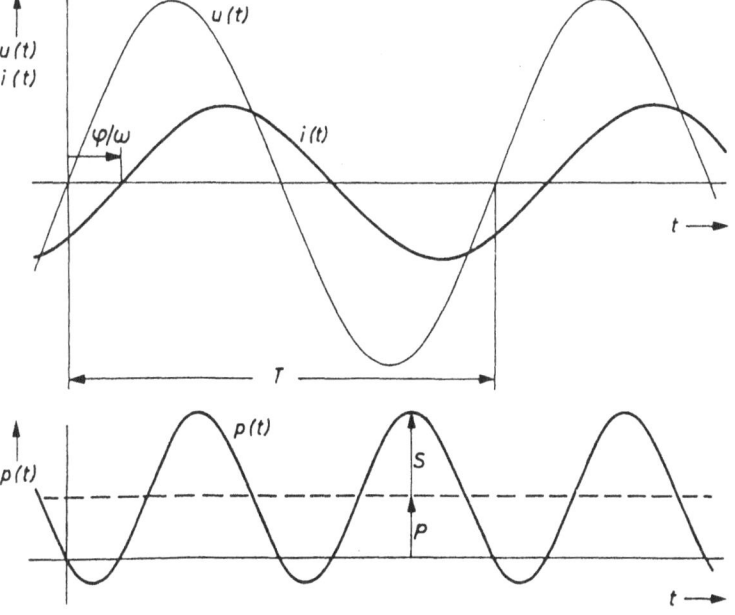

Fig. 11.2. Waveform of power p against time with sinusoidal phase-displaced current

The mean power (active power) is

$$P = UI \cos \varphi \qquad (11.3)$$

where U and I are the rms values of the voltage u and current i. The momentary value p of power is given by

$$p = ui = P - S \cos(2\omega t - \varphi). \qquad (11.4)$$

In converters generally non-sinusoidal electrical quantities occur but these are usually periodic. The currents generated in the single-phase power system by the periodic switching processes are only sinusoidal in the special case of ideal filters. In systems with an impressed sinusoidal voltage, non-sinusoidal currents also cause non-sinusoidal voltage drops across the internal impedance (see Sect. 9.3). The waveform of power against time as the product of voltage and current is likewise not sinusoidal. The usual definitions of active power, apparent power, and reactive power for sinusoidal voltage and current curves are not, therefore, adequate for the description of the energy conditions in the case of converters. It is necessary to consider the waveform of power against time

$$p(t) = u(t)i(t) \qquad (11.5)$$

with the resultant energy

$$w(t) = \int p(t)dt = \int u(t)i(t)dt. \qquad (11.6)$$

Presupposing *sinusoidal* voltage and *non-sinusoidal* current the following definitions apply

active power $P = UI_1 \cos \varphi_1$ \qquad (11.7)

where U is the rms value of the voltage, I_1 the rms value of the fundamental component of the current and φ_1 the phase displacement between the voltage and the fundamental component of the current.

Further definitions which have already been introduced in Sect. 7.1.7, are

apparent power $S = UI$ \qquad (11.8)

as the product of the rms values of voltage and current,

reactive power $Q = \sqrt{S^2 - P^2},$ \qquad (11.9)

fundamental frequency reactive power $Q_1 = Q_1 = UI_1 \sin \varphi_1,$ \qquad (11.10)

distortion power $D = U\sqrt{I_2^2 + I_3^2 \ldots},$ \qquad (11.11)

and

fundamental frequency content $g_i = \dfrac{I_1}{I}.$ \qquad (11.12)

The apparent power S, active power P, fundamental frequency reactive power Q_1, and distortion power D are related as follows:

$$S^2 = P^2 + Q_1^2 + D^2. \qquad (11.13)$$

This can be illustrated graphically by the power tetrahedron (see Fig. 7.16).

The relationship between active and apparent power has special significance. It is known as the

$$\text{power factor } \lambda = \frac{P}{S} = g_i \cos \varphi_1. \tag{11.14}$$

The fundamental power factor $\cos \varphi_1$ which is usually known as the displacement factor therefore produces the power factor λ only by multiplication with the fundamental frequency content g_1 i.e. with non-sinusoidal currents λ is smaller than the displacement factor $\cos \varphi_1$. The definitions listed above apply only when a sinusoidal voltage is presupposed. With a non-sinusoidal voltage they are no longer valid.

Other power definitions have been suggested [11.3, 11.4, 11.5] that can also be employed with non-sinusoidal periodic quantities. The relationship (11.1) always applies for the active power P.

The apparent power S can also be defined as the product of the rms value of voltage and current (Eq. (11.2)). In addition to this, Tröger [11.2, 11.20] suggested the terms

$$\text{regenerative power } P_r = \frac{1}{2T} \int_0^T [|u(t)i(t)| - u(t)i(t)] dt, \tag{11.15}$$

$$\text{transmitted power } P_d = \frac{1}{T} \int_0^T |u(t)i(t)| dt = P + 2P_r, \tag{11.16}$$

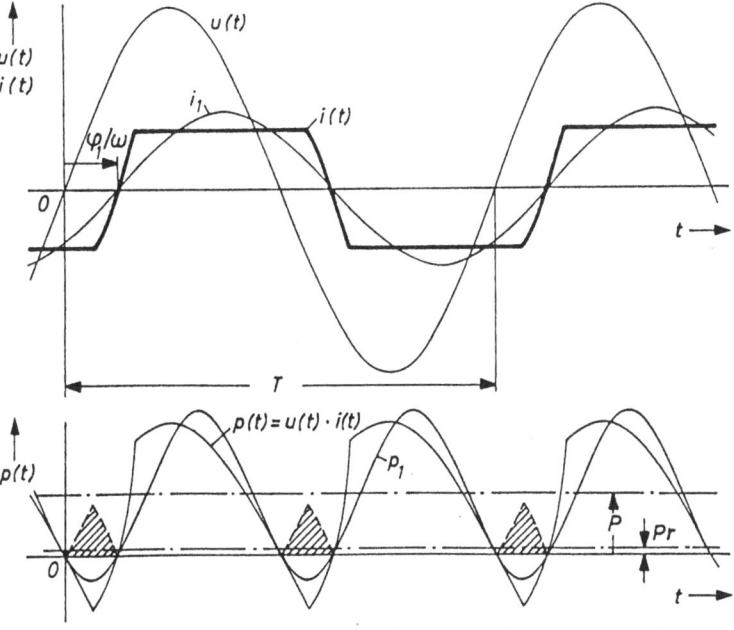

Fig. 11.3. Waveform of power p against time with a line-commutated converter (single-phase bridge connection)

and

advance power $P_v = P + P_r$.   (11.17)

The power quantities are mean values and, for systems loaded with harmonics, lead to unambiguous and physically understandable terms. They can, moreover, be measured.

Figure 11.3 illustrates the waveform of power p for a line-commutated converter. In this example a sinusoidal voltage waveform $u(t)$ was assumed. The ac current $i(t)$ is sinusoidal during commutation but is otherwise constant and lags behind the voltage by the angle φ_1. The waveform of $p(t)$ is obtained by multiplying voltage and current, the active power P as the mean value of $p(t)$, and the regenerative power P_r as the mean value of the areas under the negative power/time curve.

Summarizing we reiterate that only the waveform of $p(t)$ and the mean power resulting from it have any physical significance. The other terms such as reactive power, distortion power or apparent power are calculated quantities resulting from mathematical analysis of voltage and current into their fundamental and harmonics and active and reactive components.

11.3 Types of Converter

In Sect. 5.5, converters were classified according to their internal mode of operation, namely, the type of commutation, rather than according to the basic function performed (rectification, inversion, dc-to-dc conversion, and ac-to-ac conversion), i.e. their external mode of operation. According to this classification there are three different types of converter, namely semiconductor switches and power controllers (without commutation), externally commutated converters, and self-commutated converters which were dealt with the detail in Chaps. 6, 7, and 9.

The internal mode of operation of converters can be classified not only according to natural commutation and forced commutation as already described, but also according to whether commutation takes place on the ac side or the dc side [5.1, 5.2].

11.3.1 Converters with Commutation on the AC Side

In the case of converters with commutation on the ac side, ac current is converted into dc current or, when energy is supplied from the dc side, ac current into ac current where in both cases the current commutates on the ac side. This characteristic assumes several conditions. On the ac side owing to the commutating process occurring there only a moderate reactance is permissible. On the other hand on the dc side there can be any amount of smoothing inductance. The semiconductor switches for this type of converter only have to carry current in one direction.

In Fig. 11.4 an equivalent commutator is shown which switches periodically at the system frequency f_N between the states drawn in continuous and broken lines

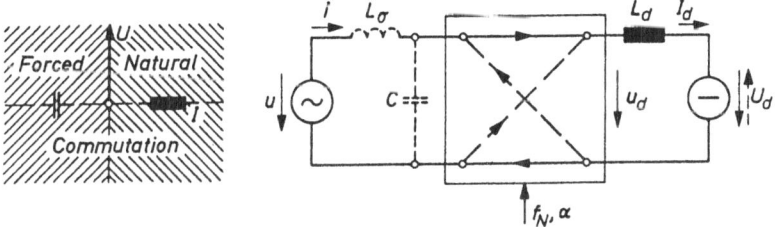

Fig. 11.4. Converter with commutation on the ac side (drawn with equivalent commutator)

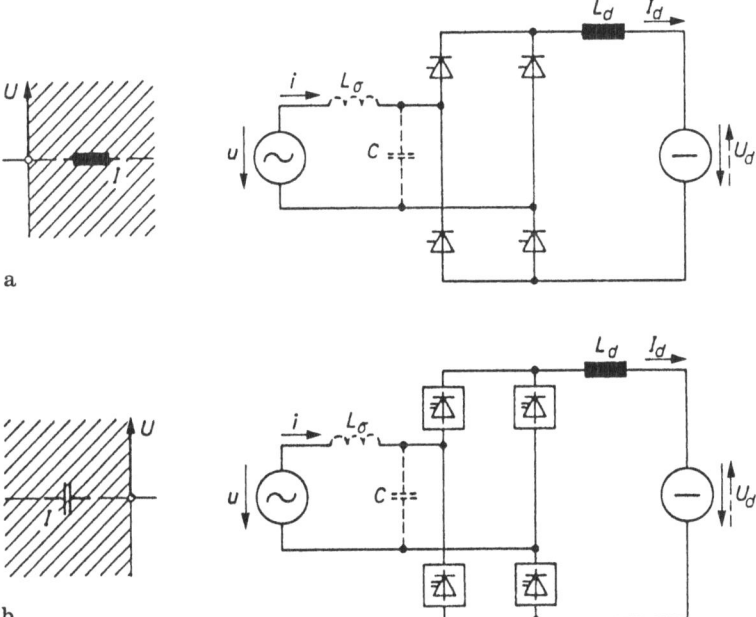

Fig. 11.5a,b. Converter with commutation on the ac side. **a** Natural commutation; **b** forced cummutation

and which corresponds to a single-phase bridge connection. The arrows shown mark the direction of current. It should be assumed that the commutator can operate not only in the range of natural commutation but also in that of forced commutation. In the range of natural commutation the ac voltage source is loaded with inductive reactive power and in the range of forced commutation with capacitive reactive power.

Figure 11.5 shows designs of converters with commutation on the ac side. These are converters in single-phase bridge connection, the equivalent commutator of Fig. 11.4 being realized by semiconductor switches. The most usual type of converter is illustrated in Fig. 11.5a. This is an externally commutated converter (with natural commutation). The semiconductor switches are rectifier diodes or thyristors. The ac voltage source is loaded with inductive reactive power over the entire operating range. Figure 11.5b illustrates a converter with commutation on the ac side but with quenchable semiconductor switches for one direction of current. This converter permits the ac voltage source to be loaded with capacitive reactive power.

In the case of converters with commutation on the ac side the mean value U_d of the dc voltage varies (owing to the solid connection with the ac voltage source via the commutator) only with the ac voltage amplitude and the phase position of the switching instants. The control angle α defines this phase displacement relative to the passage of the ac voltage through zero. The momentary value u_d of the dc voltage behind the commutator (see Fig. 11.4) can therefore assume the values $u(t)$ or $-u(t)$ according to the position of the commutator. In the special case of asymmetrical control angle the value u_d additionally equals 0 (freewheeling). The known relationship between the dc voltage U_d and the control angle α is $U_d = U_{di} \cos \alpha$. The dc voltage U_{di} on full modulation ($\alpha=0$) varies for a given converter connection only with the rms value U of the ac voltage. For the single-phase bridge connection illustrated in Figs. 11.4 and 11.5

$$U_d = \frac{2}{\pi}\sqrt{2}U \cos \alpha. \tag{11.18}$$

The control angle α can vary from 0° to 180° (range of natural commutation) and from 180° to 360° (range of forced commutation). It determines the relationship between the dc voltage U_d and the ac voltage U. The current i or I_d driven by small voltage differences between the ac and dc voltage sources is determined by a change in control angle $\Delta\alpha$. With good smoothing the relationship between the dc current I_d and the rms value U_1 of the fundamental component of the ac current is

$$I_d = \frac{\pi}{4}\sqrt{2}I_1. \tag{11.19}$$

For all converters with commutation on the ac side a fixed relationship exists between the control angle α and the phase displacement φ_1 between the voltage and the fundamental component of the current on the ac side (see Sect. 7.1.7)

$$\cos \varphi_1 = \cos \alpha. \tag{11.20}$$

As a result of overlap this relationship applies only approximately with commutating reactances on the ac side. In principle not only lagging (inductive) but also leading (capacitive) ac current is possible. Leading ac current can, however, only be realized with forced commutation in the range from $\alpha=180°$ to 360°.

11.3.2 Converters with Commutation on the DC Side

In the case of the converters with commutation on the dc side dc voltage is converted into ac voltage or — when energy is supplied from the ac to the dc side — ac voltage into dc voltage the current in both cases being commutated on the dc side. This characteristic assumes that there is only moderate reactance on the dc side. The effect of inductances L_σ being to large can be compensated for by smoothing capacitors C_d. On the ac side a reactance is required to limit current harmonics. The semiconductor switches for this type of converter should conduct current in two directions. In Fig. 11.6 an equivalent commutator is drawn which as before corresponds to a single-phase bridge connection. Converters with commutation on the dc side generally work with forced commutation. The ac voltage source is loaded with capacitive reactive power. In cases of an inductive loading of the ac voltage source natural commutation is possible. This operating range is only employed in special cases with converters with commutation on the dc side [11.25].

Figure 11.7 shows various designs of converters with commutation on the dc side in single-phase bridge connection, the equivalent commutator in Fig. 11.6 being realized by semiconductor switches. Figure 11.7b illustrates the commonest type of converter with commutation on the dc side. This works as a self-

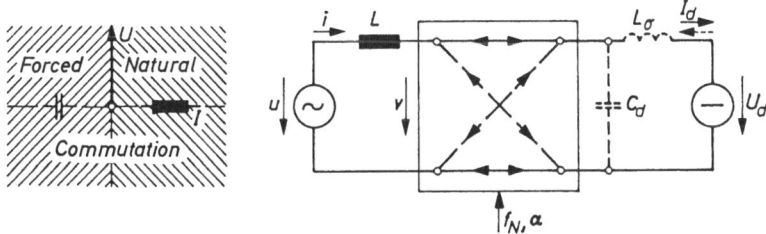

Fig. 11.6. Converter with commutation on the dc side (drawn with equivalent commutator)

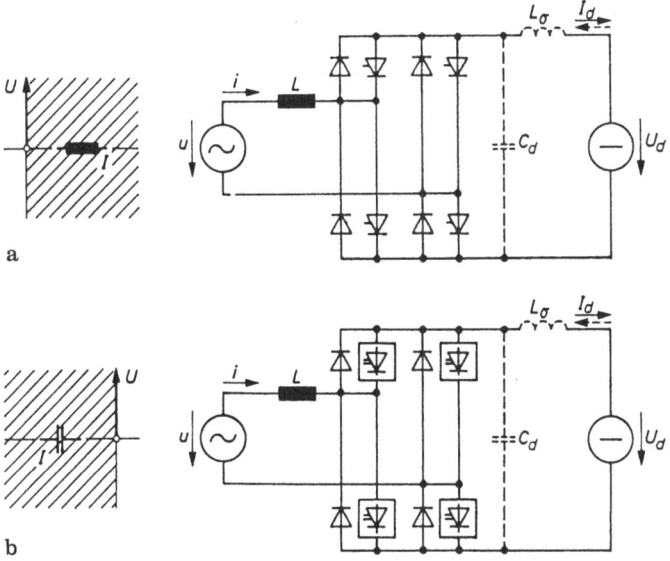

Fig. 11.7a,b. Converter with commutation on the dc side. **a** Natural commutation; **b** forced commutation

commutated converter with forced commutation. The semiconductor switches in the individual arms of the bridge are equipped for two directions of current. In one direction they can be turned off. The ac voltage source is loaded with capacitive reactive power. Figure 11.7a shows a special case of the converter with commutation on the dc side which in the range of inductive reactive power consumption from the ac voltage source, works with natural commutation, as does for example the series resonant inverter (see Fig. 7.40). It can also be employed as an infinitely adjustable reactive power converter.

Also a synchronous machine can be operated with variable frequency and voltage by this type of converter under natural commutation [11.24].

Thyristors with diodes connected in antiparallel are employed in each arm of the bridge.

In the case of converters with commutation on the dc side there is likewise a fixed relationship between the dc voltage U_d and the fundamental component V_1 of the voltage on the ac side of the commutator (see Fig. 11.6). For single-phase bridge connections with adequate smoothing of the dc voltage U_d the relationship is

$$U_d = \frac{\pi}{4}\sqrt{2}V_1. \tag{11.21}$$

The control angle α at which the commutator is driven determines the phase displacement between V_1 and U. For a given current I and inductance L this phase displacement varies with the inductance L on the ac side and approaches zero as the inductance decreases. The dc current I_d and rms value I_1 of the fundamental component of the active current are linked according to

$$I_d = \frac{2}{\pi}\sqrt{2}I_1\cos\varphi_1. \tag{11.22}$$

The current is set by small changes $\Delta\alpha$ in the control angle. The fundamental component power factor $\cos\varphi_1$ can lie in all quadrants. With this type of converter it is no longer possible to alter the relationship between the dc voltage U_d and the ac voltage U — as it is with converters with commutation on the ac side — between wide limits via the control angle α. Instead it is possible to freely exchange reactive power between the ac system and the converter [11.7].

By pulsing the commutator i.e. repeated switching on and off during a half-cycle of the ac voltage the relationship between U_d and V_1 can be made variable in the manner that for a given dc voltage U_d the fundamental oscillation U_1 of the ac voltage can be infinitely adjusted between the possible maximum value and zero (see Sects. 8.3 and 11.6).

11.4 Coupling of Power Systems

Electrical power supply systems contain ac voltage sources whose impressed voltages vary sinusoidally. It is desirable that these ac voltage sources are loaded with currents which are as sinusoidal as possible in which case the phase displacement angle φ should be small in order to prevent excess reactive current. When the voltage and current curves are sinusoidal the power in the ac voltage sources pulses sinusoidally at twice the system frequency (see Fig. 11.2).

$$p(t) = u(t)i(t) = \hat{u}\hat{i} \sin \omega t \sin(\omega t - \varphi) = UI \cos \varphi$$

$$= UI \cos \varphi \left[1 - \frac{\cos(2\omega t - \varphi)}{\cos \varphi} \right] \qquad (11.23)$$

The current in dc voltage sources and dc loads should be smoothed as much as possible. With this assumption the power on the dc side is constant

$$p_d(t) = P_d = U_d I_d. \qquad (11.24)$$

If a single-phase ac system were to be coupled to a dc system under the assumptions made there would be variations with time of the power converted in the two sources. The variation in each case is

$$p(t) - p_d(t) = -UI \cos(2\omega t - \varphi). \qquad (11.25)$$

Magnetic and electrical stores are therefore necessary to maintain the power balance.

With multi-phase symmetrically loaded ac systems with a sinusoidal current waveform, the total power of all phases is constant with respect to time. The power balance between the ac and dc sides would therefore be maintained. This condition applies only for the positive-sequence currents when multi-phase systems are asymmetrically loaded. For the negative-sequence system caused by the asymmetry, power pulsations of twice the system frequency are again produced.

Unfortunately in power conversion using non-linear semiconductor switches sinusoidal currents on the ac side cannot be attained without stores. Rather harmonic currents are produced by the periodic switching processes (see Sect. 7.1.9) which can only be suppressed by the addition of magnetic and electrical stores connected as filter circuits.

When single-phase or three-phase ac systems are coupled with dc systems or dc loads generally adequate smoothing is provided by means of a magnetic store on the dc side. In this case square-wave or stepped currents, typical for the converter load, occur on the single-phase or three-phase ac side.

In principle, magnetic and electrical stores can be combined to form filters which can absorb not only energy oscillations with twice the system frequency but other higher-frequency energy oscillations. They can, moreover, ensure a sinusoidal current waveform in the single-phase or three-phase ac system and a constant current in the dc system. In such networks consisting of voltage sources, transformers, linear energy converters (resistances), linear energy stores (see Fig. 2.1) and non-linear semiconductor switches without energy consumption capacity (see Figs. 2.2 and 2.3), the two assumptions made can be satisfied, namely that the power balance is equalized at each instant (necessary condition) and that the currents are sinusoidal or constant (desired condition). The power balance is described by the condition

$$\sum p(t) = 0. \tag{11.26}$$

The sources contribute to this balance with

$$p(t) = u(t)i(t), \tag{11.27}$$

magnetic stores with

$$p_L(t) = L i_L di_L/dt, \tag{11.28}$$

electrical stores with

$$p_C(t) = C u_C du_C/dt, \tag{11.29}$$

and resistors as energy converters with

$$u_R^2/R \text{ or } Ri_R^2. \tag{11.30}$$

11.4.1 Coupling of Single-phase AC and DC Systems

First, the energy conditions obtained when a single-phase ac voltage source is coupled to a dc voltage source via various types of converter shall be examined. In a system consisting of a single-phase ac source, filter f on the ac side, commutator (single-phase bridge connection with semiconductor switches), filter F on the dc side, and dc voltage source (Fig. 11.8), it is assumed that the ac current is sinusoidal and the dc current is constant [11.17].

Figure 11.8 illustrates the power flow between the sources and the stores. The single-phase ac voltage source supplies the power p consisting of a constant component P which flows to the dc side as dc power P_d and the power p_2 oscillating about zero at twice the system frequency. p_2 flows in the filters f and F.

$$p_2 = p_{2f} + p_{2F} \tag{11.31}$$

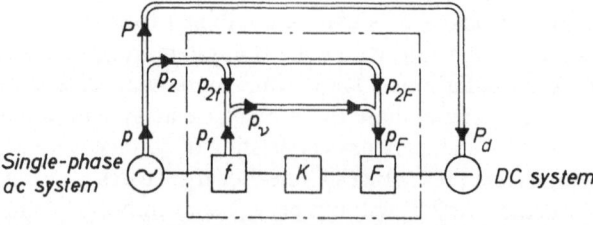

Fig. 11.8. Power flow when a single-phase as system is coupled to a dc system via a converter. *f* Filter on ac side, *K* commutator (semiconductor switch), *F* filter on dc side

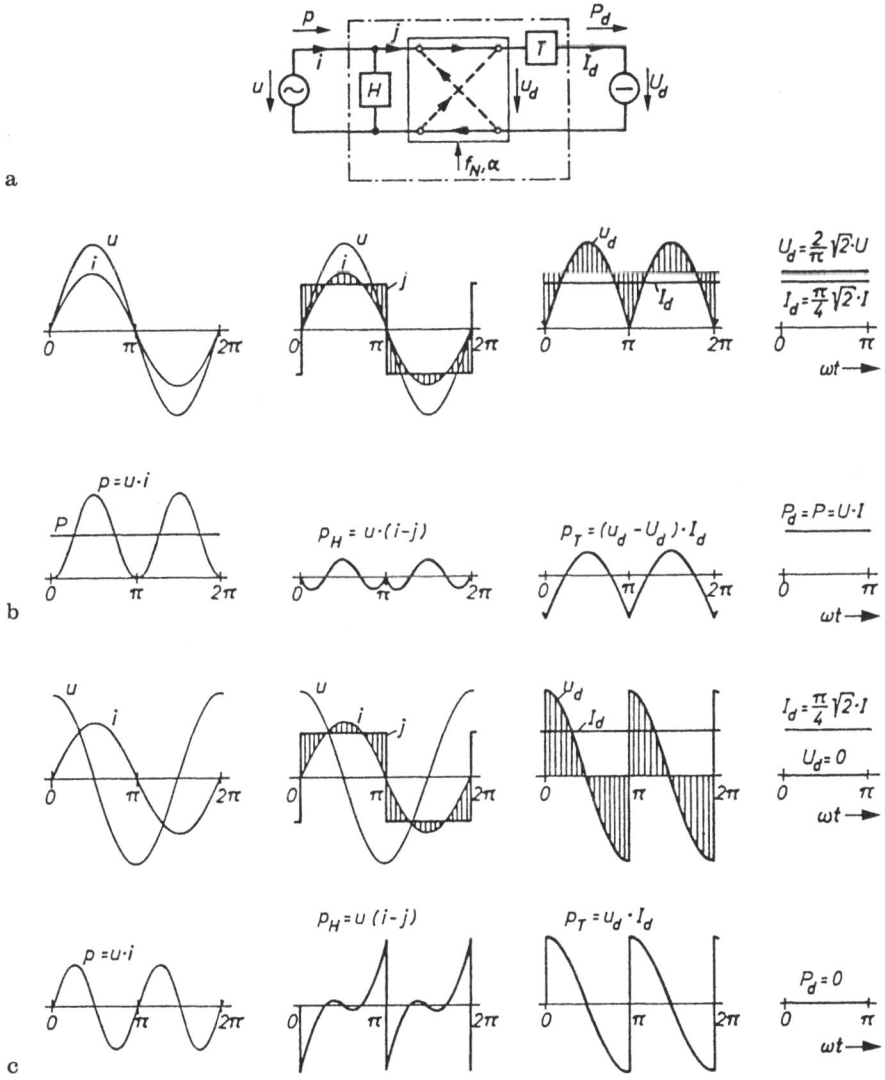

Fig. 11.9a – c. Voltage, current and power waveform for a converter with commutation on the ac side (single-phase bridge connection). **a** Circuit with equivalent commutator and ideal filters; **b** with active power ($\alpha = 0°$); **c** with inductive reactive power ($\alpha = 90°$)

In addition to P_d and p_2 an exchange harmonic power p_v occurs between the two filters.

Figure 11.9 shows the waveforms of voltage, current, and power when an ac voltage source is coupled to a dc voltage source via a converter with commutation on the ac side and two idealized filters, one with high-pass character on the ac side and the other low-pass character on the dc side. Two operating cases are illustrated i.e control angle $\alpha = 0$ (fully modulated with active load on the ac voltage source) and control angle $\alpha = 90°$ (with inductive load on the ac voltage source). Voltage and current have the familiar waveforms. Of interest is the waveforms of the powers p_H and p_T in the filters on the ac and dc sides. At any point in time

$$p = p_H + p_T + P_d. \tag{11.32}$$

The areas under the waveforms of power against time between two passages of the powers p_H and p_T through zero are a measure of the energy to be stored by the filters and also of the size of the filters. It can be seen that with a control angle $\alpha = 90°$ (reactive power operation without active power transfer), the energies to be stored by the filters are the highest. The filter T on the dc side must store more energy than the filter H on the ac side. When $\alpha = 90°$ the area under the power wave form of the ac voltage source.

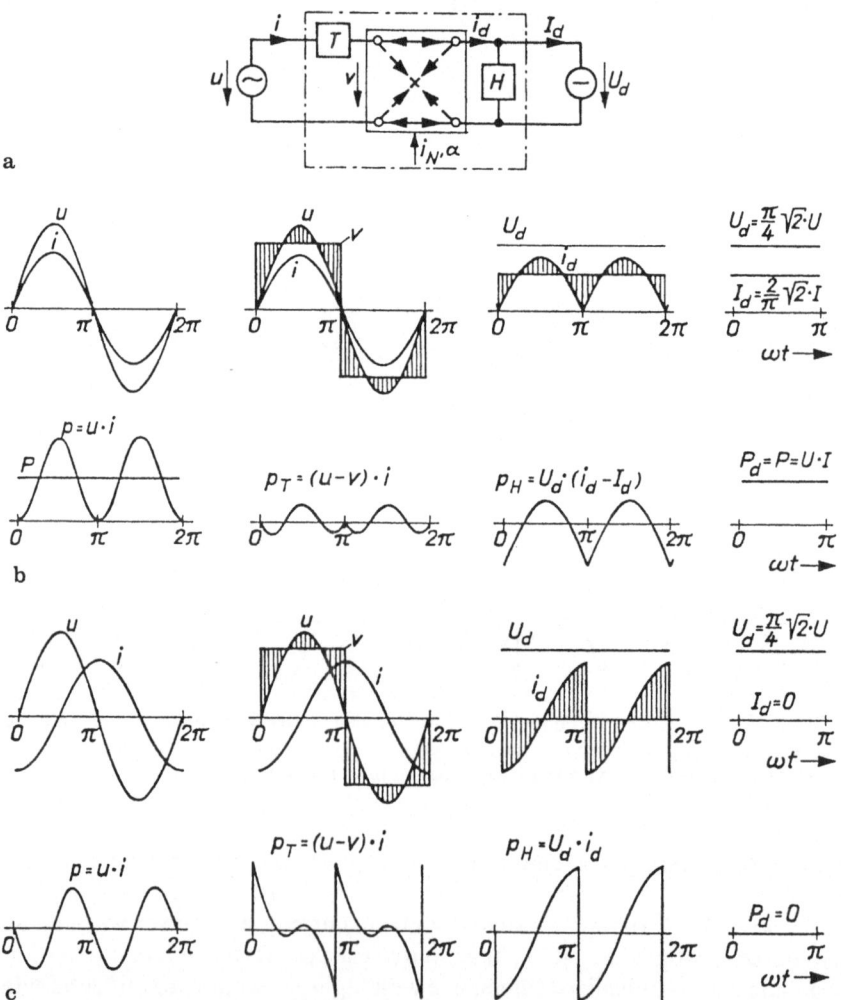

Fig. 11.10a–c. Voltage, current and power waveforms for a converter with commutation on the dc side (single-phase bridge connection). a Circuit with equivalent commutator and ideal filters; b with active power ($\alpha = 0$); c with inductive reactive power ($\alpha = 90°$)

Figure 11.10 shows the waveforms of voltage, current, and power when an ac voltage source is coupled to a dc voltage source via a converter with commutation on the dc side, a filter T with low-pass character on the ac side, and a filter H with high-pass character on the dc side. Here the operating conditions corresponding to Fig. 11.9b and c are similarly illustrated, namely active current and inductive reactive current in the ac voltage source.

Close comparison of Figs.11.9 and 11.10 indicates the dualism between sinusoidal voltage u and square-wave current j in front of the commutator, in the case of converters with commutation on the ac side, and sinusoidal current i and square-wave voltage v in front of the commutator in the case of converters with commutation on the dc side. This dualism of current and voltage leads to corresponding waveforms of with the two types of converter. Also, with converters with commutation on the dc side the filter H on the dc side has the larger energy to store.

11.4.2 Coupling of Three-phase AC and DC Systems

The conditions occuring when a multi-phase ac system is coupled to a dc system shall now be dealt with. This task is simplified by the fact that a multi-phase system with symmetrical sinusoidal currents supplies a power that is constant with time as is required on the dc side. Figure 11.11 shows the power flow. In this case three filters f_1, f_2, and f_3 are required on the ac side. The instantaneous powers p_1, p_2, and p_3 of the three ac voltage sources 1, 2, and 3 (see Fig. 11.2 and Eq. (11.23)) produce a constant power P equal to the dc power P_d. Then

$$p_1 = \frac{P}{3} + p_{21},$$ (11.33a)

$$p_2 = \frac{P}{3} + p_{22},$$ (11.33b)

$$p_3 = \frac{P}{3} + p_{23},$$ (11.33c)

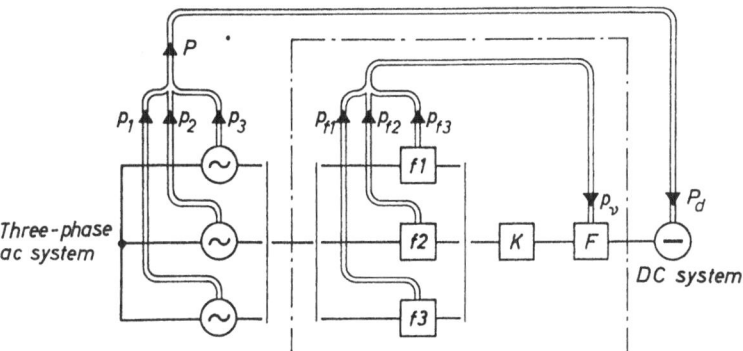

Fig. 11.11. Power flow when a three-phase ac system is coupled to a dc system via a converter. $f1...f3$ Filters on three-phase ac side, K commutator (semiconductor switch), F filter on dc side

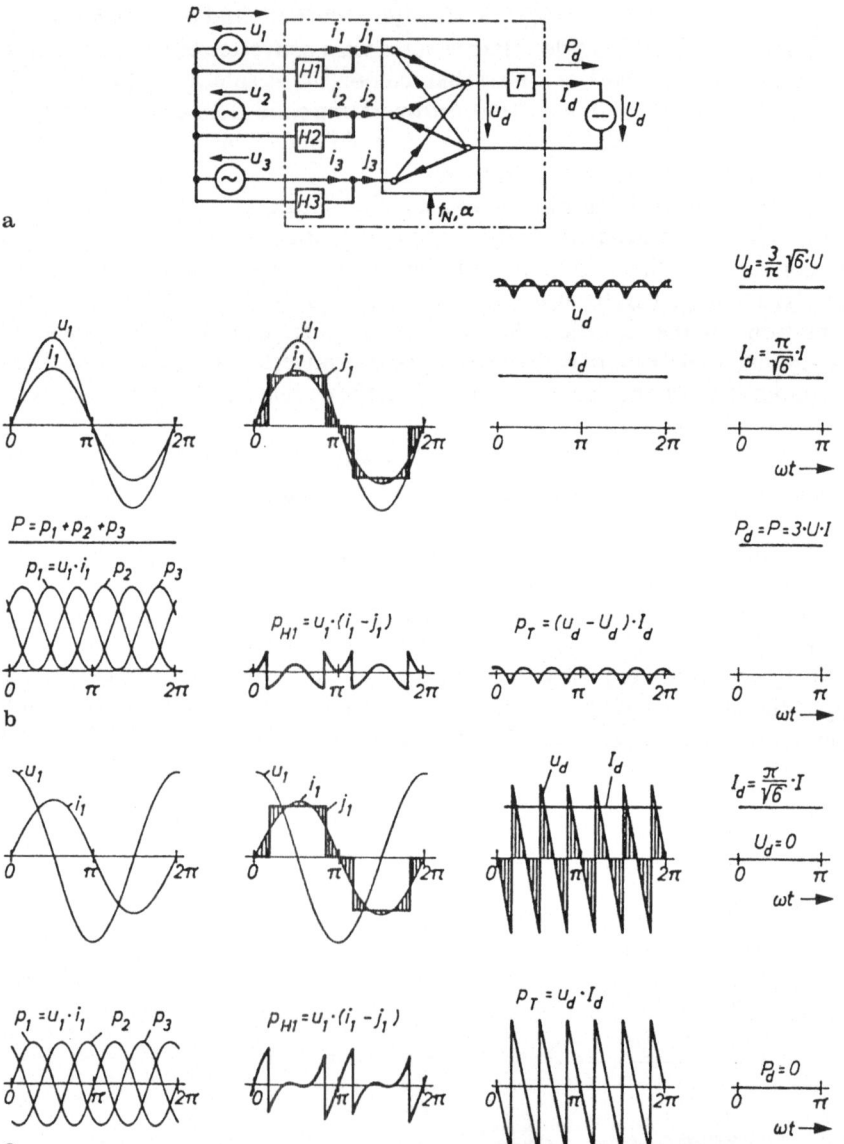

Fig. 11.12a – c. Voltage, current and power waveforms for a converter with commutation on the dc side (three-phase bridge connection). **a** Circuit with equivalent commutator and ideal filters; **b** with active power ($\alpha=0$); **c** with inductive reactive power ($\alpha=90°$)

with the sum of the partial powers pulsating at twice the system frequency

$$p_{21}+p_{22}+p_{23}=0. \qquad (11.34)$$

Only the harmonic power p_v is still exchanged between the filters on the ac and dc sides.

Figure 11.12 shows the waveforms of voltage, current, and power with a converter in three-phase bridge connection with commutation on the ac side for two characteristic operating conditions, namely control angle $\alpha = 0°$ (fully modulated with active current in the three-phase ac system) and control angle $\alpha = 90°$ (inductive reactive current in the three-phase ac system).

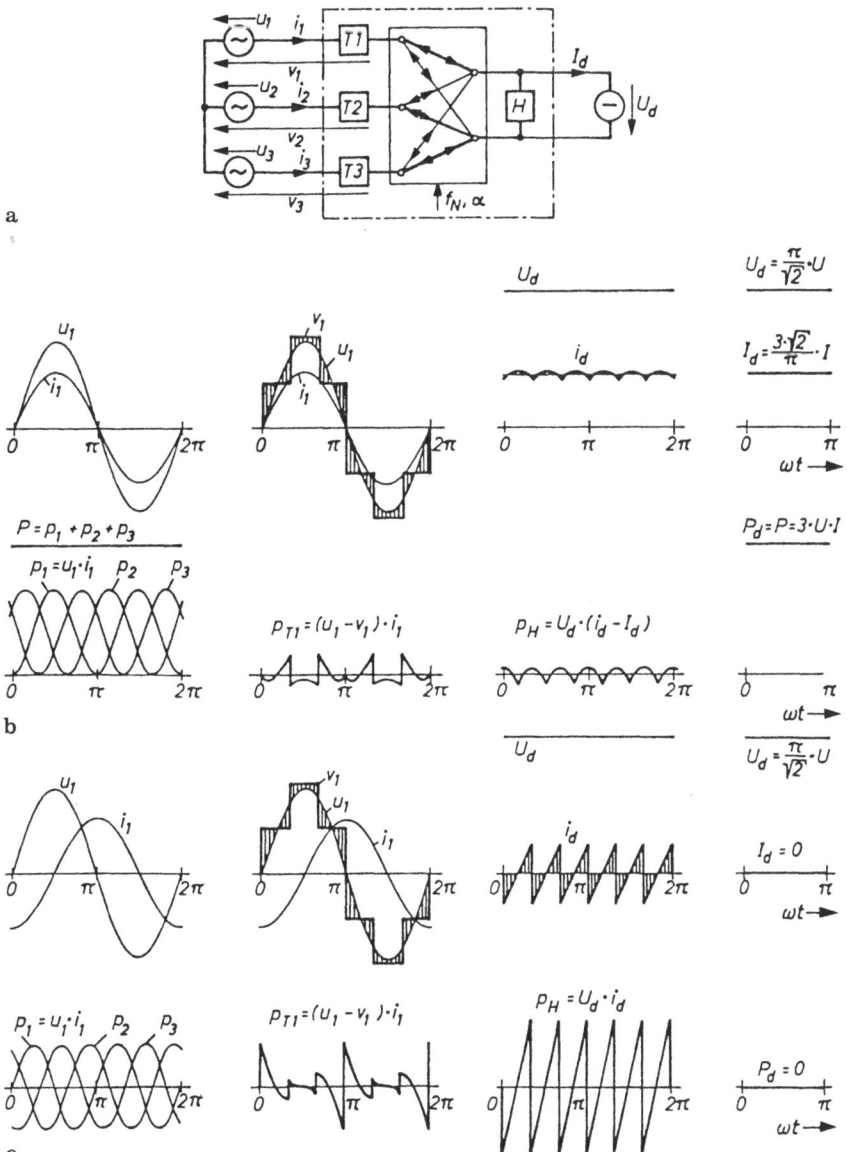

Fig. 11.13a − c. Voltage, current and power waveforms for converter with commutation on the dc side (three-phase bridge connection). **a** Circuit with equivalent commutator and ideal filters; **b** with active power ($\alpha = 0$); **c** with inductive reactive power ($\alpha = 90°$)

Figure 11.13 shows the corresponding waveforms for the case of converters in three-phase bridge connection with commutation on the dc side also for the two operating states of active current and inductive reactive current in the three-phase ac system. Compared with the single-phase converters considered previously it is characteristic that the filters no longer have to supply or absorb any power pulsations at twice the system frequency. In the filter on the dc side the power pulses with the frequency $2mf_N$ and higher harmonics (m = number of phases, f_N = system frequency) [11.28].

Also in the case of multi-phase systems a dualism exists between voltages and currents for converters with commuation on the ac side and currents and voltages for converters with commutation of the dc side. Owing to the balanced three-phase ac system and the higher frequency of the power pulsation, the energies to be stored by the filters are smaller than with single-phase converters. The greatest demands on these filters occur at a control angle of $\alpha = 90°$ (reactive power operation).

11.5 Pulse Number

Among many possible connections only two typical converter connections have been examined in the examples so far: the single-phase bridge connection and the three-phase bridge connection. The three-phase bridge connection is the most common connection with semiconductor components. Besides these a multitude of other converter connections are used in power electronics (see Sect. 7.1.5) not all of which, however, will be dealt with here. Attention should, however, again be drawn to the general circuit principle that by increasing the number of pulses of a converter connection, the voltage, and current harmonics on the ac and dc sides can be considerably reduced. The number of pulses p is defined as the total number of non-simultaneous current transfers by principal arms of a converter connection during one cycle (Sect. 7.1.1). An increase in the number of pulses therefore implies an increase in the number of principal arms of a converter i.e. the minimum number of principal arms of a converter. This leads to a reduction in the harmonics i.e. to smoother dc voltages and more sinusoidal ac currents.

The harmonics of externally commutated converters have already been dealt with in Sect. 7.1.9. It was found that with externally commutated converters, harmonic of the order $v = kp$, where $k = 1, 2, 3...$, occur in the dc voltage u_d. When fully modulated, the vth harmonic of the dc voltage is

$$U_{vi} = \frac{\sqrt{2}}{v^2 - 1} U_{di}.$$

(11.35)

(U_{di} = ideal direct voltage). The harmonic currents in the ac system have the ordinal number $v = kp \pm 1$, where $k = 1, 2, 3...$. Their amplitudes decrease as the ordinal number rises. Then

$$I_{vi} = \frac{I_{1i}}{v}$$

(11.36)

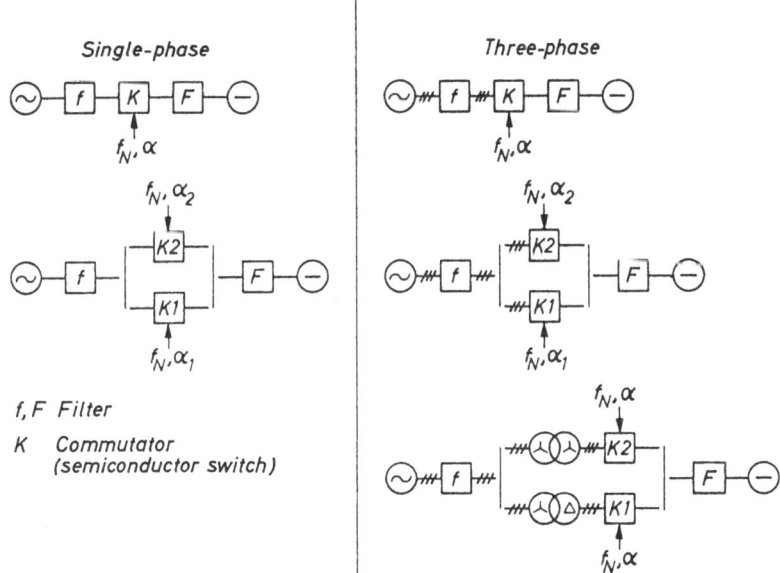

Fig. 11.14. Methods of increasing the number of pulses of rectifiers and inverters

(I_{1i} = fundamental component of the ideal system current). The same applies to the relationship between the harmonics and the number of pulses in self-commutated converters.

In Fig. 11.14 general techniques are enumerated for increasing the number of pulses of dc rectifiers and inverters. These include using multi-phase ac systems and providing additional phase displacement of the ac voltages by the use of converter transformers. Besides these a phase-displaced modulation of partial converters is also possible. In addition, the control angles α_1 and α_2 of the partial converters K1 and K2 must lead or lag by the same angle. Nevertheless, a leading control angle can be achieved only by forced commutation.

Figure 11.15 illustrates the better approximation of the system current to the desired sine wave made by increasing the number of pulses p. With a converter in three-phase bridge connection the current waveform curve shown with 120° square-wave blocks of alternate polarity is produced on the system side; this being in the ideal case when the converter transformer is connected in star on the primary and secondary sides. With the converter transformer connected in delta on the primary side the stepped current waveform is obtained in the supply system with two different values of current of alternate polarity for the same converter connection. Both current waveforms have harmonic currents of exactly the same amplitude. The 5th and 7th, 17th and 19th etc. current harmonics are, however, phase-displaced from one another in such a way that they cancel each other out when added. The summated current in the supply system has the stepped waveform drawn with three different values of current of alternate polarity. Besides the fundamental component the lowest current harmonics arising are the 11th and 13th (for p = 12).

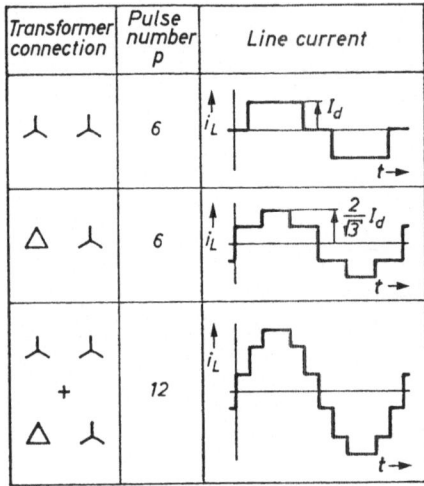

Transformer connection	Pulse number p	Line current
入 入	6	
△ 入	6	
入 入 + △ 入	12	

Fig. 11.15. Waveforms of the system currents for six- and twelve-pulse converters

So, increasing the number of pulses leads to a reduction in the current and voltage harmonics and hence to a reduction in the size of the filters on the ac or dc sides. This can be illustrated by comparing Figs. 11.9 or 11.10 with Figs.11.12 or 11.13 in which the conditions of the single-phase bridge connection (number of pulses $p=2$) are plotted against those of the three-phase bridge connection (pulse number $p=6$).

11.6 Pulse Frequency

With all the converter connections examined in this chapter it was implicitly assumed that the semiconductor switches in the converter arms or in the equivalent commutator switch periodically with the frequency f_N of the ac voltage source i.e. a semiconductor switch performs an on/off switching cycle every cycle of the supply system.

Self-commutated converters with forced commutation permit this limiting switching condition to be lifted and hence allow for repeated switching on and off per cycle of the supply system (see also Sects. 8.2 and 8.3). Thus, without increasing the number of semiconductor switches (as when increasing the number of pulses using the conventional technique described in Sect. 11.5), by repeated switching the total number of non-simultaneous commutations during a cycle of the supply system is increased, and another way to reduce current and voltage harmonics is obtained. The frequency with which the semiconductor switches switch is designated the pulse frequency f_p. In principle the pulse frequency f_p can be any (even fractional) multiple of the supply system frequency. Nor does it need to be constant with time. In the following considerations the special case of a constant pulse frequency equal to a whole multiple of the supply system frequency is assumed. Converters whose semiconductor switches operate at pulse frequency will be called pulse converters or PWM converters.

Converters of this type open up novel possibilities in the conversion of electrical energy. The most important are the conversion of the mean values of currents and voltages, the creation of any current waveform as well as the reduction of current and voltage harmonics and hence the reduction of the expense of filters and smoothing devices. Amongst other things the characteristics of pulse converters depend greatly upon the pulse frequency. With regard to harmonics and small filters the highest possible pulse frequencies should be used; however, there are limits to this owing amongst other things to the switching characteristics of the semiconductor components (see Chap. 3).

11.6.1 Pulse Converters with Commutation on the DC Side

The behavior of a single-phase ac system coupled to a dc system via a pulse converter with commutation on the dc side shall now be examined more closely [11.14, 11.15, 11.18, 11.30].

Figure 11.16a shows the circuit considered. It is assumed that the commutator is driven at constant pulse frequency f_p. The relative duty factor $\lambda = T_1/T_p$ is variable. To generate sinusoidal variables on the ac side λ must vary sinusoidally with the system frequency (see Sects. 8.3.3 and 8.3.4).

To study the limiting values of very high pulse frequency ($f_p \rightarrow \infty$) the inductance L on the ac side and the smoothing capacitor C_d on the dc side can be very small ($L \rightarrow 0$ and $C_d \rightarrow 0$). From the power balance between the ac and dc sides it follows that

$$p(t) = u(t)i(t) = p_d(t) = u_d(t)i_d(t). \tag{11.37}$$

The assumed characteristics of the commutator permit three states for the voltage $v(t)$ on the ac side of the commutator

$$v(t) = \begin{vmatrix} u_d(t) \\ 0 \\ -u_d(t) \end{vmatrix}. \tag{11.38}$$

Accordingly for the current $i_d(t)$ on the dc side of the commutator

$$i_d(t) = \begin{vmatrix} i(t) \\ 0 \\ -i(t) \end{vmatrix}. \tag{11.39}$$

Assuming very high pulse frequency and hence any size of store L and C_d where $U_d \geq \sqrt{2}U$ the current $i(t)$ can be set to any value. The only actuating variable necessary for this is the variable duty factor of the commutator. Generally a sinusoidal waveform of current $i(t)$ is desired, $i(t) = \hat{\imath} \sin(\omega t - \varphi)$. In this case the power $p(t)$ on the ac side is

$$p(t) = \frac{\hat{u}\hat{\imath}}{2} [\cos \varphi - \cos(2\omega t - \varphi)] = P_d \left[1 - \frac{\cos(2\omega t - \varphi)}{\cos \varphi} \right]. \tag{11.40}$$

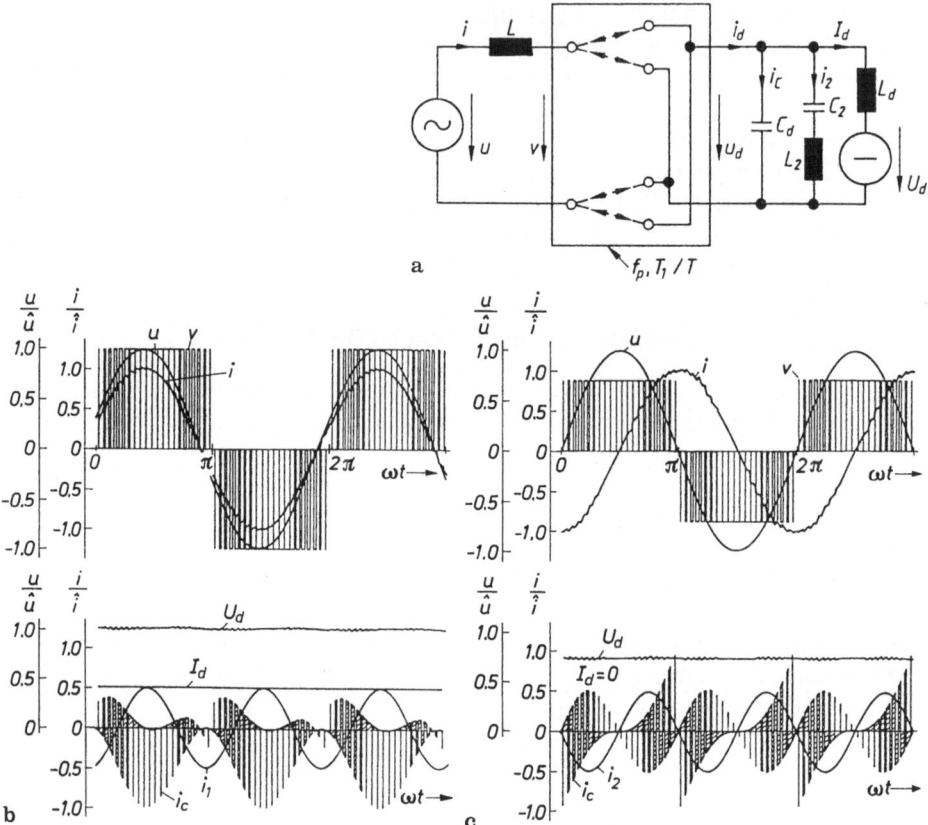

Fig. 11.16a–c. Voltage and current waveforms for a pulse converter with commutation on the dc side. **a** Circuit with equivalent commutator and filter; **b** with active power; **c** with inductive reactive power

The power $p(t)$ therefore oscillates about the mean value P_d at twice the system frequency and with an amplitude of $P_d/\cos\varphi$ (see Eq. (11.23)).

From the power balance between the ac and dc sides it follows that at constant dc voltage U_d the current $i_d(t)$ at the dc side output of the commutator

$$i_d(t) = \frac{p(t)}{U_d}.\qquad(11.41)$$

Under the assumed conditions this current therefore pulses about the mean value I_d as does the power on the ac side at twice the system frequency. In order that only the desired constant current flows in the dc voltage source U_d a filter circuit tuned to twice the system frequency must be added in parallel with the dc voltage source. This absorbs the ac portion of $i_d(t)$. This filter circuit must absorb the power pulsations of the single-phase ac voltage source independent of the pulse frequency.

Figures 11.16b and c illustrate the voltage and current waveforms when active power and inductive reactive power are transmitted via a pulse converter in single-phase bridge connection. The current and voltage waveforms were determined by simulation. A pulse frequency $f_p = 42f_N$, sinusoidal variation of the duty factor λ between the values 0 and 1, and a reactance factor $u_k = \omega L I_1 / U$ on the ac side of 30% are assumed. Making these quantitative assumptions the current i on the ac side is approximately sinusoidal. The remaining harmonics are determined by the areas under the curve of the difference between the voltage v and its fundamental component v_1 and by the magnitude of the inductance L on the ac side.

On the dc side the ac current i_2 superimposed on I_d and pulsating at twice the system frequency is absorbed by the filter circuit. The magnitude of this current is independent of whether active current or reactive current is being transmitted. The ripple of the dc voltage u_d is determined by the area under the waveforms of the current in the smoothing capacitor C_d.

The coupling of a multi-phase ac system with a dc system via a pulse converter with commutation on the dc side shall also be considered. Such a pulse converter can for example be realized in three-phase bridge connection with semiconductor switches for both directions of current (capable of being turned off in one direction and uncontrollable in the other).

Figure 11.17a shows the principle connection with equivalent commutator. The commutator works at pulse frequency f_p and connects each ac terminal alternately to the positive and negative dc terminal, the duty factor λ determining the actual positions of the three commutator switches. First ideal conditions shall be assumed, namely that the commutator operates with a very high pulse frequency $(f_p \rightarrow \infty)$. In this case the inductances on the ac side and the smoothing capacitor on the dc side can shrink to any size as they can with the single-phase bridge connection [11.13, 11.19, 11.20, 11.21, 11.22].

With sinusoidal and symmetrical currents on the three-phase ac side the sum of the power flow from there is constant.

$$p(t) = u_1(t)i_1(t) + u_2(t)i_2(t) + u_3(t)i_3(t) = 3UI \cos \varphi = \text{const} = p_d$$

$$(11.42)$$

The power balance is therefore satisfied at any instant in this case without supplementary energy stores. The filter circuit on the dc side is omitted. The amplitude and phase displacement of the current i(t) in the ac voltage sources are, in principle, continuously adjustable under the above assumptions so long as the condition $U_d \geq 2\sqrt{2}U$ is satisfied.

In Figs. 11.17b, c, and d, the current and voltage waveforms are reproduced again for operation with active power and with inductive and capacitive reactive power. The waveforms were determined quantitively by simulation. The pulse frequency f_p is again 42 f_N, the duty factor λ varies sinusoidally with the system frequency between the values 0 and 1, the reactance factor n_k on the ac side is 30%. The current i_1 on the ac side is approximately sinusoidal, the remaining current harmonics are determined by the area under the curve of the difference between the converter phase voltage v_1 and its fundamental component and by the size of the inductance L on the ac side.

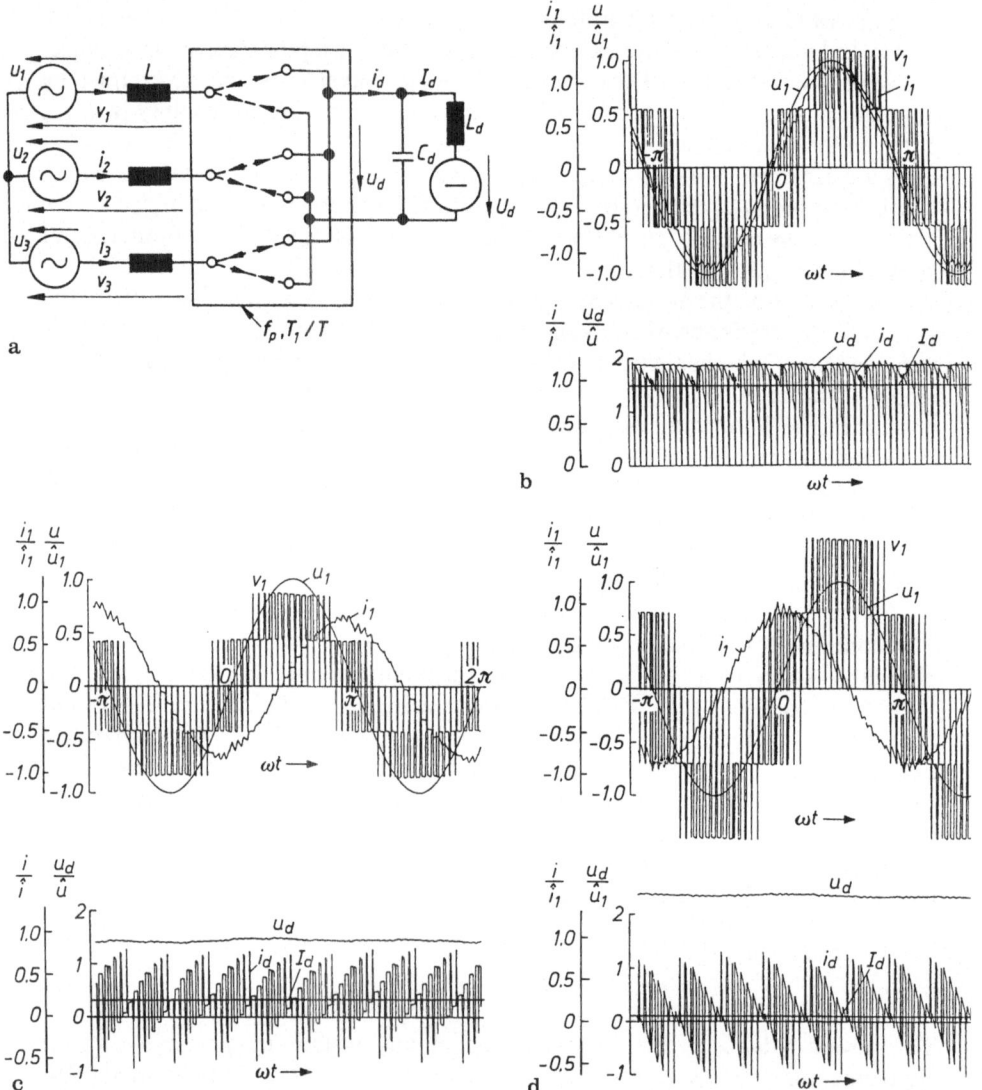

Fig. 11.17a – d. Voltage and current waveforms for a pulse converter with commutation on the dc side (three-phase bridge connection). **a** Circuit with equivalent commutator and filter; **b** with active power; **c** with inductive reactive power; **d** with capacitive reactive power

On the dc side there is still only a high-frequency power pulsation at the pulse frequency which is absorbed by the smoothing capacitor C_d. On reactive power operation the mean value of the dc current I_d disappears. The mean value of the dc voltage U_d drops with inductive reactive power in the three-phase system and rises with capacitive reactive power. Approximately

$$U_d \approx 2\sqrt{2}U(1 \pm u_k). \tag{11.43}$$

Since the harmonics superimposed on the ac current vary with the magnitude of the dc voltage U_d, they attain their maximum value with capacitive reactive power in the three-phase ac system.

11.6.2 Pulse Converters with Commutation on the AC Side

The pulse converters dealt with up to now work with commutation on the dc side. They correspond to the type of converter dealt with the Sect. 11.3.2 with a smoothing capacitor C_d on the dc side and inductance L on the ac side. This is the type of converter which has always been employed for pulse converters.

In principle, however, a converter with commutation on the ac side can also operate as a pulse converter. For this the valve arms must be capable of being turned off. The basic circuit of a pulse converter with commutation on the ac side corresponds, therefore, to the connection shown in Fig. 11.5b [11.1, 11.16].

In Fig. 11.18a this connection is represented by a mechanical equivalent commutator which only needs to be provided for the direction of current indicated. The assumed characteristics of the commutator permit three states for the voltage $u_d(t)$ on the dc side of the commutator

$$u_d(t) = \begin{vmatrix} u_C(t) \\ 0 \\ -u_C(t) \end{vmatrix}. \tag{11.44}$$

Accordingly for the current $i(t)$ on the ac side of the commutator

$$j(t) = \begin{vmatrix} i_d(t) \\ 0 \\ -i_d(t) \end{vmatrix}. \tag{11.45}$$

With a sinusoidal current i on the ac side and constant current I_d on the dc side a power balance is obtained only by introducing a blocking circuit consisting of L_2 and C_2 on the dc side. Across this blocking circuit there is a voltage u_2 pulsating at twice system frequency.

Figures 11.18c and d illustrate the current and voltage states on active power, inductive reactive power, and with capacitive reactive power with simultaneous energy absorption by the ac source. The current and voltage waveforms were determined by simulation with an assumed pulse frequeny $f_p = 42f_N$. The duty factor $\lambda = T_1/T_p$ varies sinusoidally between 0 and 1. The voltage $u_d(t)$ at the commutator output on the dc side with constant direct current I_d is given by

$$u_d(t) = \frac{p(t)}{I_d}. \tag{11.46}$$

Under the assumed conditions this voltage pulses about the mean value U_d like the power on the ac side at twice the system frequency. In order that only the desired constant ac voltage U_d appears across the dc voltage source the blocking circuit tuned to twice the system frequency must be connected in series. This blocking

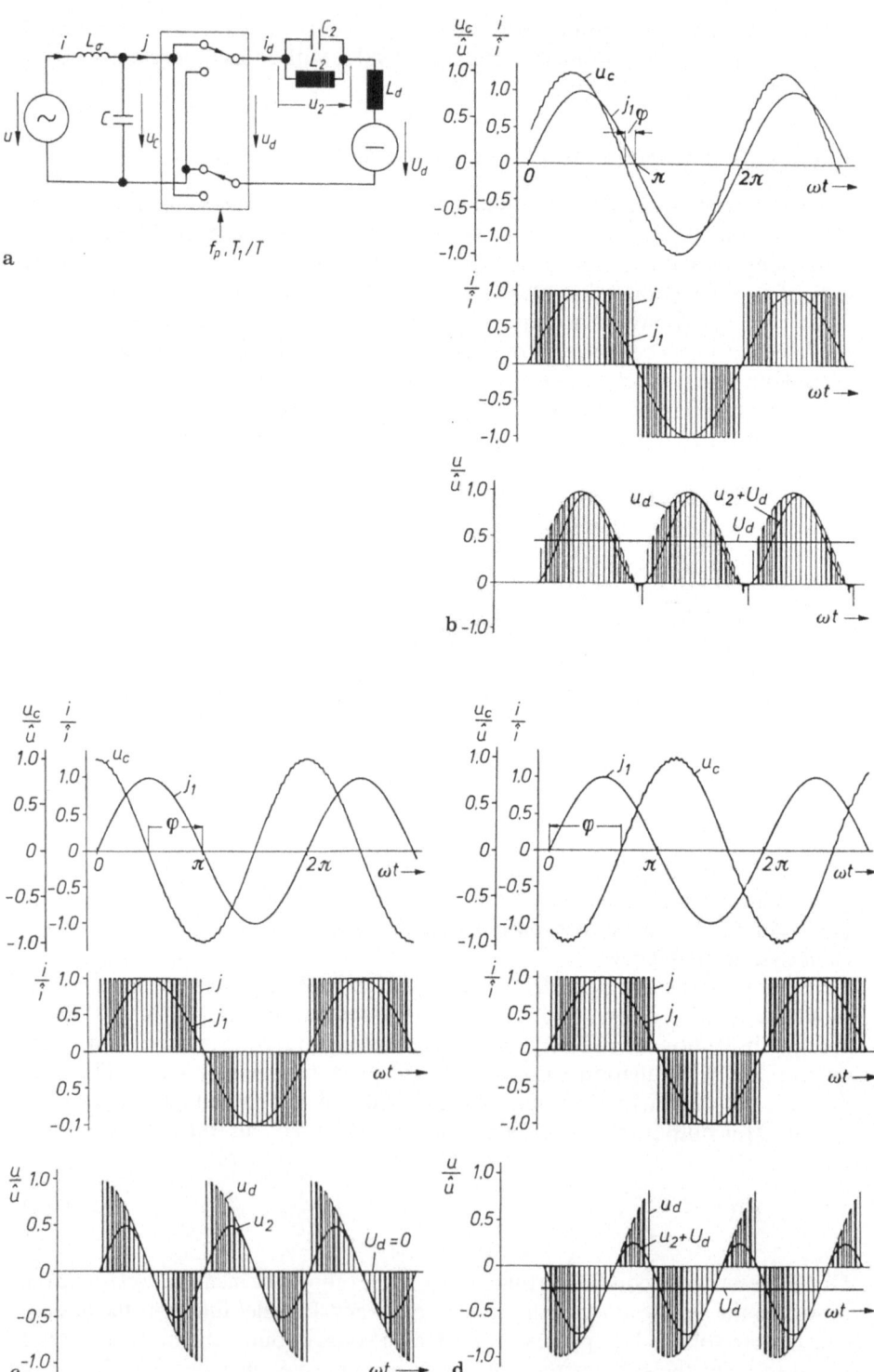

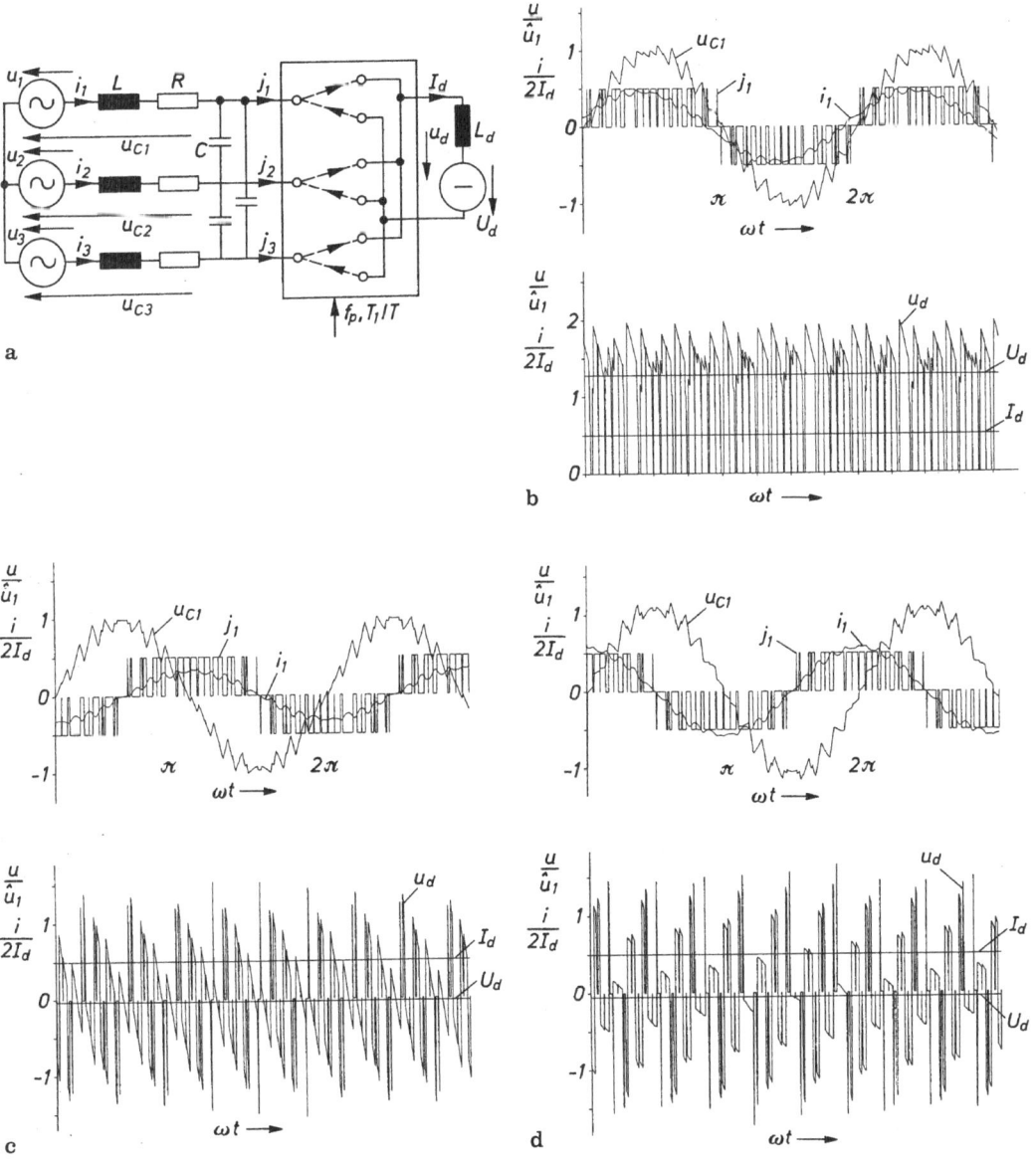

Fig. 11.19a – d. Voltage and current waveforms for a pulse converter with commutation on the ac side (three-phase bridge connection). **a** Circuit with equivalent commutator and filter; **b** with active power; **c** with inductive reactive power; **d** with capacitive reactive power

Fig. 11.18a – d. Voltage and current waveforms for a pulse converter with commutation on the ac side (single-phase bridge connection). **a** Circuit with equivalent commutator and filter; **b** with active power; **c** with inductive reactive power; **d** with capacitive reactive power with energy absorption by the ac voltage source

circuit absorbs the power pulsations of the single-phase ac voltage source independent of the pulse frequency.

Between the voltage and currents of the pulse converters with commutation on the dc side in Fig. 11.16 and those for the converter with commutation on the ac side in Fig. 11.18 there is a dualism similar to that which arose between converters illustrated in Figs. 11.9 and 11.10.

A pulse converter with commutation on the ac side can also be constructed as a multi-phase unit. This then produces dual current and voltage conditions for the pulse converter illustrated in Fig. 11.17 with commuation on the dc side in three-phase bridge connection (see Fig. 11.19). Pulse converters with commutation on the ac side have so far not been employed in practice. They tend to have oscillations between the leakage inductance L_σ of the ac side and the parallel capacitor C.

11.7 Reactive Power Compensation and Balancing of Unbalanced Load

The definitions of reactive power dealt with in Sects. 7.1.7 and 11.2 can be augmented by a general statement. So long as the voltage and current waveforms are in phase the reactive power is zero. Phase displacements between voltage and current act as reactive power even when pure resistors are supplied via the converter.

This applies generally even with non-sinusoidal ac voltage [11.1]. Reactive power always occurs with converters since the instantaneous values of voltage and current normally are not linearly related. However, this reactive power can, if necessary, be kept low by appropriate methods (e.g. by the sector control dealt with in Sect. 8.3.5). It is desirable that the power factor, determined by the ratio of transmitted active power to apparent power arising should be as high as possible. The optimum limiting value attainable for the power factor λ is therefore 1 [11.9, 11.10].

11.7.1 Reactive Power Compensation

The reactive power in single-phase and three-phase ac circuits can be compensated by supplementary reactances (usually capacitive). Such reactances can be contactlessly switched by semiconductor switches (see Sect. 6.1.4). Using semiconductor switches inductive rectances can also be continuously adjusted by phase control (see Sect. 6.2). In addition converters can be employed as pure reactive power converters whereby their inductive or capacitive reactive power consumption can be continuously varied [11.6, 11.12].

Besides the undesired loading of systems and installations by reactive power reactive currents generate voltage drops across the system reactances. The voltage drops caused by uneven loads such as welding machines are furnaces or drives with surge-like current consumption lead to system voltage fluctuations called "flicker". When the pulsations of reactive power occur in the range of several Hz, the light fluctuations caused by flicker are tiring to the human eye.

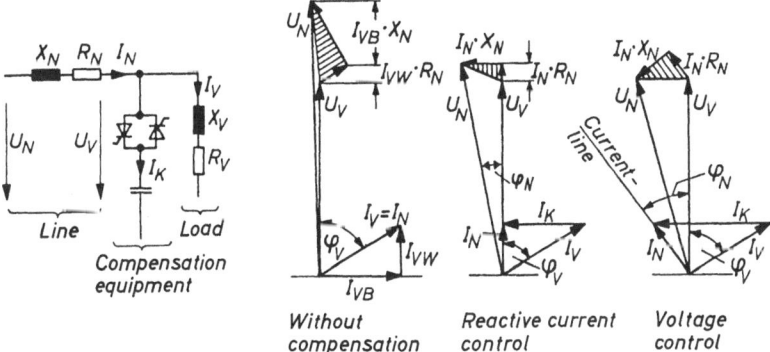

Fig. 11.20. Circuit diagram and phasor diagram of a resistive-inductive load with capacitive power factor compensation

Reactive power fluctuations can be compensated by converters or by reactances switched by power semiconductors. Not only inductive but also capacitive reactances can be switched by power semiconductors connected in antiparallel. The maximum switching frequency causing no transient oscillations equals twice the system frequency. This result in good dynamics [11.26, 11.27, 11.29].

Figure 11.20 shows the principle of parallel compensation of the voltage drop in a single-phase ac system with inductive and resistive internal impedance by a thyristor-switched compensating capacitor. With pure power factor compensation the resistive voltage drop remains. Compensation of the system current to the system current line inclined at an angle of $\varphi_N = \arc\tan(X_N/R_N)$ also reduces the resistive voltage drop.

If the compensating capacitors are binary stepped, four capacitor stages can produce 15 different steps of reactive power compensation (Fig. 11.21). In this case capacitors are connected via thyristors with diodes in antiparallel. When semicontrolled switches are employed, the maximum switching frequency drops to the system frequency. Nevertheless, good dynamic voltage stability can be attained by this method.

Instead of with compensating capacitors it is also possible to work with switched compensating inductances. In this case stepped reactors are switched in

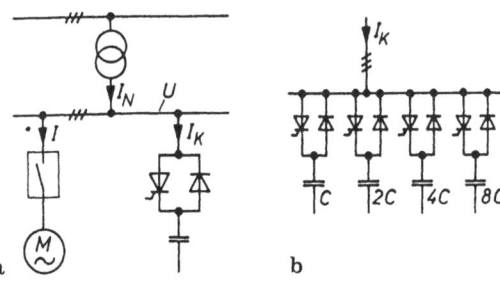

Fig. 11.21a,b. Dynamic voltage stabilization with thyristor-switched power capacitors. **a** Plant; **b** power factor compensation equipment

and out via thyristors in antiparallel. Since the switching in of inductances increases the inductive reactive current, supplementary fixed compensating capacitors must be provided.

11.7.2 Balancing of Unbalanced Load

When multi-phase ac systems are loaded unbalanced load can occur i.e. different currents in the individual phases. In this case the individual phases are unsymmetrically loaded and power pulsations occur even on otherwise balanced multi-phase systems.

According to Steinmetz a single-phase resistive load between two conductors of a three-phase system can be balanced by means of capacitive and inductive reactances. Figure 11.22 shows the conditions with load balancing in a three-phase ac system.

The capacitive and inductive reactances required for balancing (Fig. 11.22a) can be calculated in accordance with equation

$$\omega C_{23} = \frac{1}{\omega L_{31}} = \frac{1}{\sqrt{3} R_{12}}. \tag{11.47}$$

With these reactances symmetrical resistive currents I_1, I_2, and I_3 arise in the three-phase ac system.

When in addition inductive loads are to be compensated (Fig. 11.22b) the required compensating capacitors can be calculated in accordance with the

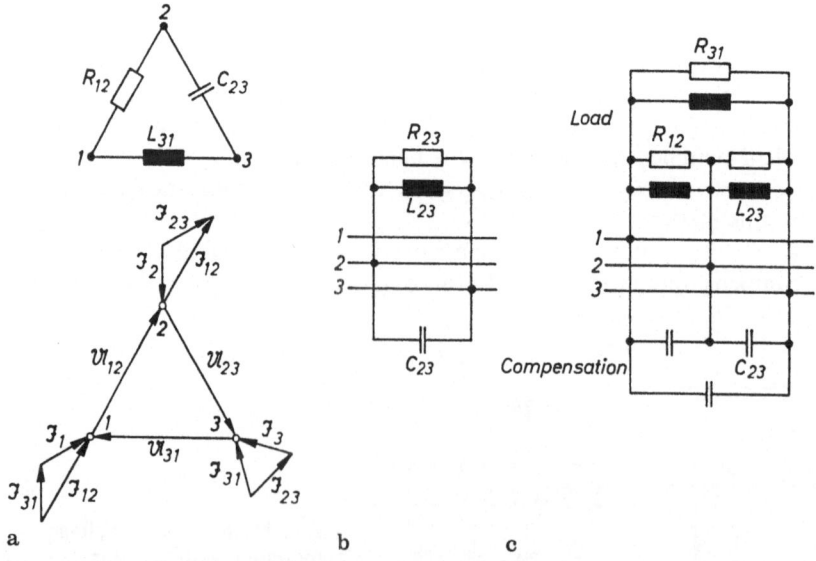

Fig. 11.22a – c. Complete reactive power compensation on unsymmetrical load. **a** Load balancing in accordance with Steinmetz; **b** reactive power compensation; **c** reactive power compensation with load balancing

equation

$$\omega C_{23} = \frac{1}{\omega L_{23}}. \qquad (11.48)$$

If reactive power compensation with load balancing is to be carried out for all three phases (Fig. 11.22c), the capacitors required can be calculated in accordance with the equation

$$\omega C_{23} = \frac{1}{\omega L_{23}} + \frac{1}{\sqrt{3}} \left(\frac{1}{R_{12}} - \frac{1}{R_{31}} \right). \qquad (11.49)$$

The capacitors for the other phases are obtained by rotating the indices. Under certain conditions Eq. (11.49) can produce a negative value. In this case instead of a capacitor an additional inductance is required.

Figure 11.22c shows how by combination of compensating capacitors between the three conductors of a three-phase ac system an unbalanced resistive-inductive load can be balanced and completely power factor corrected.

The coupling of ac systems which have different number of phases represents a similar problem. Figure 11.23 shows how the unbalanced loading of the three-phase ac system illustrated in part a can be balanced by appropriately arranged inductive and capacitive reactances whereby an inductive component of the current I_{12} in the single-phase system can be eliminated by a supplementary capacitor C_K. When the current in the single-phase system is variable the

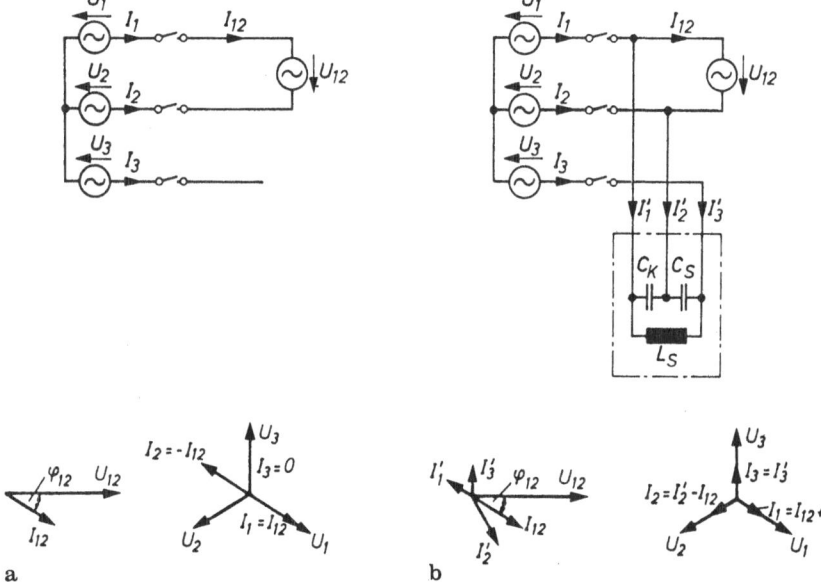

Fig. 11.23a,b. Reactive power compensation and balancing when out of phase ac systems are coupled at constant frequency. **a** Unsymmetrical loading; **b** reactive power compensation and balancing

reactances must be adjusted accordingly. This can be done with the aid of semiconductor switches. The reactances can in such a case again be binary graded and switched in and out via thyristor switches in synchronism with the supply without transient phenomena.

Instead of with reactances the balancing and power factor correction of a three-phase system can also be carried out using a reactive power converter.

11.8 Losses and Efficiency

The efficiency η of a converter is the ratio of the output power P_0 to the input power P_I. The efficiency can be calculated from the following equation:

$$\eta = \frac{P_0}{P_I} = \frac{P_0}{P_0 + \Sigma P_{Loss}} = 1 - \frac{\Sigma P_{Loss}}{P_0 + \Sigma P_{Loss}} = 1 - \frac{\Sigma P_{Loss}}{P_I} \qquad (11.50)$$

where ΣP_{Loss} is the sum of all the relevant losses.

Losses in converters arise in the power semiconductors themselves because they have a forward voltage drop (of the order of 1 V to more than 2 V) in the on-state (see Chap. 3). In addition, off-state and switching losses occur but these can generally be ignored in the case of line-commutated converters. Power must likewise be applied for triggering and cooling (see Chap. 4) and be added to the loss balance. Electrical energy is converted into heat in the smoothing elements. Series fuses also have losses owing to their internal resistance (of the order of 1 mΩ).

Besides these, losses occur due to the current in the connecting wiring within the converter. A considerable portion of the total loss consists of the copper and iron losses in the converter transformer and current-limiting and current-equalizing reactors or filters where present. To this can be added the powers of auxiliary drives (pumps, fans etc.). With self-commutated converters additional losses occur in the commutating devices.

Power Terms. Table 11.1 lists power terms on the ac and dc sides (see also Sect. 11.2). The active power P_L is the arithmetic mean over the time period of the instantaneous power values on the ac side of the converter (see Eq. (11.1)):

$$P_L = \frac{1}{T} \int_0^T u i \, dt. \qquad (11.51)$$

The fundamental power P_{1L} is that part of the active power formed by the fundamental component of current and voltage. The harmonic power is that part of the active power formed by the harmonic component of current and voltage. The reactive power Q_L is the sum of the fundamental and harmonic reactive power. The apparent power S_L is the product of the effective values of voltage and current on the ac side of the converter: the active power P_d on the dc side is the arithmetic mean over the time period of the instantaneous power values on the dc

Table 11.1. Power and efficiency terms

Power terms
on the ac side
active power P_L: fundamental power P_{1L} and harmonic power
reactive power Q_L
apparent power S_L
on the dc side:
active power P_d
dc current power S_d: product of the arithmetic mean of dc voltage and
dc current
input power P_I: active power absorbed
fundamental input power P_{1I}
output power P_0: active power delivered
fundamental output power P_{1O}
Efficiency terms

efficiency $\eta = \dfrac{P_0}{P_I}$

rectification degree $\eta_{d1} = \dfrac{S_d}{P_{iL}}$

inversion degree $\eta_{1d} = \dfrac{P_{1L}}{S_d}$

ac conversion degree $\eta_{11} = \dfrac{P_{1O}}{P_{1I}}$

---

side of the converter:

$$P_d = \frac{1}{T} \int_0^T u_d i_d dt. \tag{11.52}$$

The direct current power S_d is the product of the arithmetic mean of the direct voltage and the direct current:

$$S_d = U_d I_d. \tag{11.53}$$

Efficiency Terms. Terms of efficiency are also listed in Table 11.1.
Determination of Efficiency. It is generally accepted that the efficiency of a line commutated converter is the efficiency resulting from an almost sinusoidal operating voltage on the ac side and an almost smooth dc current on the dc side (whose ac content is $\leq 5\%$). The efficiency is usually determined for rated values of input voltage, output voltage, and output current. It is also often interesting to consider the efficiency as a function of the converter load. Since the no-load losses are constant the efficiency generally falls off with partial load.

The efficiency of a converter can be determined either indirectly or by direct measurement. With direct measurement the input and output powers are measured with wattmeters or by precision instruments for measuring voltages and currents. On the dc side additional measuring instruments can be employed to indicate the arithmetical mean when the smoothing is possibly so good that the error caused by neglecting the ac content is less than half the tolerance for determining the efficiency. On the single-phase or three-phase ac side harmonics must be taken into consideration. Owing to the distorted voltage and current waveforms the magnitude of the measuring accuracy must be watched.

With indirect determination the efficiency is determined from the sum of the losses in the individual parts of the apparatus. These individual losses can be determined partly by measurement and partly by calculation. The efficiency is then calculated in accordance with Eq. (11.50). In the case of multi-phase converters of high output (above 300 kW or above 5000 A rated ac current) this indirect method is always employed.

In the following the relevant and irrelevant losses in determining the efficiency of converters are discussed.

The losses in the cables outside the converter and in the switchgear are not taken into consideration when determining the efficiency. The losses in smoothing devices on the ac side and in smoothing inductances on the dc side are also not taken into consideration when determining the efficiency of a converter. The same applies for losses in auxiliary devices which operate only for short periods. The losses in base load resistors are taken into consideration when these devices remain continuously switched on.

It has already been mentioned in Sect. 4.3 that the main portion of the losses in the power semiconductors is the on-state loss. A decisive influence, therefore, on the proportion of losses in the power semiconductors to the total losses of a converter is the ratio of reverse voltage to on-state voltage of any power

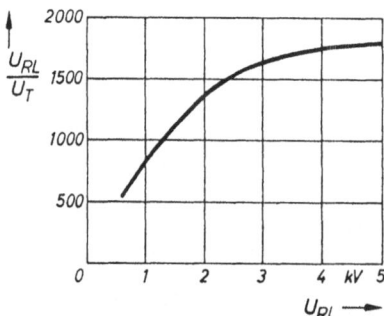

Fig. 11.24. Ratio of reverse voltage/on-state voltage of N-type thyristors

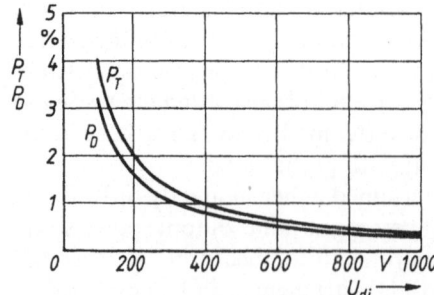

Fig. 11.25. Referred thyristor losses p_T or diode losses p_D with line-commutated converters in three-phase bridge connection as a function of the ideal dc voltage U_{di} (1 power semiconductor in series)

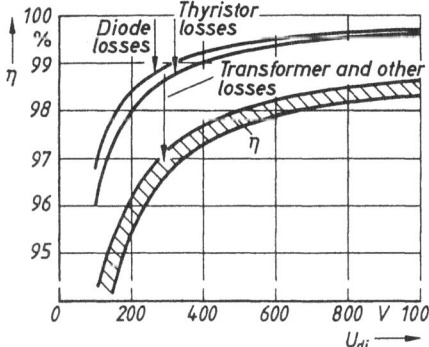

Fig. 11.26. Efficiency η, as a function of the ideal dc voltage U_{di} ($P_d > 100$ kW), for line-commutated converters in three-phase bridge connection

semiconductors employed. In Fig. 11.24 this ratio of repetitive peak reverse voltage to on-state voltage is plotted against the peak reverse voltage for N-type thyristors. The curve indicates the values attainable with optimum design of the N-type thyristors. Above 3 kV the ratio U_{RL}/U_T reaches an only slightly inclined plateau with values between 1500 and 2000.

In Fig. 11.25 the referred thyristor losses p_T or diode losses p_D are plotted against the ideal dc voltage U_{di}. These are estimated values for a line-commutated converter in three-phase bridge connection with one semiconductor valve per arm in series. These loss curves are determined by the ratio U_{RL}/U_T shown in Fig. 11.24 on the one hand and by the necessary safety factors on the other. At dc voltages above 600 V the referred power semiconductor losses are only around 0.5%.

Figure 11.26 similarly shows estimated values for the efficiency η of line-commutated converters as a factor of the ideal dc voltage U_{di}. The curves apply for converters in the power range above 100 kW. The transformer and other losses dominate the diode and thyristor losses in the upper voltage range. Typical efficiencies for such converter are 97% to 98%.

The efficiencies attainable with static converters therefore lie considerably above the values that can be attained with other converters (e.g. rotary converters). This is an essential advantage of power electronics.

12 Control Conditions

Owing to their good controllability and their output power being easily matched to the requirements of the application, static converters are particularly suitable as correcting units for open-loop and closed-loop electrical controls. The control units of converters can be directly triggered by electronic controllers. Control unit and controller are compatible i.e. they are constructed from the same components and have the same power level in the information-processing section.

The control conditions with converters shall now be dealt with briefly [11, 16, 12.1, 12.2].

Treatment is limited to statement of the most important terms for an open-loop or closed-loop control system and their representation in the signal flow diagram and configuration diagram.

An exact mathematical investigation on the converter as a correcting unit in open-loop and closed-loop controls would have to take into consideration the fact that the control angle is not a continuous function and that the relationship between the control angle and the converter output is not linear. Up to now there is no exact general theory. In practice the converter is approximated by a simplified model, for which a dead time between the change in the input quantity and in the output quantity is assumed and the control characteristics is linearized at least in partial ranges.

Since the beginning of the eighties microprocessors are more and more employed in control equipment. Signal processing becomes digital instead of analogue [12.7, 12.8, 12.9, 12.13, 12.15, 12.17]. Also adaptive control is applied with converters [12.10].

12.1 Terms and Designations

The most important terms and designations of open-loop and closed-loop control technology will be introduced here shortly.

12.1.1 Open-loop Control

In the case of an open-loop control one or more quantities act as inputs and other quantities as outputs. A characteristic of an open-loop control is the open sequence of action with which for the purpose of automatic correction there is neither detection nor feedback of the output quantity.

Figure 12.1 shows the signal flow diagram of an open-loop control.

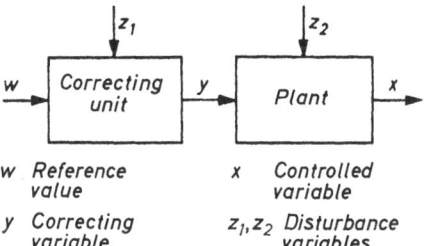

w Reference   x Controlled
   value   variable

y Correcting z_1, z_2 Disturbance
   variable   variables

Fig. 12.1. Signal flow diagram of an open-loop control

Such a signal flow diagram describes the relationships between the quantities. The various equations of the system are represented in the form of a block with input and output quantities, the quantities being represented by directional arrows.

The reference value w is supplied to the control from outside. It should be followed by the output quantity (controlled variable x) in a preset relationship. The correcting unit at the input intervenes in the mass flow or energy flow. In the case of a control with converters it comprises a trigger pulse unit and a power section (see Fig. 12.4). The correcting variable y transmits the controlling action of the control device to the plant that is that part of the path of action which represents the area of the plant it is desired to manipulate. Disturbances z_1 and z_2 are quantities acting from the outside in so far as they impair the intended manipulation in a control.

Along the so-called open-loop control chain consisting of the correcting unit and the plant there is a fixed direction of action which is indicated by the arrows in the signal flow diagram. A change in the reference value w effects a change in the correcting variable y and this produces a change in the output quantity x of the plant. The correction speed is then the speed at which the correcting condition is changed. The transient response of the plant is specified by the totality of the appropriate characteristics of its elements.

12.1.2 Closed-loop Control

In the case of a closed-loop control the quantity to be controlled (controlled variable) is continuously detected compared with the reference input and depending upon the result of this comparison adjusted to the reference input. This produces a sequence of action in a closed circuit the control system.

Figure 12.2 shows the signal flow diagram of a closed-loop control. The controlled variable x is detected as the quantity to be manipulated and compared with a preset reference input w at the input of the control device. So long as there is a difference between the reference input and the controlled variable, the correcting variable y acting on the plant is manipulated with the aid of the correcting unit of the control device such that the difference between the controlled variable and the reference input is reduced.

The input quantities of the control device are the reference input w, the controlled variable x and where applicable a disturbance z_1. The output quantity of the control device is the correcting variable y. The control device effects the

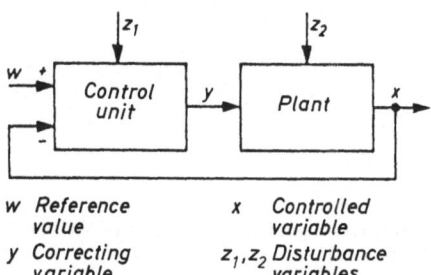

w Reference   x   Controlled
   value   variable
y Correcting z_1, z_2 Disturbance
   variable   variables

Fig. 12.2. Signal flow diagram of a closed-loop control

desired manipulation of the plant via the correcting unit. It contains elements for the conversion of the reference input and controlled variable, their comparison, dynamic correction elements, and amplifiers as well as the correcting unit (see Fig. 12.5). The plant is the area of the installation in which the manipulation of the controlled variable takes place. A characteristic of the plant is that it is permeated by the main energy flow.

The quality of a closed-loop control is determined by its steady-state and dynamic response. The accuracy of a closed-loop control is indicated by the maximum residual deviation of the controlled condition from the reference input under the action of the most unfavourable combination of disturbances. The constancy of a closed-loop control indicates the maximum residual deviation of the controlled variable with constant reference input under the action of the most unfavourable combination of disturbances.

The dynamic response of a closed-loop control is described by its reaction to sudden changes in reference input or to a sudden disturbance. The controlled variable reacts to both with a transient oscillation. The control response time is the time that elapses between a step in the reference input or a disturbance and the time the controlled variable again enters the preset tolerance range. The correction time beings when the controlled variable leaves a preset tolerance range after a step in the reference input or a disturbance. It ends when the controlled variable again enters this range to remain there permanently.

Figure 12.3 shows the waveform of the controlled variable after a sudden change in the reference input and after a sudden change in the load.

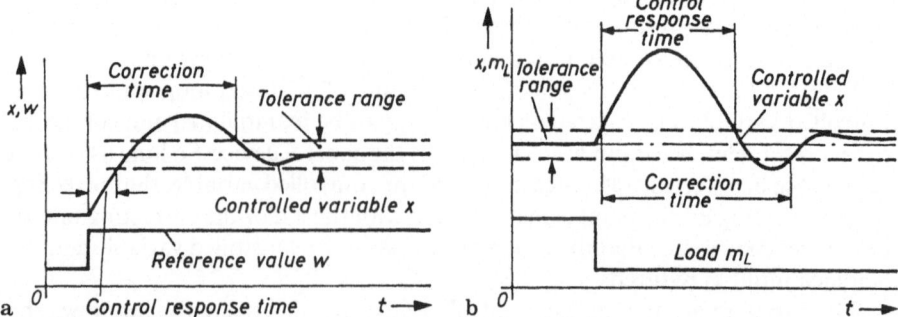

Fig. 12.3a,b. Waveform of the controlled variable after a sudden change of reference input (**a**) and load (**b**)

12.2 Converters as Correcting Unit

Converters are employed as correcting units in open-loop and closed-loop control systems. They operate as power amplifiers. Their output voltage or output current acts as the correcting variable on the open-loop or closed-loop control system. The output voltage or current is adjusted by changing the instants of triggering the controlled converters valves. Generation of the trigger pulses for the controlled converter valves is carried out in the control unit (see Fig. 4.13), the pulse displacement by a signal to the input amplifier of the control unit.

The response of the converter as a correcting unit is determined by three things: first by its steady-state characteristic which is for example the dependence of the output voltage on the pulse displacement signal; second by its internal impedance which alters the output voltage depending upon the current loading; and third by its dynamic response. A characteristic of converters is that the possible instants of triggering of the valves are discrete which produces dead times and that the functions of the output quantities of the input signals are generally not linear. A further feature is the directional dependence of the converter valves.

Analysis of the dynamic processes in the converter leads to non-linear differential equations which cannot be completely solved [12.3, 12.4]. By means of simplifying assumptions and linearizations in partial ranges models are, however, arrived at which make possible the mathematical treatment of converters as the correcting unit in open-loop and closed-loop controls and which lead to satisfactory results in practice [12.6, 12.11, 12.18, 12.19, 12.20].

12.2.1 Open-loop Control with Converters as Correcting Unit

Figure 12.4 shows the speed control of a dc machine via a line-commutated converter in six-pulse bridge connection. The reference input to the control unit is an adjustable dc voltage U_S. The correcting unit comprises the control unit and the

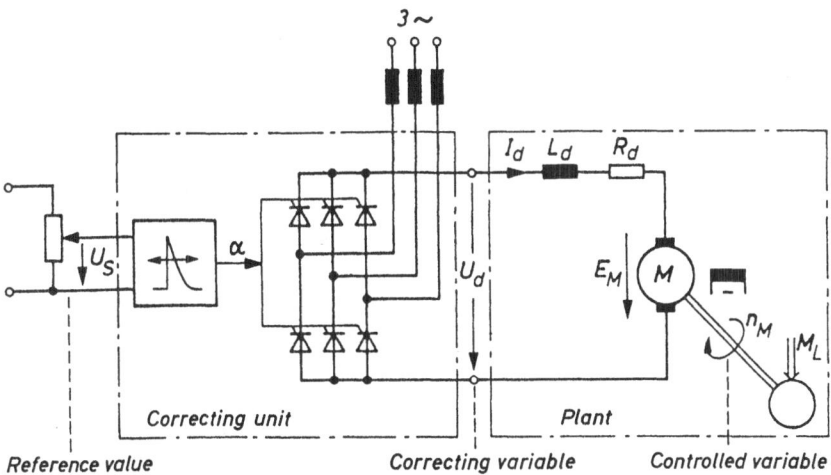

Fig. 12.4. Speed control (open-loop) of a dc machine

three-phase bridge connection i.e. the power circuit of the converter. The correcting variable is the dc voltage U_d at the converter output. The plant includes the connected dc machine including a load with the load torque M_L. In this case the controlled variable is the machine speed n_M. By changing the reference input the speed of the drive can be manipulated, via the correcting unit and plant to match a desired reference input. An example of such a reference input is the value which a quantity (in this case, the speed n_M) should have under steady state conditions at the instant under consideration.

Characteristic of the open-loop control is an open sequence of action. No feedback of the controlled variable occurs. Supply voltage fluctuations or load changes can occur as disturbances which lead to undesired changes in the controlled variable (see Fig. 12.1).

12.2.2 Closed-loop Control with Converters as Correcting Unit

Figure 12.5 shows a converter as the power-correcting unit in a closed-loop speed control system for a dc machine by the inner current loop method. The desired quantity of the closed-loop control i.e. the quantity which it is desired to manipulate is speed n_M of the drive. The speed is detected by a tachometer and compared with the reference input n_W.

The speed control is superimposed upon an armature current control which has prevailed, in the case of line-commutated converter drives because, on one hand, it protects the converter from overload and, on the other hand, it leads to a good dynamic response for the current control.

The current for the subordinated current control system is detected on the three-phase ac side and converted into a dc quantity proportional to the dc current I_d. The superimposed speed control system supplies the reference input for the armature current control system.

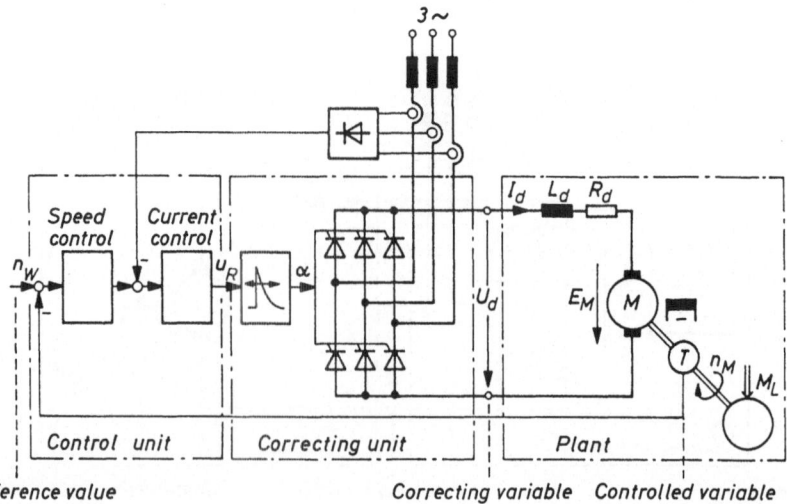

Fig. 12.5. Speed control (closed-loop) of a dc machine by the inner current loop method

12.3 Control System Elements

Open-loop and closed-loop controls can be subdivided into elements along the path of action. The most important control system elements are listed in the following.

12.3.1 Linear Control System Elements

Control circuit elements can be characterized by the transient response, their equation in the time domain and the circle diagram of the frequency response. They can be represented in block diagram form as symbols with transient response.

The transient response is the waveform of the output signal as the result of a step function at the input, the change in the output signal being referred to the size of step at the input.

The equation in the time domain describes the relationship between the input quantity x_i and output quantity x_o as a function of the time.

The frequency response of a linear system is the ratio of the inhomogeneous partial solution of the output quantity to the input quantity, the input quantity

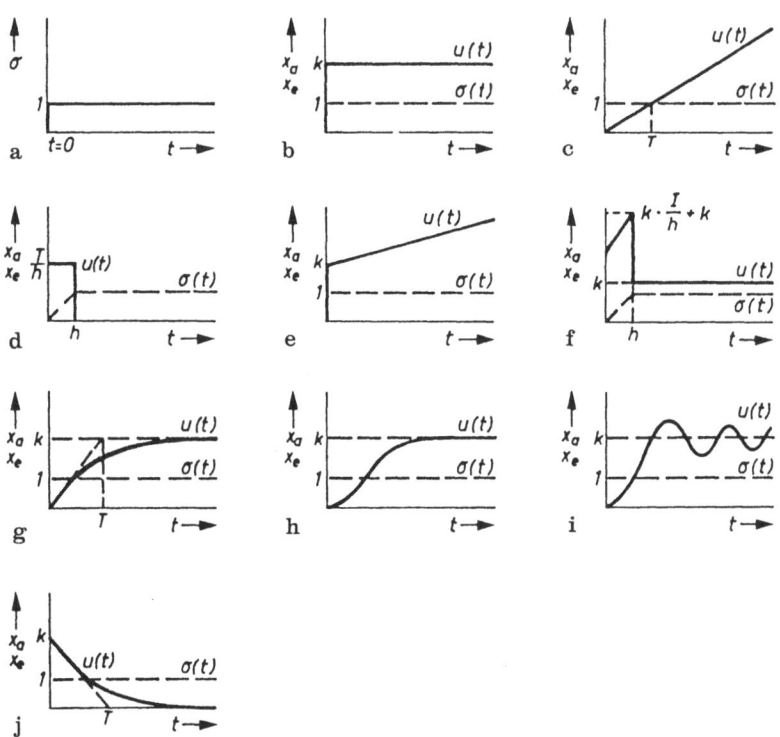

Fig. 12.6a–j. Transient response of linear control system elements for a unit step at $t=0$.
a Unity step; **b** P-element; **c** I-element; **d** D-element; **e** PI-element; **f** PD-element; **g** delay element (1st order); **h** delay element (2nd order) for $d>1$; **i** delay element (2nd order) for $d<1$; **j** delay differentiating element

varying sinusoidally. The frequency response equation which describes the transient response of a system for a sinusoidal input quantity, is obtained by application of the Laplace transformation to the differential equation of the system. The Laplace transformation of a function is an image function. If the original function has the form of a linear differential equation, the image function is a linear algebraic equation. The solution of the differential equation is therefore obtained by applying the inverse of the Laplace transformation to the solution of the algebraic equation which first, has been derived by Laplace transformation of the differential equation [10].

Elementary linear elements are

- P elements (proportional elements) with proportional action,
- I elements (integrating elements) with integrating action,
- D elements (differentiating elements) with differentiating action.

Class of element	Equation in the time domain	Frequency response	Symbol with transfer function
P	$x_o = k \cdot x_i$	k	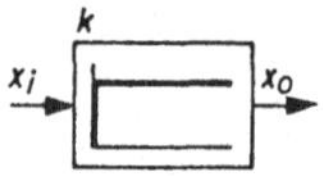
I	$T\dot{x}_o = x_i$	$\dfrac{1}{pT}$	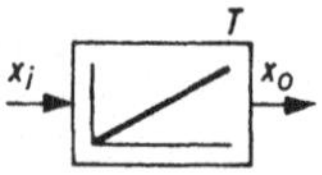
D	$x_o = T \cdot \dot{x}_i$	pT	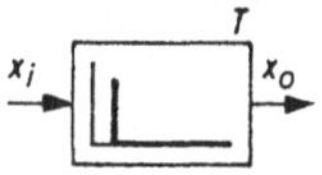
PI	$T\dot{x}_o = k(x_i + T\dot{x}_i)$	$\dfrac{k(1+pT)}{pT}$	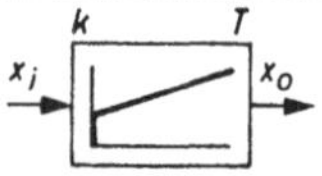
PD	$x_o = k \cdot x_i + kT\dot{x}_i$	$k(1+pT)$	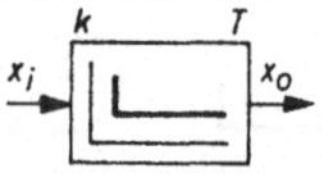
Delay 1st order	$T\dot{x}_o + x_o = k \cdot x_i$	$k\,\dfrac{1}{1+pT}$	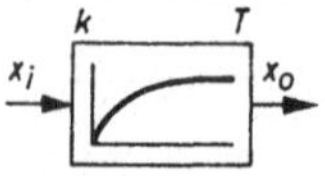
Delay 2nd order	$T^2\ddot{x}_o + 2dT\dot{x}_o + x_o = kx_i$	$\dfrac{k}{1+2dpT+p^2T^2}$	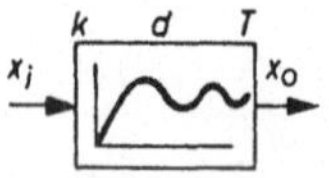
Delay differentiating	$T\dot{x}_o + x_o = kT\dot{x}_i$	$k\,\dfrac{pT}{1+pT}$	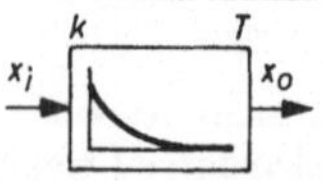

Fig. 12.7. Identification of linear control system elements

The elementary types of action can be combined additively, thus producing PI elements, PD elements, and PID elements.

Further elements are delay elements of the 1st and 2nd order and a delay differentiating element.

In Figure 12.6 the transient responses of the elements listed are given. The transient response $u(t)$ is the waveform of the output quantity for a unit step $\sigma(t)$ at the input. Solution of the equation in the time domain (of a differential equation) produces the transient response $u(t)$ when the input quantity $x_i = \sigma(t) = 1$ is set for $t > 0$. With D and PD elements a unit step with finite slope must be assumed since otherwise the derivative of the unit step for $t = 0$ is infinite.

In Fig. 12.7 linear control system elements are characterized by their equation in the time domain their frequency response or their transient response.

12.3.2 Dead Time Element

There are elements of a closed loop which cannot be described by differential equations. Figure 12.8 shows the transient response of one known as a dead time element. Between the input quantity and the output quantity there is a time delay equal to the dead time T_t. The characterization of a dead time element is given in Fig. 12.9.

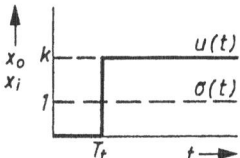

Fig. 12.8. Transient response of a dead time element

T_t	Equation in the time domain	Frequency response	Symbol with transfer function
T_t	$x_o = \sigma$ for $t < T_t$ $x_o = k \cdot x_i$ for $t \geq T_t$	ke^{-pT_t}	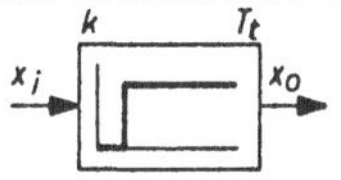 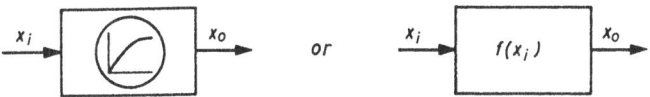

Fig. 12.9. Identification of a dead time element

12.3.3 Characteristic Element

The elements listed up to now including the dead time element are linear elements with a linear relationship between the input and output quantities. To designate a non-linear relationship characteristic elements (NL elements) are employed. Non-linear relationships occur for example in magnetizing characteristics.

Fig. 12.10. Characteristic element

Figure 12.10 shows the block representation of a characteristic element. Often the characteristic is linearized for particular working ranges so that a NL element can be replaced by a P element in the linearized ranges.

12.3.4 Configuration Diagram

The symbolic representation of a system of quantities that are interlinked in their action is called a configuration diagram. Other descriptions are block diagram and schematic diagram. A configuraton diagram characterizes not only the steady-state but also the dynamic action of the system represented.

Closed-loop control systems are often represented in a configuration diagram. The configuration diagram is a signal flow diagram (see Figs. 12.1 and 12.2). In the case of extensive systems it makes for greater clarity of the mutual interdependence of the quantities.

Figure 12.11 shows for example the configuration diagram of the speed control of a dc machine using the inner current loop method as is produced with a closed-loop control system in accordance with Fig. 12.5. The blocks can be specified either by transient response (Fig. 12.11a) or by the frequency response (Fig. 12.11b).

The example shows a meshed control system with control loops and the dc machine represented as a counter e.m.f. loop in the configuration diagram being enlarged.

The converter is introduced into the configuration diagram as a linear element with dead time action. The input quantity is the output voltage U_R of the series-connected controller; the output quantity is the dc voltage u_d i.e. the armature voltage of the dc machine. All quantities are given in normalized form. The mathematical treatment of configuration diagrams is not gone into further here but reference is made to relevant literature [10, 16, 17].

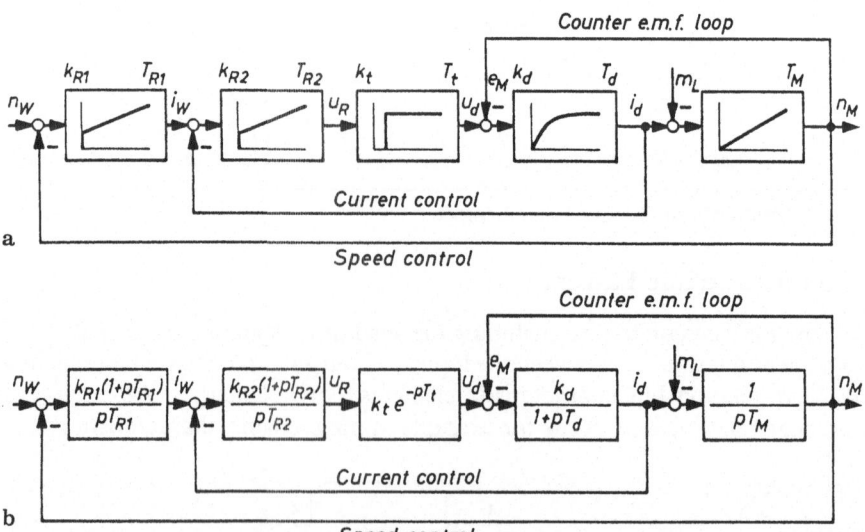

Fig. 12.11a,b. Configuration diagram of the speed control (closed-loop) of a dc machine by the inner current loop method. **a** Blocks with transient response; **b** blocks with frequency response

12.4 Internal Closed-loop Controls

To improve their operational behaviour, to increase their operational reliability or for protection reasons converter can contain internal closed-loop controls.

The subordinated current control in the case of a speed control using the inner current loop method (see Figs. 12.5 or 12.11) is in this sense an internal closed-loop control that protects the converter valves from overload and moreover improves the dynamic response.

Further internal controlled quantities are for example voltages or the margin angle γ in the case of line-commutated or load-commutated inverters (see Sect. 7.1.2). In the case of two-way converters with circulating current (see Sect. 7.2.1) the circulating current can be controlled whereby the rectifier groups can be mutually modulated without time delay and the circulating current is suppressed as the load current rises [12.5, 12.12, 12.14, 12.16].

13 Semiconductor Converter Applications

Power electronics are employed wherever the task is to convert, control, and contactlessly switch electrical energy. Reasons for the steadily increasing application of static converters are to be found in their technical properties and operating characterstics.

Their most important technical properties are the capacity to convert electrical energy according to voltage, frequency or number of phases, their infinite and rapid adjustability, and their high efficiency.

Concerning the operating characteristics of static converters, their reliability, modest maintenance requirements, and low wear should be mentioned. In many applications these advantages must be weighed against higher investment costs.

In Sect. 1.1 the development of the converter valves and the first applications for battery charging and supplying dc loads via dc substantions were described. This was soon followed by the supply of electrolytic plants with dc current as well as the operation of dc railways. In the course of the decade-long development the number of applications has increased considerably. Even today the great majority of static converters are employed for the generation of dc current from the single-phase or three-phase ac supply system, the dc voltage or dc current generally being controlled. To this can be added inverters for the generation of single-phase or three-phase ac when driving induction machines with variable frequency. Inverters for medium frequency are employed for induction heating and hardening. The main applications for power electronics are listed below.

13.1 Main Applications

Power electronics are employed in the entire field of electrical power engineering. The main applications are industrial drives, electrical power generation and distribution, electrical heating, electrochemistry, electric traction, and to an increasing extent electrical domestic equipment. Moreover, there are interesting applications in special fields such as particle accelerators and other physical apparatus.

13.1.1 Industrial Drives

In motor drives the converter-fed drives have prevailed to a considerable extent when speed adjustment is required [13.6, 13.10, 13.19, 13.26, 13.29, 13.32, 13.64, 13.66, 13.67, 13.70].

In industrial installations many drives must be continuously matched to the varying working conditions by speed adjustment. Pole-changing motors permit matching in only two or three stages. Mechanical variable-speed drives have only a limited speed of adjustment. Gearing converts not only the speed, but also the torque, and thus gives the drive a high torque in the lower speed range. However, mechanical drives are subject to wear and need maintenance.

Converter-fed drives can be equipped not only with dc machines but also with induction machines. For both types of machine, however, it is true that the torque output by an electrical machine is proportional to the product of the magnetic flux times the current, i.e. at a given current an approximately constant torque can be output over the entire speed range as the magnetic flux can be maintained constant (see Figs. 10.7 and 10.8). For increased starting torque an increase in current or flux is required, contrary to adjustable gearing.

A distinction is made between dc drives and ac drives according to the type of electrical machines employed. The most important converter-fed drives are listed below.

Converter-fed dc drives are employed the most frequently. There are numerous variants of dc drives. Fig. 13.1 illustrates operation with a single converter and four-quadrant operation with a double converter (two-way converter) in the armature circuit.

Four torque/speed ranges can be distinguished (see Fig. 10.6). Quadrants I and III indicate operation in the motor mode (the electrical machine supplies mechanical power) quadrants II and IV indicate generator or braking operation

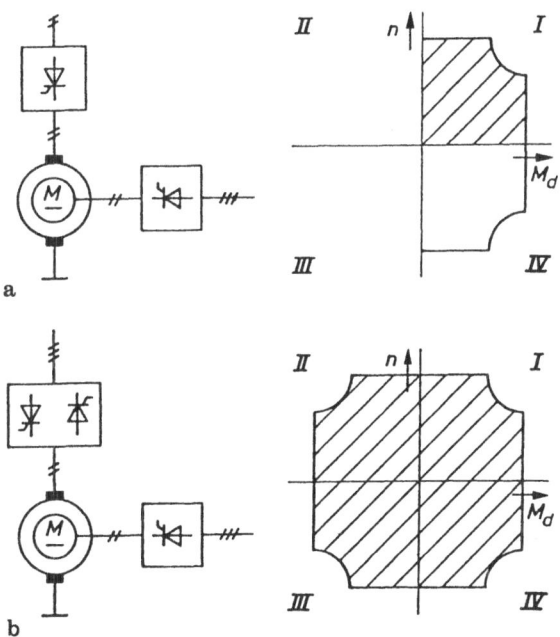

Fig. 13.1a,b. DC drive with line-commutated converter (separately excited dc motor). **a** Single converter; **b** double converter (for four-quadrant operation)

(the electrical machine consumes mechanical power). Transition from driving to braking means reversal of the direction of energy. With a single converter in the armature circuit it is only possible to operate in quadrants I and IV. With a double converter four-quadrant operation is possible. Four-quadrant operation can also be achieved by mechanical armature switching or by field reversal.

DC drives as in Fig. 13.1 are adjusted by changing the stator voltage (armature setting range) and the excitation current (field weakening range). The torque is reversed by reversing the stator or excitation current. The converters are line-commutated. A universal controllable drive is achieved which can be built for a wide power and speed range (power range: 1 kW to 10000 kW, speed range: up to 6000 rpm).

Figure 13.2 shows dc drives with direct dc converters as found on electrically driven vehicles.

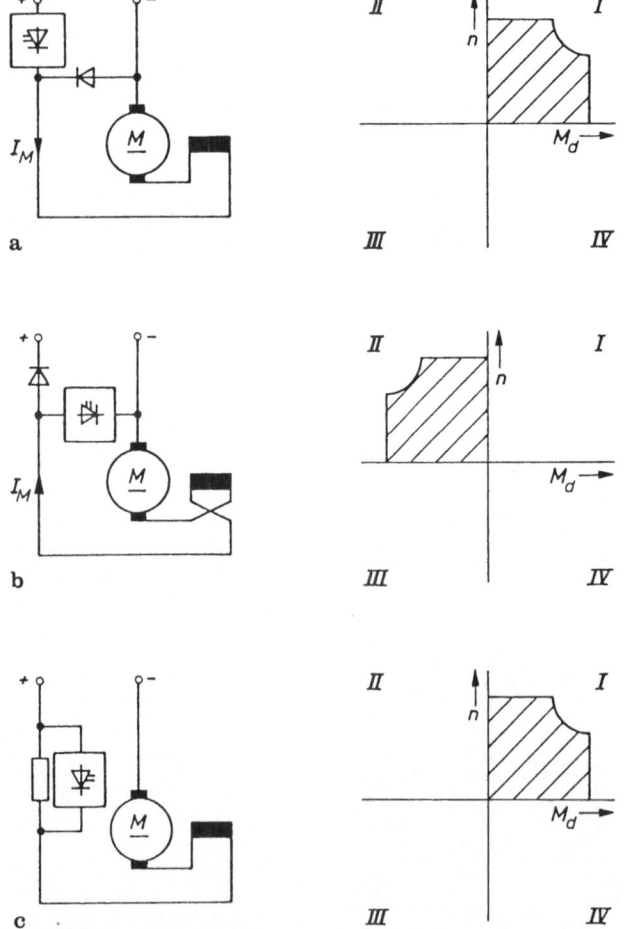

Fig. 13.2a – c. DC drive with dc power controller (series motor). **a** Driving circuit; **b** braking circuit; **c** pulse-controlled resistor for starting

The mean value of the constant dc voltage can be infinitely varied by pulsed operation with an adjustable pulse control factor (see Sect. 8.2). Figure 13.2a shows the circuit for driving, Fig. 13.2b a circuit for braking whereby the series winding of the dc machine must be reversed at the terminals. Also, starting is possible using a pulse-controlled resistor (Fig. 13.2c).

DC power controllers are built in a range from about 1 kW to more than 1 MW for speed-controlled dc motors operating in both the driving and braking modes (electric traction). Transition from driving to braking is generally performed by mechanical switchgear.

Besides the converter-fed dc machines converter-fed three-phase machines have various application. Compared to dc drives they have certain advantages: no commutator, robust construction i.e. to a great extent they are free from maintenance, lower weight and smaller dimensions, higher output limits, and higher maximum speeds.

Three-phase ac drives generally make higher demands upon the converter than dc drives for which adjustment of the armature voltage is sufficient. Therefore, converter-fed three-phase drives are generally more expensive than dc drives [13.39, 13.43].

Speed adjustment with converter-fed drives is in the simplest case performed by means of a voltage control. However, this is only economical in the lower power range and for a limited setting range. For slip-ring motors the slip power can be adjusted via converters. Use is made of this possibility preferably in applications which need only a limited range of speed adjustment. Three-phase drives for applications needing the full range of speed adjustment require the provision of variable frequency and voltage.

Figures 13.3 to 13.6 show converter-fed three-phase drives for limited requirements for which the slip power is controlled (slip control) [13.46]. In Figs. 13.7 to 13.11 converter-fed three-phase drives with variable frequency supply are illustrated. Figure 13.12 shows the use of a resonant-circuit frequency converter for high-speed three-phase drives.

For the drive illustrated in Fig. 13.3 the stator voltage of the induction motor is adjusted via a three-phase ac controller [13.41]. High-resistance squirrel-cage motors or slip-ring motors with external resistors are needed for this. Consider-

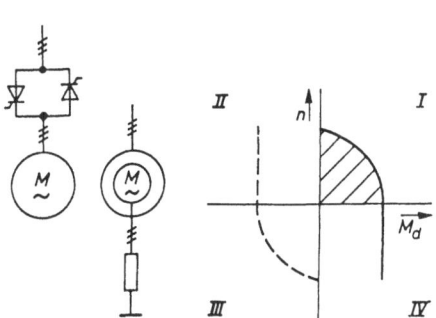

Fig. 13.3. Squirrel-cage or slip-ring motor with three-phase power controller

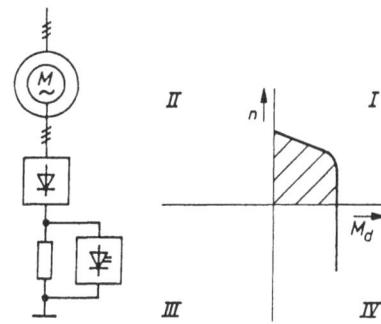

Fig. 13.4. Slip-ring motor with pulse-controlled resistor in rotor circuit

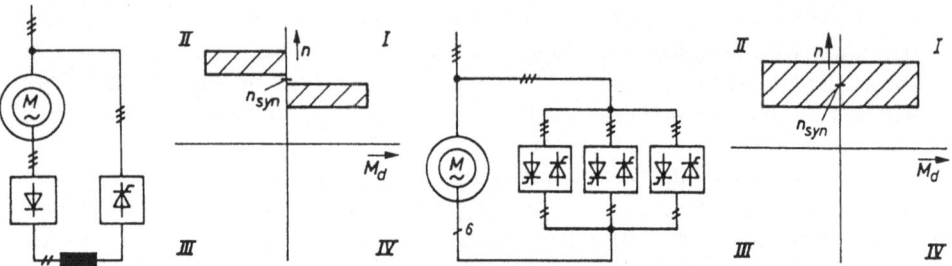

Fig. 13.5. Subsynchronous converter cascade with slip-ring motor

Fig. 13.6. Doubly fed slip-ring machine with cycloconverter in rotor circuit

able losses arise in the rotor circuit at partial speeds because the slip power is converted into heat. Braking is possible by reversing the rotating field.

Figure 13.4 shows a drive with slip-ring motor with which the rotor resistance can be varied by pulse control. Again, with this drive the slip power is converted into heat in the rotor circuit [13.45].

In Fig. 13.5 the subsynchronous converter cascade is illustrated with which the slip power of the slip-ring motor is first rectified via a controllable rectifier, then smoothed and fed back into the three-phase supply system via a line-commutated inverter. The subsynchronous converter cascade is employed for drives of high output (up to 20 MW) with a limited range of speed adjustment. Braking is possible in the supersynchronous speed range [13.25].

Figure 13.6 shows a slip-ring machine with dual supply, the rotor being connected to a cycloconverter. Via the rotor slip rings energy can be added to or subtracted from the machine. Motor and generator operation are possible as well below synchronism as above synchronism (with respect to the mains frequency) [13.60].

Figure 13.7 shows the feeding of an ac machine from a line commutated cycloconverter whereby the stator frequency and voltage are matched to the actual speed at any moment.

The three-phase machine can be not only a squirrel-cage induction motor but also a synchronous machine. The maximum frequency of the line commutated cycloconverter is limited to about 40% of the supply system frequency. The main application is for low-speed drives of high output (up to 10000 kW).

Figures 13.8 and 13.9 show three-phase drives with variable frequency and voltage with which the machine voltage is impressed via dc link converters [13.2].

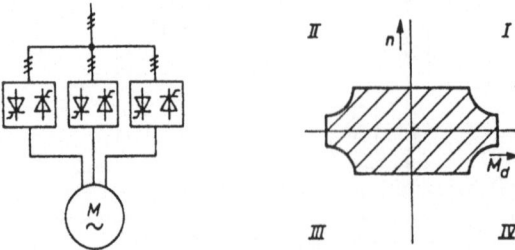

Fig. 13.7. Three-phase drive with cycloconverter in stator circuit

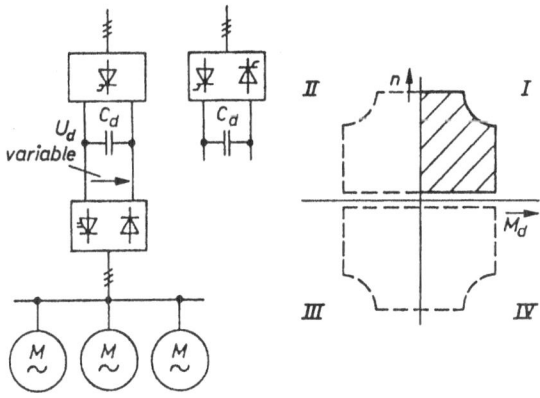

Fig. 13.8. DC link circuit converter with impressed voltage (variable link circuit voltage)

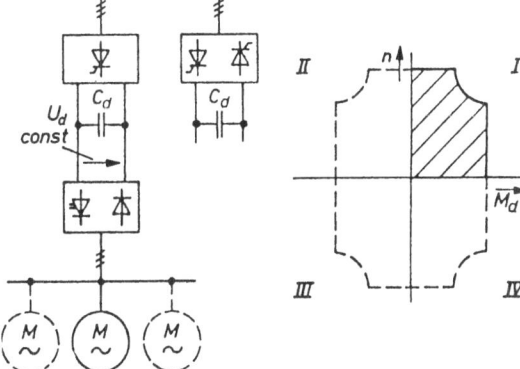

Fig. 13.9. DC link circuit converter with impressed voltage (constant link circuit voltage), pulse converter

With the dc link converter in Fig. 13.8 the voltage U_d in the link circuit is variable. Voltage adjustment is performed by phase control of the converter on the line side. For four-quadrant operation this must be constructed as a double converter.

With the three-phase drive in Fig. 13.9 the link circuit voltage U_d is constant. Both alteration of the frequency and adjustment of the voltage is carried out in the inverter next to the machine by pulse technique (see Sect. 8.3.4). DC link converters with impressed voltage are employed for group drives with severe requirements concerning synchronous operation. In the case of converters with variable link circuit voltage the frequency range goes up to about 600 Hz and even above 1000 Hz with reduced power. In the case of pulse converters the upper frequency lies at about 200 Hz because the pulse frequency must be a multiple of the output frequency. Pulse converters can be employed for the dynamic four-quadrant operation of squirrel-cage induction machines [13.1]. Their power has been extended into the megawatt range.

Figures 13.10 and 13.11 show dc link converters with impressed current. These have been developed for single-motor drives. The inductive energy store in the link circuit impresses the motor current which is periodically switched to the phase windings of the machine by the inverter. The voltage at the machine terminals then sets itself automatically.

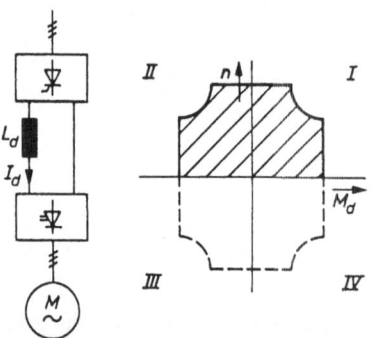

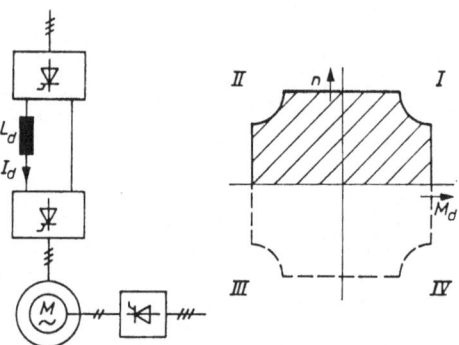

Fig. 13.10. DC link circuit converter with impressed current for induction machine

Fig. 13.11. Circuit converter with impressed current for synchronous machine (converter motor)

Figure 13.10 shows the dc link converter with impressed current for the drive of an induction motor. The converter on the machine side operates with forced commutation, preferably with phase-sequence quenching (see Sect. 8.3.2).

Figure 13.11 shows a dc link converter with impressed current to drive a synchronous machine (see Sect. 7.3.3). The synchronous machine is overexcited and supplies the reactive power required for the inverter [13.13]. The trigger pulses of the inverter can be derived from the machine via a rotor position sensor. The drive is thus given a response similar to a separately excited dc machine and cannot drop out on load surges. For the run-up from standstill special measures are required (e.g. pulsing of the converter on the supply side) because at standstill no commutating power can be made available by the synchronous machine. Converter-fed drives with synchronous machines (converter motors) are built for ratings exceeding the limiting power range of dc machines (over 20 MW) [13.49].

Figure 13.12 shows a three-phase parallel resonant circuit converter for the drive of high-speed three-phase machines. The three-phase machines (e.g.

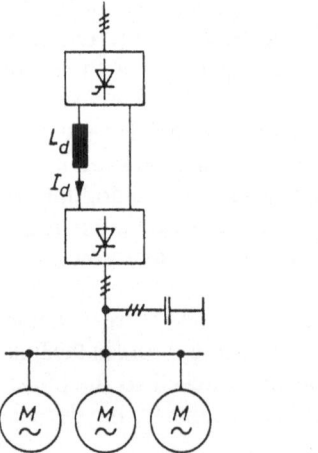

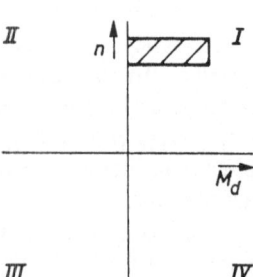

Fig. 13.12. Multi-phase parallel-resonant circuit converter for high-speed three-phase drives

hysteresis motors) are extended by parallel capacitors into a resonant circuit with impressed current. The converter next to the load draws its commutating reactive power from the load (see Sect. 7.3.1). The frequency can be more than 1000 Hz.

13.1.2 Power Generation

In power generation systems converters are employed to drive pumps, blowers, fans, and auxiliaries, the drives being predominantly three-phase (subsynchronous converter cascades, converter motors in the medium and high power ranges, three-phase controllers in the lower power range) [13.8, 13.20, 13.59].

Moreover, diode and thyristor converters are employed for the excitation of synchronous machines [13.11, 13.34, 13.44, 13.63]. Rotating diode converters can be used for this, the excitation power being supplied via a shaft-mounted generator whose outer poles are energized via a thyristor converter. Such an arrangement works without slip-rings. Exciters with rotating thyristor converters have also been developed.

Besides these there are static exciters with which the excitation power is controlled and supplied to the synchronous generator via thyristor converters.

13.1.3 Power Distribution

In power distribution systems there are applications for static converters for coupling supply systems either of the same frequency and phase relationship or of different frequencies [13.61]. In high-voltage direct-current (HVDC) transmission they are employed for the conversion of three-phase ac into direct current (rectifier station) and of direct current into three-phase ac (inverter station) [13.48, 13.58]. Moreover, in substations, three-phase ac is converted into direct current to feed dc power supply systems. An important sphere of application is the maintenance of a guaranteed power supply by means of converters for sensitive loads for which in the event of loss of the mains power supply no interruption of the power supply can be permitted [13.7, 13.47, 13.65]. Further applications of converters are to be found in power distribution as ripple control transmitters. A special field is power factor compensation by thyristor-switched reactances or other reactive power converters. They are for example employed for the elimination of mains flicker. Besides this, semiconductor switches are increasingly being used in power distribution where high numbers of switching cycles are demanded.

In HVDC transmission which until the middle of the sixties had been the last domain of the mercury-arc rectifiers, thyristor valves are now used exclusively. The mechanical design of these systems employ either air cooling or liquid cooling [13.18, 13.41].

Figure 13.13 shows the block circuit diagram of an HVDC station (Cabora Bassa) where in the second development stage 8 thyristor bridges (three-phase bridge connection B6) connected in series generate a dc voltage of ± 533 kV on the 1350 km long overhead line. To obtain a twelve-pulse reaction on the supply system, the converter transformers are alternately connected in star and delta on

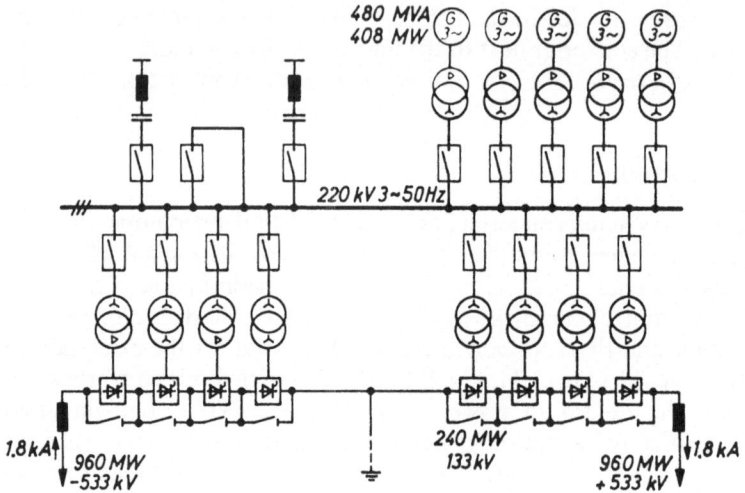

Fig. 13.13. Block circuit diagram of an HVDC station

the secondary side. Five hydroelectric generators (output: 480 MVA each) feed via transformers the 220 kV busbars to which compensating capacitors formed into filter circuits are also connected.

Each converter bridge produces an output of 240 MW at 133 kV and 1800 A. The thyristor valves are built into the valve tank under oil (280 thyristors per valve in series, 2 in parallel). The outer valve tanks at the overhead line must be

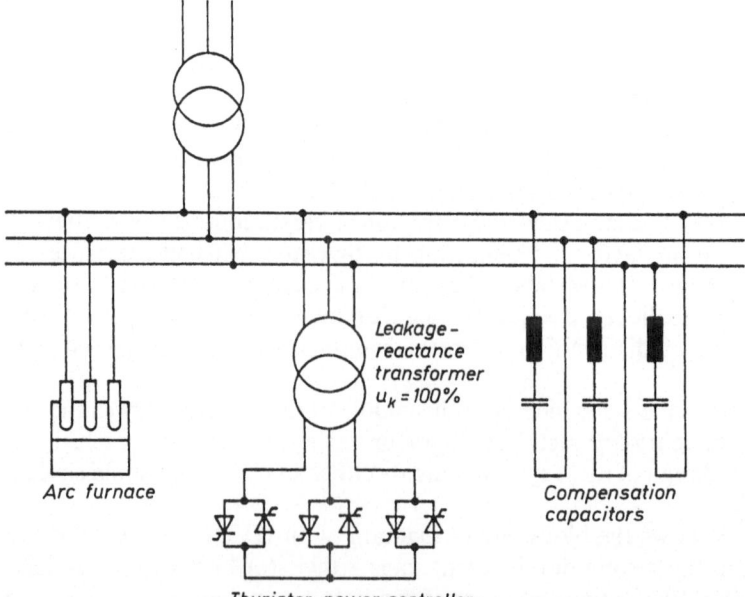

Fig. 13.14. Dynamic power factor correction via leakage transformer and thyristor power controller

insulated from earth on an insulating table for a dc voltage of 533 kV. Transmission of the signals from the control room to the valve potential is carried out via light signal transmitters.

A special application of thyristor switches is the synchronous switching of reactances (capacitors or inductances). This is employed for the stepped correction of the power factor in three-phase power supply systems where mains flicker is caused by uneven loads (see Sect. 11.7) [13.14, 13.21].

Figure 13.14 shows one method of power factor correction using a leakage transformer and thyristor power controller. The reactive power consumption of the transformer designed with extremely high leakage ($u_k = 100\%$) can be continuously varied via the three-phase controller on the secondary side. Since with this method only the inductive reactive power consumption of the leakage transformer can be infinitely adjusted between rated power and zero, it is first necessary to overcorrect using fixed compensating capacitors. For reasons of protection, inductances lie in series with these compensating capacitors for current limitation in the event of disturbances.

In dynamically regulating reactive power fluctuations in three-phase lines, reactive power converters can also be used. Figure 13.15 shows the power section of a reactive power converter with capacitive storage and commutation on the dc voltage side with pulse number $p = 6$ (see Fig. 8.30). This enables the capacitive or inductive reactive power to be quickly adjusted. By connecting phase-differing six-phase units a twelve-pulse reaction can be obtained in a three-phase line.

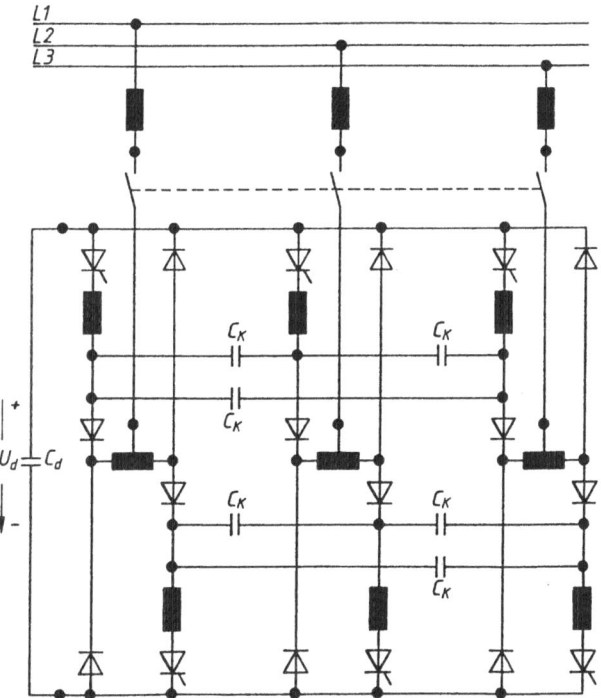

Fig. 13.15. Power circuit of a self-commutated reactive power converter

13.1.4 Electric Heating

In the field of electric heating converters are employed to switch and control heating equipment, for welding control, to supply arc furnaces, and as inverters for induction hardening, heating and smelting.

Resonant circuit converters are particularly suitable for generation of medium frequencies. These are load-commutated converters where the reactive power necessary for commutation is provided by the load (see Sect. 7.3). For this, resistive-inductive loads must have capacitors added to create series or parallel resonant circuits [13.15, 13.33].

Figure 13.16 shows series correction of the load power factor (see Sect. 7.3.2). With this series resonant converter the voltage in the link circuit is constant. The inverter has reverse-current diodes (double converter). Series resonant circuit converters are used to supply induction smelting furnace installations with ratings of up to several MW. Their frequency range extends from 200 Hz to 3 kHz. The efficiency lies around 95%.

Figure 13.17 shows parallel correction of the load power factor (see Sect. 7.3.1). With the parallel resonant circuit converter the voltage in the link circuit is adjusted via the controlled line-side rectifier. The current in the link circuit is impressed by the smoothing inductance. The inverter next to the load only needs valves for one direction of current (single converter).

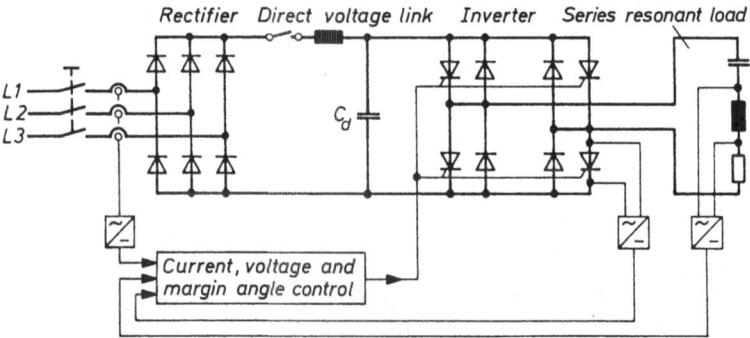

Fig. 13.16. Series-resonant circuit converter

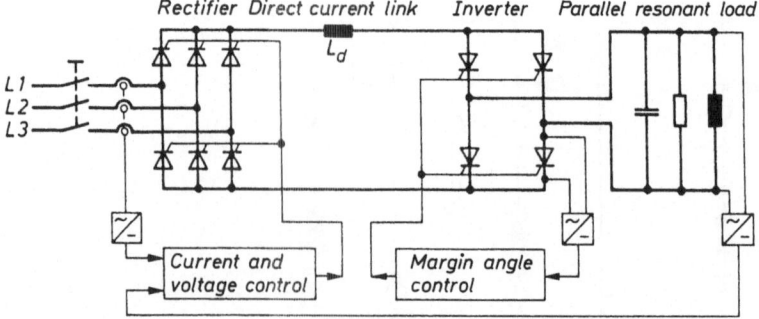

Fig. 13.17. Parallel-resonant circuit converter

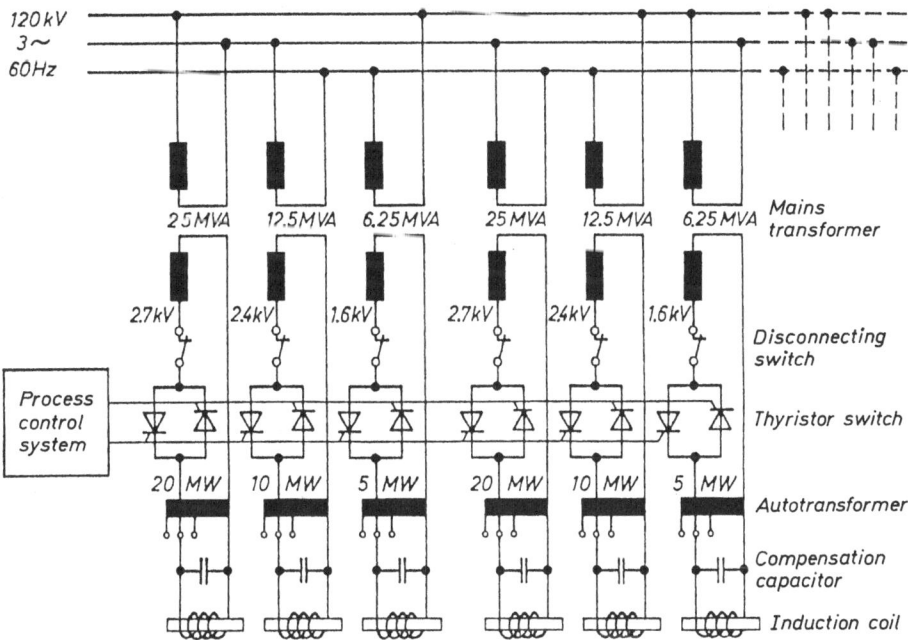

Fig. 13.18. Induction heating of large steel ingots (General Electric/Ajax 1968)

Parallel resonant circuit converters are employed for all applications in induction electric heating (induction heating for forging, hardening and induction smelting) up to frequencies of 10 kHz. The full-load efficiency is 95% whereas rotary converters achieve a maximum of 90%. Outputs of up to 1800 kW per converter unit are attained at frequencies up to 3 kHz. With 10 kHz converters the power is reduced owing to the high stressing of the thyristors and components.

Water cooling is prefered to dissipate the losses in resonant circuit converters. Special steps are taken for the oscillation build-up. To achieve still higher frequencies, special forms of construction of resonant circuit inverters have been developed.

A special application of line-commutated cycloconverters is electric slag resmelting. A multi-phase cycloconverter is employed to generate low-frequency ac current of variable frequency (0 to 10 Hz) [13.30]. With the low-frequency current resmelting occurs under a layer of slag.

Figure 13.18 shows the induction heating of large steel ingots using mains frequency. This plant commissioned in 1968 has a total rating of 210 MW. The induction coils in which the steel ingots are heated to rolling temperature are connected via antiparallel thyristors and compensated by parallel capacitors. Phase control is used temporarily for switching on (soft start-up) [13.9].

13.1.5 Electrochemistry

In electrochemistry static converters are employed as electrolysis and electro-plating rectifiers as well as for battery charging and plate forming. A further field of applications is electrophoretic paint spraying.

Next to HVDC transmission electrolysis needs the highest converter outputs. Of all the consumer of electric power, the electrochemical industry is by far the largest consumer of dc current. It uses very high dc currents for electrolytic plants for the manufacture of chlorine, acetone, hydrogen, oxygen as well as aluminium, magnesium, zinc, and pure copper. As approximate values for electrochemical installations, currents of 170 kA with dc voltages of 1100 V (rating: 165 MW), are used for aluminium electrolytic plants and currents of over 200 kA at voltages of 250 to 500 V for electrolytic chlorine plants [13.31, 13.35].

The demands made on an electrolysis rectifier installation are: highest possible efficiency owing to the considerable portion of the production costs represented by the current, controllability within the setting range required by the electrolysis process, reliability and low maintenance requirements, and, moreover, acceptable reaction on the supply system as well as lowest possible investment costs.

For electrolytic plants liquid-cooled rectifiers have prevailed to a great extent as they have the highest specific output. Figure 13.19 shows as an example the

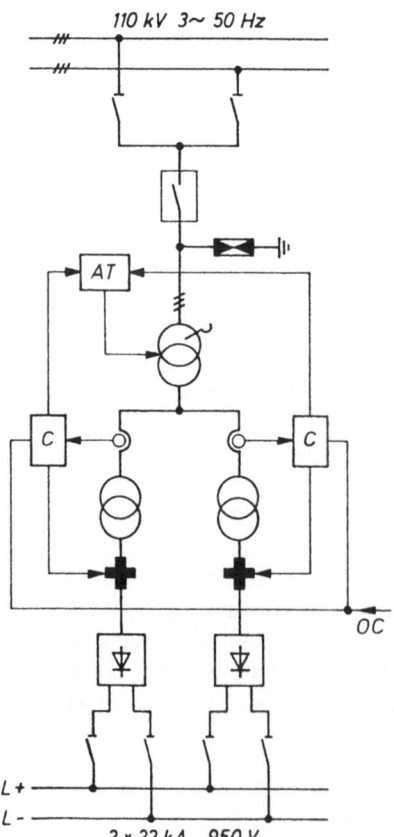

Fig. 13.19. Twelve-pulse rectifier group in an aluminium electrolytic plant for 110 kV supply. *C* Controller, *AT* automatic tap-changing, *OC* overall control of a row of furnaces

block circuit diagram of a twelve-pulse rectifier group for direct 110 kV connection. The energy flows from the 110 kV duplicate busbar system via a circuit-breaker into a regulating transformer and two rectifier transformers connected in series. The windings of the rectifier transformers are mutually displaced by 30° to achieve a twelve-phase reaction on the supply system. The series-connected transductor reactors provide fine control between the steps of the on-load tap changer.

They can be included in the transformer tank or, when space permits, they can form the interconnection between transformer and rectifier (single-winding reactors). One rectifier group here supplies 2×22 kA at 950 V dc. In total the plant rating is 2×176 kA at 950 V. The two groups of electrolytic baths are each supplied by four 44 kA rectifier groups.

Electrolysis rectifiers can also be constructed with thyristors instead of with silicon diodes [13.27]. In this case the open-loop or closed-loop control of the output dc voltage is by phase control whereby a regulating transformer or saturable ironcored reactors are omitted. The reaction on the supply system of such a phase-controlled large rectifier is problematic because with partial phase control considerable reactive power arises in the three-phase supply system which has to be partially eliminated by power factor correction equipment.

Electro-plating plants require lower dc powers. The voltage must be able to be continuously adjusted from zero to the rated value. In addition thyristor correcting units can be used in place of the previously usual regulating transformers or transductors.

13.1.6 Traction

Converters are employed to a considerable extent in electric traction, as rectifiers in railway substations, for coupling systems (railway converters) [13.51], to supply universal and three-phase motors, for trains' supply systems, as chargers, and as heating converters. Moreover, antiparallel thyristors are used as the change-over device for tap changers on locomotive transformers in order to switch without drawing an arc [13.3].

Converter-fed drives are replacing other solutions in traction to an increasing extent [13.50, 13.54, 13.55]. In the case of tramways and underground railways dc power controllers are used in place of mechanical switching mechanism and resistors [13.69], on motor coaches and locomotives half-controllable bridge connections and universal motors [13.5]. Locomotives with pulse converters and induction traction motors are already in service [13.36, 13.40, 13.42, 13.52, 13.53, 13.56, 13.57, 13.62].

Converters are also employed for system coupling e.g. for coupling three-phase systems to single-phase railway systems with different frequencies. Figure 13.20 shows a variable-frequency coupling between the three-phase 50 Hz national grid and the single-phase 16 2/3 Hz system of the German Federal Railways [13.28]. For this application a rotary frequency converter is used with an induction motor and a single-phase synchronous generator. The variable slip power of the slip-ring motor is fed back into the three-phase system via cycloconverters (see Fig. 13.6). The antiparallel thyristor converters in three-phase bridge connection are supplied

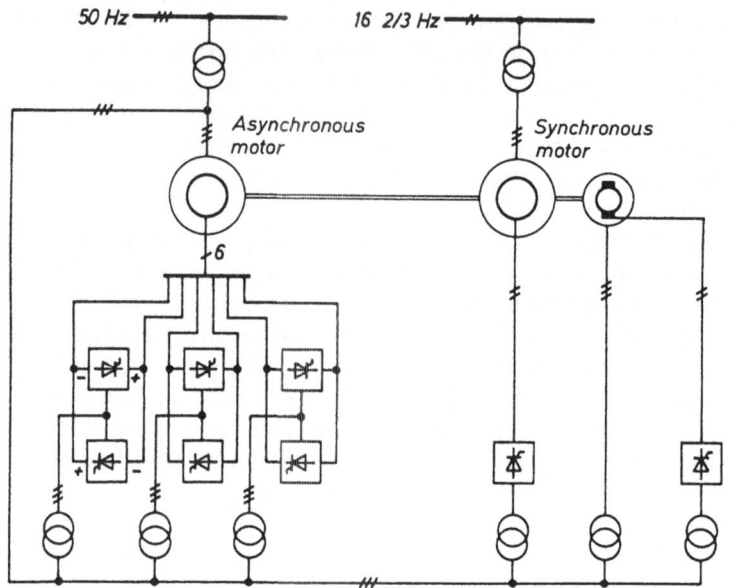

Fig. 13.20. Flexible coupling of power supply systems 50 Hz 3ph/16 2/3 Hz 1ph

by their own three transformers (each 1250 kVA) which are connected to the 50 Hz national grid via the machine transformer. The single-phase synchronous generator is excited by means of converter excitation using thyristors as is the shaft-mounted generator. The shaft-mounted generator allows the synchronous generator to operate as a phase-shifter independent of the 50 Hz grid.

Figure 13.21 shows, as a further example of the use of static converters in traction, the equipment of an unsymmetrically half-controllable bridge connec-

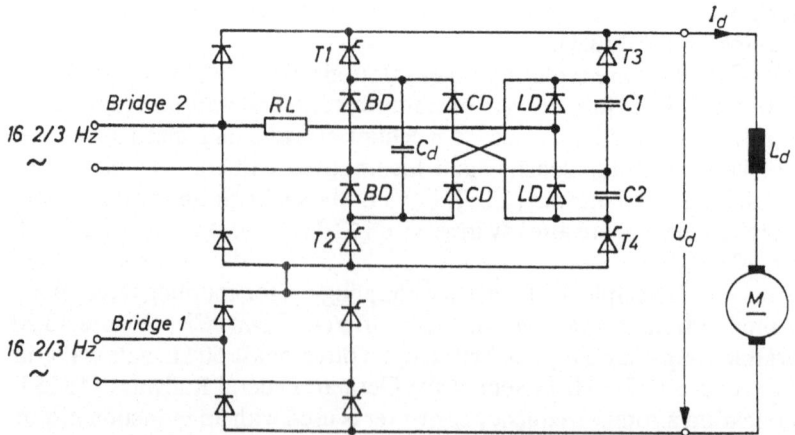

Fig. 13.21. Two asymmetrically half-controllable bridges in sequential connection with quenching device for bridge 2. *T1, T2* Main thyristors, *BD* blocking biodes, *T3, T4* quenching thyristors, *C1, C2* quenching capacitors, *CD* commutating diodes, *LD* charging diodes, *RL* charging resistor, C_d DC voltage capacitor

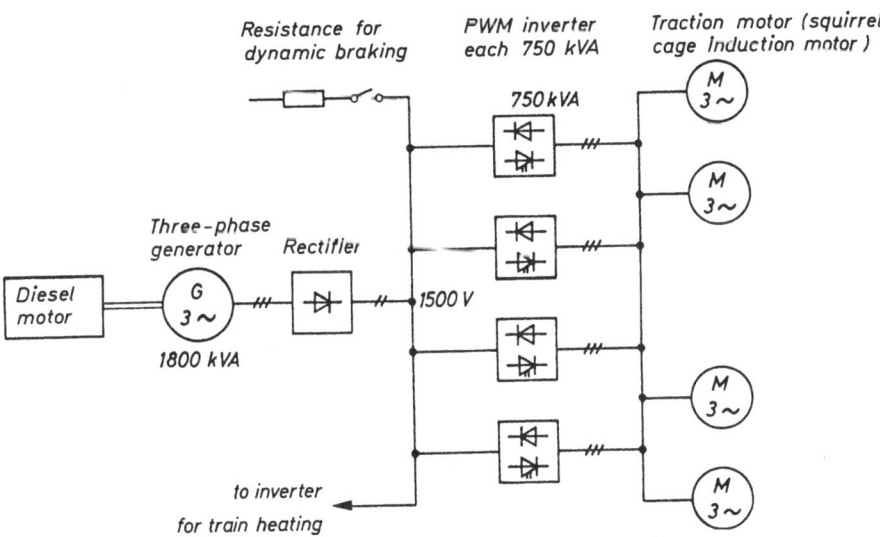

Fig. 13.22. Dieselelectric locomotive with pulse inverters and induction motors

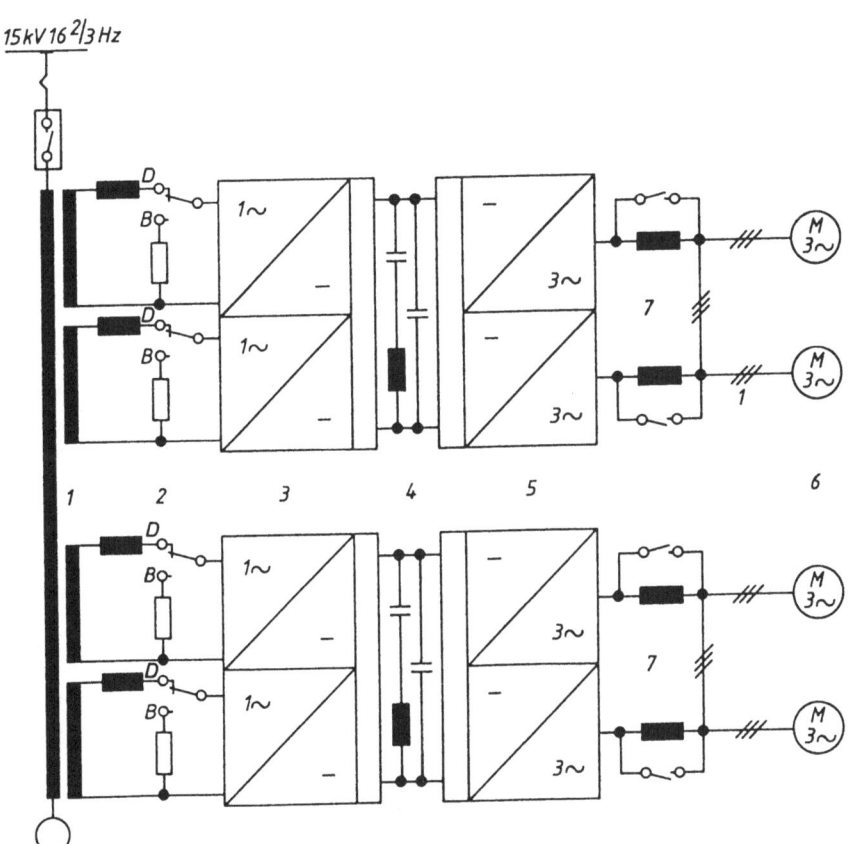

Fig. 13.23. Block circuit diagram of the universal locomotive E 120 with ac traction motors. *D* Drive, *B* brake, *1* transformer, *2* brake resistors, *3* four-quadrant controller, *4* link circuit, *5* inverter, *6* traction motors, *7* series inductance

tion with a quenching device (see Sect. 8.3.5). With the two bridges 1 and 2 in phase sequence connection the current in the principal thyristors T1 T2 of bridge 2 can be interrupted early via the quenching thyristors T3 and T4 and the quenching capacitor C1 or C2. The magnetic energy of the system inductances is temporarily stored as electrical energy in the dc voltage capacitur C_d and discharged to the dc load after completion of the quenching process.

This connection (quenchable asymmetrical bridge connection) facilitates the elimination of the control reactive power otherwise occuring with phase control by means of so-called termination control. The commutating reactive power can also be eliminated. The power factor λ attains values of over 0.95 [13.16].

Figure 13.22 shows the block circuit diagram of a diesel-electric locomotive with pulse inverters and induction traction motors. The three-phase generator driven by the diesel motor generates the electrical energy that is supplied to the dc busbar via a rectifier. In total, four pulse inverters (see Sect. 8.3.4) generate a three-phase system of variable frequency and voltage to which the traction motors are connected. The drive corresponds to the dc link converter illustrated in Fig. 13.9.

When such a locomotive with induction motors is to be supplied from the single-phase ac supply system, this ac system must be coupled to the dc link circuit. Pulse converters in single-phase bridge connection can be used for this (see Fig. 8.20 and 11.16) [13.37, 13.38, 13.68]. Figure 13.23 gives the principal connection diagram of a three-phase locomotive BR 120 of the West German Railways with asynchronous traction motors (power 5.6 MW, max. vel. 200 km/h). This universal locomotive can be used with both fast passenger trains and heavy goods trains.

13.1.7 Domestic Equipment

Power electronics are also finding an increasing number of applications in domestic equipment for open-loop or closed-loop electronic power control, for example small drives with transistor inverters or thyristor power controllers, semiconductor switches and controllers for automatic heating controls, and brightness control using triacs for stepless adjustment of lighting (light dimmers) [13.12].

13.2 Power Range

Converters are employed in a power range extending from a few VA to over 1000 MVA. The most important converter power ranges in which they have a wide application are given below. For this, the static converters were again characterized in accordance with their internal mode of operation, namely the type of commutation into line-commutated converters (see Sects. 7.1 and 7.2), load-commutated converters (see Sect. 7.3), self-commutated converters (see Chap. 8), and semiconductor switches and power controllers for ac (see Chap. 6).

13.2.1 Limiting Specifications of Power Semiconductor Devices

The development of static converters and their advance into new spheres of application is closely linked with the power capacity of the semiconductor components (silicon diodes, thyristors, and power transistors, see Chap. 3) used for the power section. In Fig. 13.24 the highest listed values of power semiconductors are plotted on a log-log scale. Silicon diodes attain the highest peak inverse voltage U_M and the highest maximum mean forward currents I_M. The limiting values of N-type thyristors are only slightly less. With F-type thyristors a maximum peak reverse voltage of 2500 V is reached. Because of the switching losses and the divided cathode surface, the maximum mean on-state current is lower for F-type than for N-type thyristors. Above a few kHz the current-carrying capacity decreases sharply as the frequency rises.

The voltage rating of thyristors has been considerably increased by the introduction of special techniques for grinding bevels on the boundary zones. The increase in current-carrying capacity was achieved by constantly increasing the diameter of the silicon slice and by structural improvement of the crystal diameter. The highest values of current are achieved with thyristors in the disc cell form of construction (see Fig. 3.2) with cooling on both sides. N-type thyristor development is aiming for peak reverse voltages of 4 to 5 kV and crystal diameters exceeding 100 mm. For F-type thyristors the development aim is to further

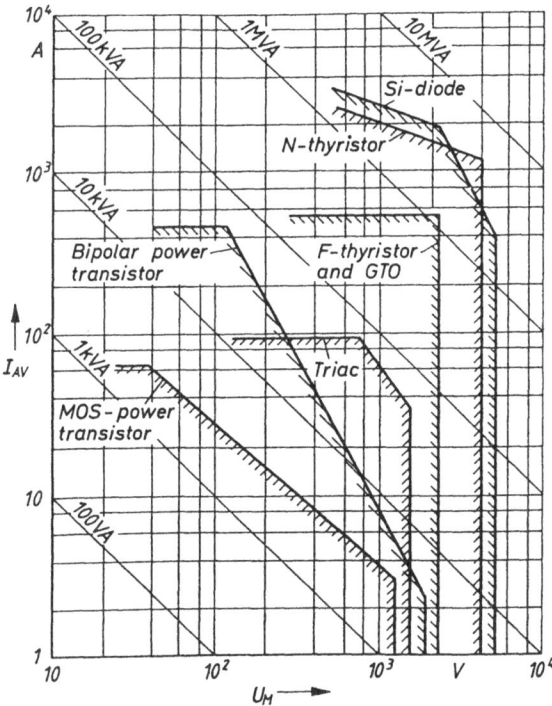

Fig. 13.24. Maximum repetitive peak reverse voltage and maximum mean on-state current of power semiconductors (maximum specified values)

improve the switching characteristics and circuit-commutated turn-off times which is generally achieved by subdivision of the cathode surface (similar to the construction of a transistor). Extremely short turn-off times (less than 10 µs) are achieved by means of negative gate currents during the turning-off process and the subsequent hold-off interval.

Thyristors which can be switched off (GTOs) can withstand voltages of more than 2500 V and currents in excess of 1000 A (turn-off time < 10 µs).

Bidirectional triode thyristor (triacs), which are employed for single-phase and three-phase switches and power controllers, attain peak reverse voltages of between 1000 and 1500 V at currents up to 100 A.

Power transistors have made considerable advances. High voltage transistors have maximum voltages of 1 to 2 kV at currents above 50 A. High current transistors switch currens up to 500 A at maximum voltages of over 100 V. Using these power transistors, converter equipment can be made with ratings of more than 50 kW. To increase the output parallel connection of power transistors in a converter arm is possible.

In the lower power range field effect transistors (MOSFETs) can be used. Power MOSFETs reach voltages above 1000 V at currents of 5 A or voltages of more than 50 V at currents of 50 A. They require only very little control power and switch in a time of < 100 ns (suitable for frequencies above 50 kHz).

13.2.2 Line-commutated Converters

Line-commutated converters are the most widespread in their application. In their case the commutating power is made available by the single-phase or three-phase supply system. For 16 2/3, 50 or 60 Hz systems they are equipped with N-type thyristors because at these switching frequencies sufficiently long hold-off intervals (several 100 µs) are available. The freedom to connect silicon diodes or N-type thyristors in series or parallel enables any converter rating to be produced.

Figure 13.25 shows the main applications of line commutated converters. The rated dc current is plotted against the rated dc voltage U_{dN} on a log-log scale. The highest converter ratings are employed for HVDC transmission (up to more than 1000 MVA). To supply electrolytic plants with stabilized dc current, plant ratings of up to several 100 MW are needed. Of all consumers of electrical energy the electrochemical industry is by far the largest consumer of dc current. Further applications of line-commutated rectifiers are electro-plating plants and battery charging. DC systems e.g. for traction are supplied via rectifier substations.

The converter-fed dc drives extend over a very wide power range that is limited at the top by the output limit of the dc machine. If possible, the dc voltage is chosen so that it is handled by one thyristor per arm of the bridge in series.

In the range of low rated dc voltages the increasingly important on-state voltage drop of the power semiconductors impairs the efficiency of the converter equipment (see Figs. 11.24 and 11.25).

13.2.3 Load-commutated Converters

In the case of load-commutated converters the reactive power needed for commutation is made available by the load. For this, resistive-inductive loads

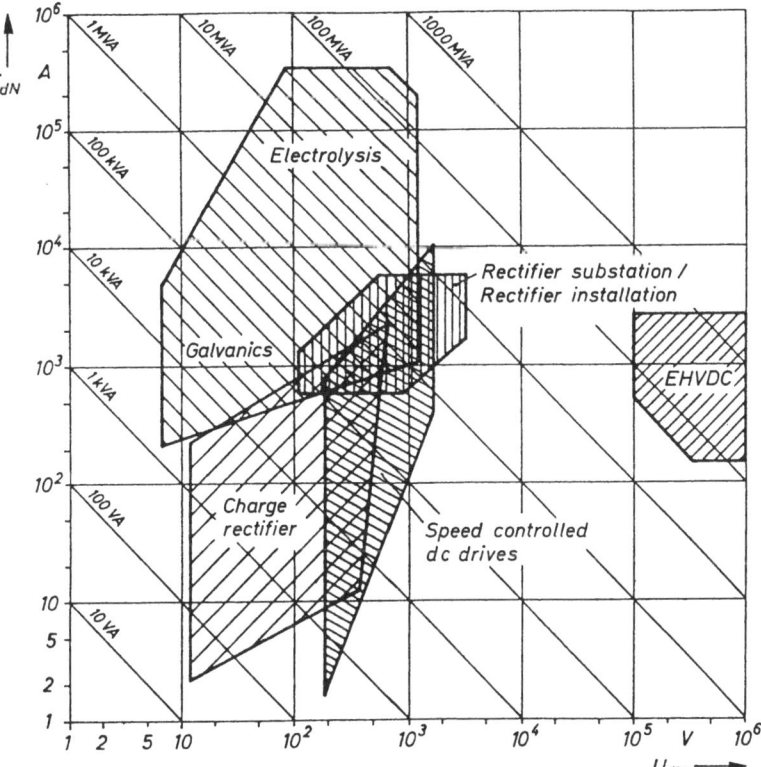

Fig. 13.25. Main applications of line-commutated converters (plant ratings)

must be augmented by capacitors into series or parallel resonant circuits (see Sect. 7.3). Resonant circuit converters are particularly suitable for the generation of medium frequency. Synchronous machines can be excited so that they absorb a current leading the voltage and so can supply the energy needed for the load-commutated converter (see Sect. 7.3.3).

Figure 13.26 shows plant ratings of load-commutated converters. The most important applications are the converter-fed synchronous machines (see Fig. 13.11) and the medium-frequency converter for induction heating, hardening and smelting. A special application is formed by multi-phase resonant circuit inverters for the supply of extremely high-speed drives for example gas ultra-centrifuges (see Fig. 13.12). The TV deflection circuit with thyristor and silicon diode can likewise be considered a load-commutated converter.

Converter-fed synchronous machines are employed for pumps and fan drives in the power range from 100 kVA up to several MVA and as the run-up device for gas turbines whereby the turbo-generator is temporarily driven as a motor via a converter of approximately 5% the rated power of the turbo-generator [13.23]. Also, in hydroelectric power stations the generators can be temporarily driven via converters as motors for pumping.

In the case of resonant circuit inverters for induction heating the plant ratings lie between 10 kVA in the package form of construction for plugging into normal

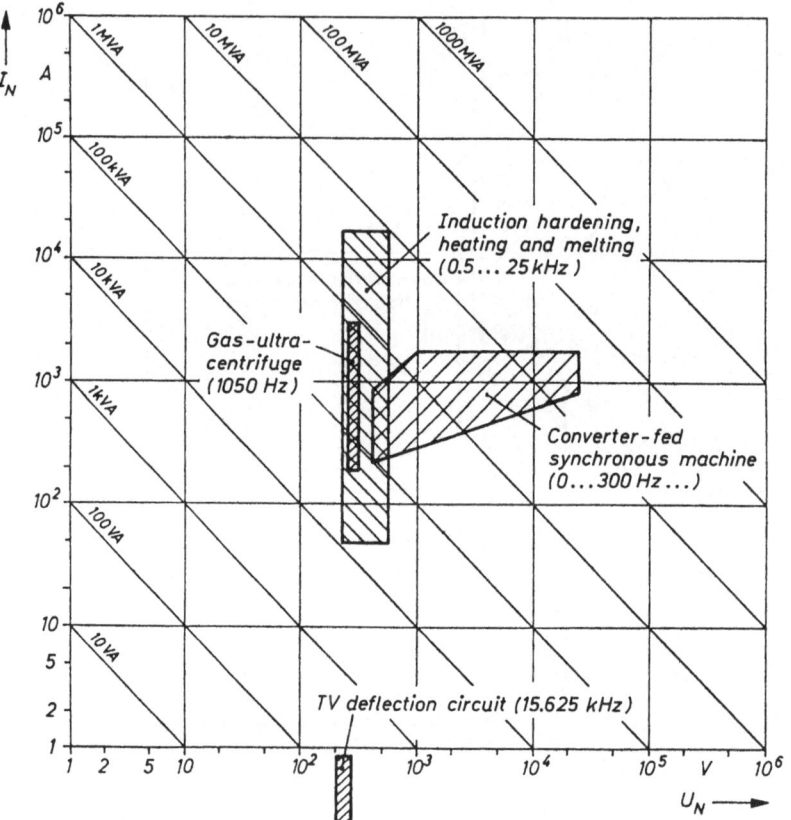

Fig. 13.26. Applications of load-commutated converters (plant ratings)

socket outlets, up to 10 MVA installations whereby frequencies ranging from several 100 Hz up to more than 10 kHz are possible. The main applications are at 500 Hz, 1000 Hz, 2000 Hz, 3000 Hz, and 10 kHz.

13.2.4 Self-commutated Converters

With self-commutated converters commutation is performed either by the quenching capacitors in the converter or by converter valves capable of being turned off (see Chap. 8). They have found many areas of application. This development was promoted by the improvement of the F-type thyristors and associated components such as the quenching and smoothing capacitors and, in the lower power range, the improvement of the power transistor.

Figure 13.27 shows applications of self-commutated converters. In motor drives they are increasingly used to supply induction machines (preferably squirrel-cage induction machines) with variable frequency and voltage where use of the robust and commutatorless induction machines justifies the extra cost of the more expensive converter. Self-commutated inverters are also used for uninter-

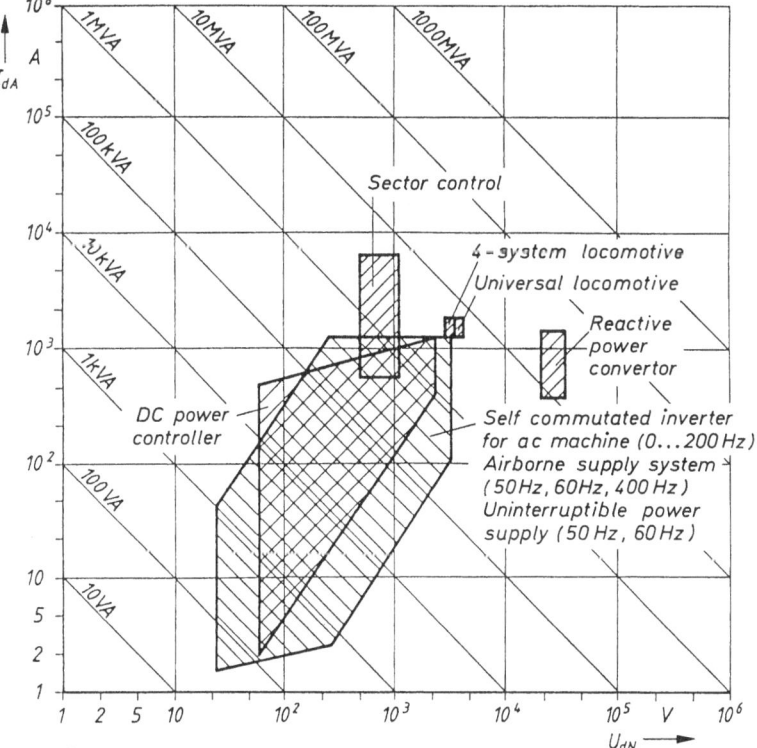

Fig. 13.27. Applications of self-commutated converters (plant ratings)

ruptible power supplies to feed vehicles' supply systems and as ripple control transmitters.

DC power controllers for the conversion of mean dc voltage and dc current values are employed in battery-powered vehicles for instance as well as on railway motor coaches supplied from a dc overhead contact wire. DC chopper converters can be built for voltages up to several kV and ratings of several MVA.

The 4-system locomotive is a locomotive intended for border traffic between four different overhead contact-wire systems applied in Europe, namely single-phase ac 16 2/3 Hz, 15 kV or 50 Hz, 25 kV and dc voltage 1.5 kV or 3 kV. On operation under dc overhead contact wire the electrical energy is first converted into single-phase ac in a self-commutated 100 Hz inverter [13.17]. The sector control (see Sect. 8.3.5) works with capacitive quenching and by termination control (as opposed to phase control) permits a reduction in the undesired reactive power. It is employed on motor coaches of the German Federal Railways (see Fig. 13.21). The universal locomotive (power 5.6 MW) has four asynchronous motors.

The connection variations of self-commutated inverters are manifold and the development of standard connections is still not complete. Their field of application is constantly expanding. It is a mark of the reliability of self-commutated inverters that important loads requiring an uninterruptible power supply are supplied not from the grid but continuously via these inverters [13.22].

With the application of GTOs dc choppers and self-commutated inverters are considerably simplified because of the abolition of auxiliary quenching circuits. Thereby a reduction of volume and weight of up to 50% an improvement of efficiency by about 0.5% and a decrease in audible noise by 10 dB are obtained.

Reactive power converters with capacitive energy storage (reactive power range ± 10 MVA) can be used to stabilize medium voltage networks (30 kV).

13.2.5 Semiconductor Switches and Power Controllers

Figure 13.28 shows applications of power semiconductors to single-phase and three-phase switches and power controllers. Where there is a requirement for high numbers of switching cycles the electronic contactor (preferably using triacs) starts to prevail [13.24]. Triac contactors can be built for terminal voltages of up to 500 V. Compared to mechanical contactors they have the advantage of fast

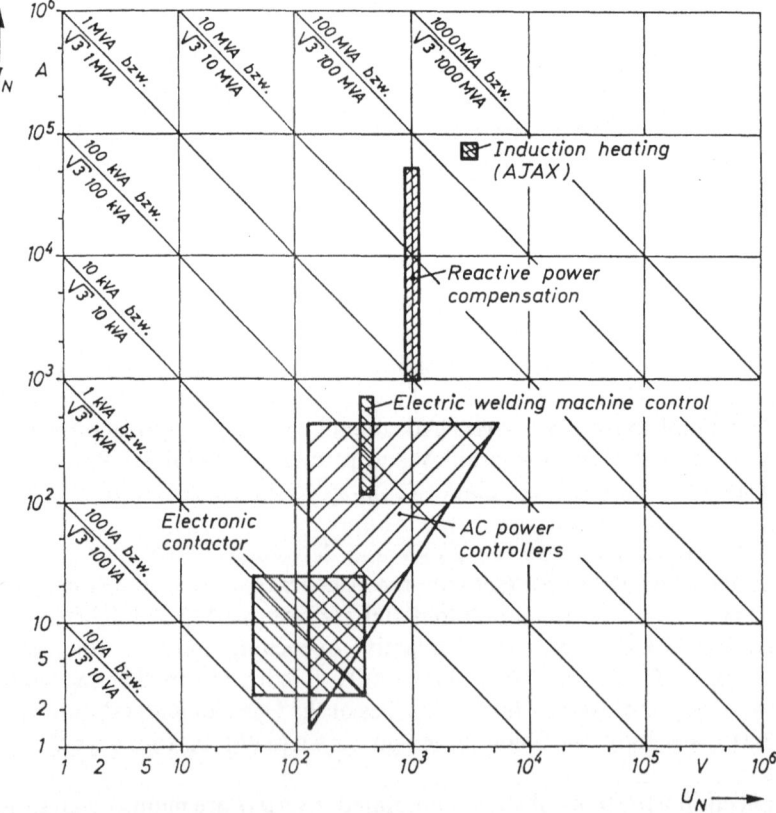

Fig. 13.28. Applications of single and three-phase switches and power controllers with power semiconductors (plant ratings)

switching without arcing and are readily synchronizable. They are practically free from wear and maintenance, quiet, but still considerably more expensive than mechanical contactors. For welding control antiparallel thyristors are increasingly replacing ignitron contactors.

A special area of application of thyristor switches is the synchronous switching of reactances (capacitors and inductances) which can be employed to eliminate mains flicker in the event of uneven loads such as arc furnaces or welding machines (see Sect. 11.7).

13.3 Frequency Range

In converters the valves switch with a generally periodic switching frequency depending upon the type and connection of the converter. With line-commutated converters this is the power supply frequency of 16 2/3, 50 or 60 Hz; with load and self-commutated converters this is the preset operational frequency which can be variable over wide limits or can lie in the medium frequency range. With pulse converters, finally, the converter valves are operated with the pulse frequency f_p. This pulse frequency can be constant or variable. It is generally several 100 Hz for direct dc chopper converters. With pulse inverters the valve arms are switched on and at pulse frequency according to the connection only during the current flow or duration of pulse operation of individual converter arms.

Higher switching frequencies of power semiconductors are, of course, limited. Important criteria are the turn-off time and the losses on turning on and off (see Chap. 3).

The highest switching frequencies can be attained with power transistors and power MOSFETs. F-type thyristors (inverter thyristors) permit switching frequencies up to about 10 kHz whereby hold-off intervals of less than 10 µs occur. By means of negative gate current during the turn-off process and the subsequent hold-off interval the circuit-commutated turn-off time of F-type thyristors can be further reduced. The current-loading capacity is considerably reduced as the frequency rises above several kHz.

Also, the switching frequency of the other components such as quenching capacitors and commutating inductances is limited. The losses in the snubber circuits increase considerably with the frequency.

In the resonant circuit converters using thyristors operating frequencies of over 10 kHz can be obtained. Power supply units using power transistors or high-speed thyristors work sometimes with even higher intermediate frequencies in the single-phase intermediate circuit. Recently, field effect transistors have been able to achieve switching frequencies of over 500 kHz.

In single-phase converters the frequency range is determined by the requirements of the application. When induction machines are supplied via converters with variable frequency the upper frequency is generally determined by the characteristics of the machine. According to the type of converter and induction machine the upper frequency limit lies between 200 Hz and 1000 Hz. A decisive factor is the inductive voltage drop which at constant current rises in direct proportion to the frequency.

14 Tests

The semi-conductors themselves as well as the convertes and converter installations are tested according to their designed properties (electrical, thermal, etc.)
[14.1, 14.2]. A distinction is made between type tests and individual tests (routine
tests).

Type tests are meant to determine whether or not the component type or device
type has the properties claimed by the manufacturer and satisfies the appropriate
specifications. For low power converters (≤ 10 kW) manufactured in small
numbers and, in general, for separately manufactured devices, a simplified type
test suffices.

Routine tests are intended to test for uniform manufacturing quality. If large
numbers are manufactured, then in testing for a particular property individual
items can be tested at random.

Tests on Semiconductor Valves. All tests on rectifying diodes and thyristors should
be performed according to given procedures (e.g. in Germany according to the
standards DIN 41783 for diodes and DIN 41784 for thyristors). Table 14.1 lists
the properties to be tested in type and routine tests. At the very least, the properties
marked with a cross should be tested. In individual tests the values ascertained for
the properties tested should not exceed the given limits of tolerance.

Converter Tests. Table 14.2 lists the tests required for converters. The individual
elements of converter installations must be tested separately. The insulation has to
be tested when the converter installations are ready for operation.

As far as possible the tests should be performed under operational electrical
conditions. This is intended to determine whether or not every electrical part of the
converter functions properly. In the heating test there should be no non-
permissible heating of the device under the permissible load. Each part of the
device must operate faultlessly up to final operating temperatures. The control
unit (including the form, duration and symmetry of the control impulse) of
controllable converter devices must also be tested. With series-connected thyristors or diodes in converter arms and series-connected converter units the
voltage distribution must be tested. With parallel-connected thyristors or diodes
in converter arms the current distribution must be tested. The safety and control
equipment must also be tested.

Tolerances. Tolerances accomodate unavoidable irregularities in the nature of
materials and unavoidalbe production variation and inaccuracy of measurements.
Values ascertained in tests are acceptable if they fall within the tolerances in Table
14.3. They must be ascertained for rated operation and operating temperatures.
When determining efficiency the individual losses of the converter unit the
transformer and the choke do not need to be established.

Table 14.1. Type and routine test of diodes and thyristors

Property to be tested	Type test		Routine test	
	Diodes	Thyristors	Diodes	Thyristors
On-state characteristic	x	x		
Forward off-state characteristic		x		
Reverse off-state characteristic	x	x		
On-state characteristic values	x	x	x	x
Greatest permissible non-repetitive peak off-state voltage	x			
Ditto in forward and reverse directions		x		
Off-state characteristic values			x	x
Holding current		x		
Latching current		x		x[a]
Impulse current limit value	x	x		
Critical current rate of rise		x		
Critical voltage rate of rise		x		x[a]
Off-state delay charge	x	x		
circuit-commutated recovery time		x		x[a]
Thermal resistance	x	x		
Transient thermal resistance	x	x		
Firing current and firing voltage		x		x
Greatest non-firing gate voltage		x		
Firing delay		x		

[a] This test is only then routine test if minimum or maximum values are indicated by the manufacturer.

Table 14.2. Test of converters

Type of test	Type test	Simplified type test	Routine test
Function test	x	x	x
Heating test	x	x	
Monitoring of characteristics	x		
Monitoring of characteristics values		x	x
Determination of the internal voltage variation	x	x	
Determination of efficiency	x		
Determination of the fundamental frequency power factor	x		
Insulation test	x	x	x
Determination of the superimposed ac voltage	x	x	
Determination of the degree of radio interference	x	x	
Shock protection test	x	x	

Table 14.3. Tolerances of electrical quantities

Electrical quantities		Permissible variation from the required value
Losses in the converter assembly		+10%
Sum of the transformer and choke losses		+10%
Efficiency		$-0,1\ (1-\eta)$, at least $-0,002$
Fundamental frequency power factor		$-0,2\ (1-\cos\varphi_1)$
Inductive dc voltage regulation governed by the transformer		$\pm10\%$
Internal voltage regulation		$\pm15\%$
Output voltage	for $U_N \leqq 10$ V	$\pm 0{,}10\ U_N$
	for $U_N > 10$ V	$\pm\ (0{,}02\ U_N + 1\ \text{V})$

References

A. Books

1 Marti, O.K.; Winograd, H. (German from Gramisch, O.): Stromrichter, unter besonderer Berücksichtigung der Quecksilberdampf-Großgleichrichter. München, Berlin 1933
2 Glaser, A.; Müller-Lübeck, K.: Theorie der Stromrichter. Bd.1: Elektrotechnische Grundlagen. Berlin 1935
3 Lappe, R.: Stromrichter. Stuttgart 1959
4 Wasserrab, Th.: Schaltungslehre der Stromrichtertechnik. Berlin, Göttingen, Heidelberg 1962
5 Gentry, F.E.; Gutzwiller, F.W.; Holonyak, N.; Zastrow, E.E.: Semiconductor Controlled Rectifiers: Principles and Applications of p-n-p-n Devices. Englewood Cliffs, N.Y. 1964
6 Bedford, B.D.; Hoft, R.G.: Principles of Inverter Circuits. New York, London, Sydney 1964
7 Gutzwiller, F.W.: Silicon Controlled Rectifier Manual. 4th Ed. Syracuse, N.Y. 1967
8 Meyer, M.: Selbstgeführte Thyristor-Stromrichter. 3. Aufl. Siemens-Fachbuch 1974
9 Möltgen, G.: Netzgeführte Stromrichter mit Thyristoren. 3.Aufl. Siemens-Fachbuch 1974
10 Landgraf, Chr.; Schneider, G.: Elemente der Regelungstechnik. Berlin, Heidelberg, New York 1970
11 Kümmel, F.: Elektrische Antriebstechnik. Berlin, Heidelberg, New York 1971
12 BBC, Silizium Stromrichter Handbuch. Baden 1971
13 Maggetto, G.: Le Thyristor: définitions − protections − commandes. Presses Universitaires de Bruxelles 1971
14 Hilpert, H.: Halbleiterbauelemente. 3.Aufl. Stuttgart 1983. Teubner Studienskripten Bd.8
15 Heumann, K.; Stumpe, A.C.; Thyristoren − Eigenschaften und Anwendungen. 3.Aufl. Stuttgart 1974
16 Leonhard, W.: Regelung in der elektrischen Antriebstechnik. Stuttgart 1974. Teubner Studienbücher Elektrotechnik
17 Chen, C.-T.: Analysis and Synthesis of Linear Control Systems. New York 1975
18 Dewan, S.B.; Straughen, A.: Power Semiconductor Circuits. New York, London, Sydney, Toronto 1975
19 Lappe, R.; u.a.: Thyristor-Stromrichter für Antriebsregelungen. Berlin 1975
20 Jötten, R.: Leistungselektronik, Bd.1. Wiesbaden 1976
21 Gyugyi, L.; Pelly, B.R.: Static Power Frequency Changers. Theory, Performance and Application. New York, London, Sydney, Toronto 1976
22 Jäger, R.: Leistungselektronik. Grundlagen und Anwendungen. 2.Aufl. Berlin 1981
23 Hartel, W.: Stromrichterschaltungen. Einführung in die Schaltungen netzgeführter Stromrichter. Berlin, Heidelberg, New York 1977
24 VEM-Handbuch Leistungselektronik. Berlin 1978
25 Hütte. Elektrische Energietechnik, Bd.2, Geräte. Berlin, Heidelberg, New York 1978
26 Bystron, K.: Leistungselektronik. Techn. Elektronik Bd.II. München, Wien 1979
27 Gerlach, W.: Thyristoren. Berlin, Heidelberg, New York 1979
28 Wood, P.: Switching Power Converters. New York, London, Toronto, Melbourne 1981
29 Möltgen, G.: Stromrichtertechnik, Einführung in Wirkungsweise und Theorie. Siemens-Fachbuch 1983
30 Middlebrook, R.D.; Ćuk, S.: Advances in Switched-Mode Power Conversion, Vol. I − III. Pasadena 1983

31 Heumann, K.: Grundlagen der Leistungselektronik, 3.Aufl. Stuttgart 1985
32 Leonhard, W.: Control of Electrical Drives. Berlin, Heidelberg, New York, Tokyo 1985

B. Conference Records

IFAC Symposium: Control in Power Electronics and Electrical Drives, Vol. 1, 2 and Survey Papers, Düsseldorf 1974
International Semiconductor Power Converter Conference, Lake Buena Vista 1977 and Orlando 1982
International Power Electronics Conference, Vol. 1 and 2, Tokyo 1983
Power Electronics Specialists Conference. San Diego 1979, Atlanta 1980, Boulder 1981, Cambridge 1982 Albuquerque 1983, Gaithersburg 1984, Toulouse 1985.
Power Electronics and Applications, Brussels 1985

C. References

Chapter 1

1.1 Köhl, G.: Hochleistungselektronik in Deutschland. Ein Vergleich des technischen Standes. VDI-Nachrichten (1970) H. 44 – 46
1.2 Heumann, K.: Leistungselektronik. Fortschritte und Entwicklungstendenzen. ETZ – B 23, Nr. 11 (1971) 253 – 258
1.3 Schmidt, J.; Schräder, A.: Vom Quecksilberdampf-Gleichrichter zur Leistungselektronik, etz Bd. 101, Nr. 16/17 (1980) 955 – 960
1.4 Grüneberg, J.: Neue Entwicklungen bei Schaltungen in der Leistungselektronik. etz 104, H. 24 (1983) 1241 – 1245

Chapter 2

2.1 Eisenack, H.; Hofmeister, H.: Digitale Nachbildung von elektrischen Netzwerken mit Dioden und Thyristoren. Arch. f. Elektr. 32 (1972)
2.2 Lakota, J.: Simulation von stromrichtergespeisten Gleichstrom-Motorantrieben. ETZ – A 94, Nr.1 (1973) 26 – 30
2.3 Vogt, F.: Die Simulation von Stromrichtern. ETZ – A 94 Nr.8 (1973) 479 – 482
2.4 Zeiner, M.: Ein Beispiel zur digitalen Simulation von Netzwerken der Leistungselektronik. Wiss. Ber. AEG – TELEF. 47, Nr.1 (1974) 21 – 27
2.5 Hoffmann, D.: Ein Beitrag zur automatischen Simulation von Stromrichterschaltungen mittels Digitalrechner. Diss. TU Berlin 1974
2.6 Hoft, R.G.: Power Electronics Circuit Analysis Techniques. Survey Paper. IFAC Sympos. Düsseldorf 1977
2.7 Foch, H.; Réboulet, C.; Schonek, J.: A General digital computer simulation programme for thyristor static converters (programme SACSO) application examples. IFAC Sympos., Düsseldorf 1977
2.8 Gutzwiller, R.: Methode der digitalen Simulation von Stromrichterschaltungen gezeigt am Beispiel eines Gleichstromstellers. ETZ – A 99, Nr.1 (1978) 8 – 10
2.9 Mehring, P.; Jentsch, W.; John, G.; Krämer, D.: NETASIM – ein digitales Simulationssystem für die Leistungselektronik. ETZ – A 99, Nr.4 (1978) 189 – 191
2.10 Yuvarajan, S.; Bellamkonda Ramaswami; Subrahmanyam, V.: Analysis of a Current-Controlled Inverter-Fed Induction Motor Drive Using Digital Simulation. IEEE Trans. Appl. Ind., Vol. IECI – 27, No.2 (1980) 67 – 76
2.11 Jawassoglou, K.; Safacas, A.: Digitale Simulation eines selbstgeführten Drehstromwechselrichters mit Einzellöschung. E u. M 98, Heft 3 (1981) 82 – 88
2.12 Möltgen, G.: Simulationsuntersuchungen zum Stromrichter mit Phasenfolgelöschung. Siemens Forsch.- u. Entwickl.-Ber. Bd.12 (1983) Nr.3, 166 – 175

Chapter 3

3.1 Moll, J.L.; Tanenbaum, M.; Goldey, J.M.; Holonyak, N.: P-N-P-N Transistor Switches. Proc. Inst. Radio Eng. 44 (1956) S. 1174 – 1182
3.2 Stumpe, A.C.: Kennlinien der steuerbaren Siliziumzelle, ETZ – A 83, Nr.4 (1962) 81 – 87

3.3 Gerlach, W.; Seid, F.: Wirkungsweise der steuerbaren Siliziumzelle. ETZ−A 83, Nr.8 (1962) 270−277

3.4 Stumpe, A.C.: Das Schaltverhalten der steuerbaren Siliziumzelle. ETZ − A 83, Nr.9 (1962) 291−298

3.5 Bösterling, W.; Fröhlich, M.: Die dynamischen Eigenschaften von Thyristoren. AEG-Mitt. 54, Nr.5/6 (1964) 459−463

3.6 Gerlach, W.: Thyristor mit Querfeld-Emitter. Z. f. angew. Phys. 19, Nr.5 (1965) 396−400

3.7 Köhl, G.: Steuermechanismus und Aufbau bilateral schaltender Thyristoren. Scienta Elektrica 12, Nr.4 (1966) 123−132

3.8 Gerlach, W., Stumpe, A.C.: Das Schaltverhalten von Thyristoren. VDE-Buchr. Bd.11, 1966, S. 32−51

3.9 Bösterling, W.; Sonntag, A.: Ein volldiffundierter Frequenzthyristor für große Einschaltstrombelastbarkeit und hohen Dauergrenzstrom. Techn. Mitt. AEG-TELEF. 59, Nr.3/4 (1969) 238−240

3.10 Gerlach, W.; Köhl, G.: Thyristoren für hohe Spannungen. Festkörperprobleme IX.: Edit.: Madelung, O. Braunschweig 1969, S. 354−370

3.11 Peter, J.M.: Die Grenzen des di/dt von Thyristoren und Schutzmethoden. Elektrie 25 (1971) S. 266−267

3.12 Ganner, P.; Kirchner, F.: Schnelle hochsperrende Thyristoren. Siemens-Z. 46, Nr.11 (1972) 841−843

3.13 Steimel, A.: Untersuchungen über das Einschaltverhalten eines neuartigen gatelosen Thyristors. ETZ−A 95 Nr.5 (1974) 288−289

3.14 Meyer, U.: Die heutige und künftige Anwendung von Leistungstransistoren in der Antriebstechnik. Zürich 1974

3.15 Herlet, A.; Voss, P.: State of the Art in Power Semiconductor Devices. Invited Paper. IEEE/IAS Conf. Florida 1977

3.16 Ginsbach, K.-H.; Silber, D.: Fortschritte und Entwicklungstendenzen bei Silizium-Leistungshalbleitern. ETZ−A 99, Nr.1 (1978) 11−19

3.17 Brisby, K.: Thyristoren für die HGÜ. ASEA Nr.4, Jahrg. 24 (1979) 88−91

3.18 Bell, G.; Ladenhauf, W.: SIPMOS Technology, an Example of VLSI Precision Realized with Standard LSI for Power Transistors. Siemens Forsch. u. Entwickl.-Ber. Bd.9, Nr.4 (1980) 190−194

3.19 Severns, R.: MOSFETs rise to new levels of power. Electronica, May 22 (1980) 143−152

3.20 Tihanyi, J.: A Qualitative Study of the DC Performance of SIPMOS Transistors. Siemens Forsch.- u. Entwickl.-Ber. Bd.9, Nr.4 (1980) 181−189

3.21 Tihanyi, J.; Huber, P.; Stengl, J.P.: Switching Performance of SIPMOS Transistors. Siemens Forsch.- u. Entwickl.-Ber. Bd.9, Nr.4 (1980) 195−199

3.22 Hebenstreit, E.: Driving the SIPMOS Field-Effect Transistor as a Fast Power Switch. Siemens Forsch.- u. Entwickl.-Ber. Bd.9, Nr.4 (1980) 200−204

3.23 Bösterling, W.; Fröhlich, M.: Frequenzthyristoren im Schwingkreisbetrieb. ETZ 101, Nr.9 (1980) 537−538

3.24 Schröder, D.: Neue Bauelemente der Leistungselektronik. etz 102, H. 17 (1981) 906−909

3.25 Fischer, F.; Conrad, H.: Thyristormodifikationen für höhere Frequenzen, Teil I−IV, Elektrie 35 (1981) H. 2−5

3.26 Schlangenotto, H.; Silber, D.; Zeyfang, R.: Halbleiter-Leistungsbauelemente: Untersuchungen zur Physik und Technologie. Wiss. Berg. AEG-TELEF. 55, Nr.1−2 (1982) 7−24

3.27 Temple, V.A.K.: Thyristor devices for electric power systems. IEEE Transactions on Power Apparatus and Systems, Vol. PAS-101, No.7 (1982) 1186−1191

3.28 Fischer, F.; Conrad, H.: Leistungs-MOSFETs in der Energieelektronik. Teil I−IV u. Schluß. Elektrie 36 (1982) H. 3

3.29 Leipold, L.; Tihanyi, J.: Experimental Study of a SIPMOS Power Field-Effect Transistor with Integrated Input Amplifier. Siemens. Forsch.- u. Entwickl.-Ber. Bd.12, Nr.5 (1983) 327−331

3.30 Bösterling, W.; Fröhlich, M.: Thyristorarten ASCR, RLT und GTO − Technik und Grenzen ihrer Anwendung. etz 104, H. 24 (1983) 1246−1251

3.31 Braukmeier, R.: Zwischen Transistor und Thyristor der GTO-Thyristor. etz 104, H. 24 (1983) 1252−1255

3.32   Baab, J.; Fischer, F.: Rückwärtsleitende Thyristormodule für Anwendungen bis 25 kHz. etz
       104, H. 24 (1983) 1256–1258
3.33   Kaesen, K.; Tihanyi, J.: MOS-Leistungstransistoren. etz 104, H. 24 (1983) 1260–1263
3.34   Lorenz, L.: Zum Schaltverhalten von MOS-Leistungstransistoren bei ohmsch-induktiver
       Last. Diss. Hochsch. d. Bundeswehr München 1984
3.35   Stein, E.: Elektrische Modelle von Leistungshalbleitern für den Entwurf von Stromrich-
       terstellgliedern. Diss. Univ. Kaiserslautern 1984.

Chapter 4

4.1   Lim, J.S.; Wilsen, K.: Some Aspects of Thyristor Series Operation. Mullrad Techn. Comm.
       7 (1964) März, 266–270
4.2   Mulica, A.R.: How to Use Silicon Controlled Rectifiers in Series or Parallel. Control Eng.
       (1964) Mai, 95–99
4.3   Reichmann, A.; Schräder, A.: Steuergeräte für die Anschnittsteuerung von Stromrichtern.
       AEG-Mitt. 55, Nr.7 (1965) 613–620
4.4   Thiele, G.: Richtlinien für die Bemessung der Trägerspeichereffekt-Beschaltung von
       Thyristoren. ETZ–A 90, Nr.14 (1969) 347–352
4.5   Korb, F.: Die thermische Auslegung von fremdgekühlten Halbleitern bei netzgeführten
       Stromrichtern. ETZ–A 92, Nr.2 (1971) 100–107
4.6   Korb, F.: Das thermische Verhalten selbstgekühlter Halbleiter bei netzgeführten Strom-
       richtern. ETZ–A 92, Nr.4 (1971) 228–234
4.7   Depenbrock, M. (Hrsg.): Dynamische Probleme der Thyristortechnik. Berlin 1971
4.8   Herrmann, D.: Digitale Zündwinkelsteuerung für eine Drehstrombrücke zum Betrieb an
       Netzen mit starken Frequenz- und Spannungsschwankungen. ETZ–A 94, Nr.1 (1973)
       31–34
4.9   Daum, D.: Digitale Steuereinrichtung für Stromrichteranlagen. ETZ–A 94, Nr.5 (1973)
       299–301
4.10   Gammel, G.; Heidtmann, U.: Anwendung von Wärmerohren in der Leistungselektronik.
       BBC-Nachr. Nr.6/7 (1973) 143–152
4.11   Reiche, W.: Steuerung von Stromrichtern. ETZ–A 95, Nr.9 (1974) 446ff.
4.12   Otani, S.; Tanaka, O.; Suzuki, T.; Shikano, Y.; Kobayashi, G.: High-Power Semiconductor
       Rectifier Equipment Using Boiling and Condensing Heat Transfer, IAS 75 ANNUAL
4.13   Braunsteiner, F.: Ermittlung der Thyristor-Sperrschichttemperatur bei Belastung mit
       Stromblöcken, E u. M 96, Nr.12, 545–548
4.14   Howe, A.F.; Newbery, P.G.: Semiconductor fuses and their applications. IEE Proceedings,
       Vol. 127, No.3 (1980) 155–168
4.15   Stamberger, A.: Die Projektierung einer RC-Beschaltung in der Leistungselektronik.
       Elektroniker CH Nr.12 (1980)
4.16   Gupta, S.C.; Venkatesan, K.; Eapen, K.: A Generalized Firing Angle Controller Using
       Phase-Locked Loop for Thyristor Control. IEEE Trans. on Ind. Electronics and Control
       Instrumentation. Vol. IFCI–28 (1981) 46–49
4.17   Wetzel, P.: Metalloxid-Varistoren schützen Leistungshalbleiter-Bauelemente. BBC-Nachr.
       Nr.3 (1981)
4.18   Büttner, J.; Berger, G.: Steuergerät für einen nach dem Unterschwingungsverfahren
       gesteuerten Wechselrichter. Elektrie 35, H. 6 (1981) 318–320
4.19   Evans, P.D.; Saied, B.M.: Fault-current control in power-conditioning units using power
       transistors. IEEE Proc., Vol. 128, Pt. B, No.6 (1981) 335–337
4.20   Treutler, H.: Die Entwicklung von Kurzschlußstrombegrenzungseinrichtungen auf der
       Basis induktiv belasteter Stromrichter. Elektrie 35, H. 12 (1981) 638–641
4.21   Bresch, W.; Sander, J.: Thyristoren störsicher und anwendungsgerecht zünden. BBC-
       Nachr., H. 2 (1982) 43–49
4.22   Evans, P.D.; Saied, B.M.: Protection methods for power-transistor circuits. IEEE Proc.,
       Vol. 129, Pt. B, No.6 (1982) 359–362
4.23   Marquardt, R.: Untersuchungen von Stromrichterschaltungen mit GTO-Thyristoren.
       Diss. Universität Hannover 1982
4.24   Heumann, K.; Marquardt, R.: GOT-Thyristoren in selbstgeführten Stromrichtern. etz 104,
       H. 9 (1983) 328–332

Chapter 5

5.1   Abraham, L.; Koppelmann, F.: Die Zwangskommutierung, ein neuer Zweig der Stromrich-
      tertechnik. ETZ−A 87, Nr.18 (1966) 649−658
5.2   Heumann, K.: Elektrotechnische Grundlagen der Zwangskommutierung − Neue
      Möglichkeiten der Stromrichtertechnik. E u. M 84, Nr.3 (1967) 99−112
5.3   Clewing, M.: Kommutierungsvorgänge in selbstgeführten Wechselrichtern. Techn. Mitt.
      AEG-TELEF. 67, Nr.1 (1977) 61−64

Chapter 6

6.1   Storm, H.F.: A gate-controlled ac power switch. Proceedings of the Intermag Conference,
      Washington (1964) paper 4.4
6.2   Weber, J: Elektronische Wechselstrom- und Drehstromsteller. VDE-Buchr. Bd.II, 1966, S.
      200−209
6.3   Heumann, K.; Koppelmann, F.: Kontaktloses Schalten mit steuerbaren Halbleiterelemen-
      ten im Niederspannungsbereich. ETZ−A 86, Nr.17 (1965) 552−557
6.4   Michel, M.: Die Strom- und Spannungsverhältnisse bei der Steuerung von Drehstrom-
      lasten über antiparallele Ventile. Diss. TU Berlin, 1966
6.5   Sapper, M.: Triac − Leistungsstellglieder für die Temperaturregelung. BBC-Nachr. Nr.1/2
      (1973) 10−16
6.6   Unterweger, H.: Vergleichende Betrachtung zwischen mechanischem Schütz und
      Halbleiterschütz zum Schalten von Lasten im Niederspannungsbereich. Zürich 1974
6.7   Heumann, K.: Halbleiterschalter für die Energietechnik. Wiss. Ber. AEG-TELEF. 48,
      Nr.2/3 (1975) 106−118
6.8   Lukanz, W.: Halbleiterschalter und -steller für den Mittelspannungsbereich. Wiss. Ber.
      AEG-TELEF. 50, Nr.1/2 (1977) 22−31
6.9   Gyugyi, I.; Otto, R.A.; Putman, T.H.: Principles And Applications Of Static, Thyristor-
      Controlled Shunt Compensators. IEEE Trans. on Power Apparatus and Systems, Vol.
      PAS-97, No.5 (1978) 1935−1944
6.10 Hammad, A.E.; Mathur, R.M.: A New Generalized Concept For The Design of Thyristor
      Phase-Controlled VAr Compensators. Part I. Steady State Performance. IEEE Trans. on
      Power Apparatus and Systems, Vol. PAS-98, Nr.1 (1979) 219−226
6.11 Mathur, R.M.; Hammad, A.E.: A new Generalized Concept For The Design Of Thyristor
      Phase-Controlled VAr Compensators. Part II: Transient Performance. IEEE Trans. on
      Power Apparatus- and Systems, Vol. PAS-98, Nr.1 (1979) 6−13
6.12 El-Bidweihy, E.; Kadry Al-Badwaihy; Sadek Metwally, M.; El-Bedweihy, M.: Power
      Factor AC Controllers for Inductive Loads. IEEE Trans. on Ind. Electronics and Control
      Instrumentation, Vol. IECI-27, Nr.3 (1980) 210−212
6.13 Ölwegard, A.; Walve, K.; Waglund, G.; Frank, H.; Torseng, S.: Improvement of
      Transmission Capacity By Thyristor Controlled Reactive Power. IEEE Trans. on Power
      Apparatus and Systems, Vol. PAS-100, Nr.8 (1981) 3930−3937
6.14 Frank, H.; Ivner, S.: Statische Blindstromkompensation in der elektrischen Energieversor-
      gung. ASEA-Zeitschrift, Jahrg. 26, Heft 5−6 (1981) 113−119
6.15 Lemire, G.; Rajagopalan, V.; Antiparallel-Connected Thyristor Scheme Suitable for
      Feeding Highly Inductive Reversing Load. IEEE Trans. on Industrial Electronics and
      Control Instrumentation, Vol. IECI-28, No.3 (1981) 173−179
6.16 Stamberger, A.: Ein Drehstromsteller zum Herabsetzen des Wirk- und Scheinleistungs-
      bedarfs von Asynchronmaschinen bei Teillast. Elektroniker 9 (1983) 15−19
6.17 Brünnler, A.; Schmidt, H.: Hochleistungskrane für den Kombinierten Ladungsverkehr −
      Elektronische Steuerung und Thyristorantriebstechnik. EB.-Elektr. Bahnen, 81. Jahrgang,
      Heft 12 (1983) 356−361

Chapter 7

7.1   Hölters, F.: Schaltungen von Umkehrstromrichtern. AEG-Mitt. 48, Nr.11/12 (1958)
      621−629
7.2   Hölters, F.; Mikulaschek, F.: Das Blindleistungsproblem bei Stromrichter Umkehran-
      trieben. AEG-Mitt. 48, Nr.11/12 (1958) 649−659

7.3   Ostermann, H.: Der fremdgesteuerte Stromrichtersynchronmotor mit steuerbarer Dreh-
       zahl. Diss. TU Stuttgart 1961
7.4   Meyer, M.; Möltgen, G.: Kreisströme bei Umkehrstromrichtern. Siemens-Z. 37, Nr.5
       (1963) 375–379
7.5   Depenbrock, M.: Die Verknüpfungen von Frequenz, Dämpfung und Steuerwinkel beim
       Schwingkreiswechselrichter. Arch. f. Elektr. 49, H. 4 (1964) 235–239
7.6   Kanngießer, K.W.: Schwingkreisumrichter für induktive Erwärmung. BBC-Nachr. 46, H.
       12 (1964) 637–647
7.7   Weber, J.: Stromrichter in halbgesteuerter Brückenschaltung mit Freilaufventil. Siemens-
       Z. 39, Nr.4 (1965) 272–274
7.8   Golde, E.; Lehmann, G.: Schwingkreisumrichter für induktive Erwärmung. AEG-Mitt. 56,
       Nr.7 (1966) 445–450
7.9   Köllensperger, D.; Tovar, K.: Stromrichtermotoren größerer Leistung. Siemens-Z. 43, Nr.8
       (1969) 686–690
7.10   Pomper, P.: Über das statische und dynamische Verhalten von Schwingkreiswechselrich-
       tern. ETZ-A 92, Nr.4 (1971) 223–227
7.11   McMurray, W.: The Theory and Design of Cycloconverters. The M.I.T. Press, 1972
7.12   Gölz, G.; Grumbrecht, P.: Umrichtergespeiste Synchronmaschinen. Techn. Mitt. AEG-
       TELEF. 63, Nr.4 (1973) 141–148
7.13   Föhse, W.; Weis, M.: AEG-Reihe der BL-Motoren für den mittleren Leistungsbereich.
       Techn. Mitt. AEG-TELEF. 67, Nr.1 (1977) 16–19
7.14   Imai, K.: New Applications of Commutatorless Motor Systems for Starting Large
       Synchronous Motors. Invited Paper, IEEE/IAS Conf. Florida 1977
7.15   Nestler, J.: Oberschwingungsverhältnisse bei Schwingkreiswechselrichtern. ETZ-A 99,
       Nr.3 (1978) 147–151
7.16   Büttner, W.: Stromrichter in zweipaar-halbgesteuerter Zweipulsbrückenschaltung mit
       gemischter Last und ohmisch-induktivem Innenwiderstand der Wechselstromquelle.
       Archiv f. Elektrotechnik 60 (1978) 161–167
7.17   Intichar, L.: Anlaufverfahren für wechselrichtergespeiste Synchronmotoren. E u. M 96,
       Nr.9 421–424
7.18   Issa, N.A.H.; Williamson, A.C.: Control of a naturally commutated inverter-fed variable-
       speed synchronous motor. Electric Power Applications Nr.6 (1979) Vol. 2, 199–204
7.19   Steinfels, M.: Netzgeführter Direktumrichter mit erhöhter Ausgangsfrequenz. Elektrie 34,
       Nr.8 (1980)
7.20   Slonim, M.A.; Biringer, P.P.: Harmonics of Cycloconverter Voltage Waveform (New
       Method of Analysis). IEEE Trans. on Industrial Electronics and Control Instrumentation,
       Vol.-27, Nr.2 (1980) 53–56
7.21   Ericsson, H.: Stromrichter für Gleichstromantriebe. ASEA-Zeitschrift, Jahrg. 26, Heft
       5–6 (1981) 101–105
7.22   Seelig, A.: Mittelfrequenz-Wechselrichter für das induktive Kochen. Wiss. Ber. AEG-
       TELEF. 55, Nr.1–2 (1982) 80–89

Chapter 8

8.1   Alexanderson, E.F.W.: System of distribution USA Patent 1 800 002, 13. Juli 1923
8.2   Morgan, R.E.: A New Magnetic-Controlled Rectifier Power Amplifier with a Saturable
       Reactor Controlling On Time. AIEE Trans. (Communic. and Electron.) 80 (1961)
       152–155
8.3   McMurray, W.; Shattuck, D.P.: A Silicon-Controlled Rectifier with Improved Commu-
       tation. AIEE Trans. 80 (1961) Teil I, 531–542
8.4   Abraham, L.; Heumann, K.; Koppelmann, F.: Wechselrichter zur Drehzahlsteuerung von
       Käfigläufermotoren. AEG-Mitt. 54, Nr. 1/2 (1964) 89–106
8.5   Steimel, K.; Heumann, K.: Kommutatorloser Bahnmotor mit Pulswechselrichter für
       Akkumulatortriebwagen. AEG-Mitt. 55, Nr.3 (1965) 220–226
8.6   Meyer, M.: Beanspruchung von Thyristoren in selbstgeführten Stromrichtern. Siemens-Z.
       Nr.5 (1965) 495–501
8.7   Abraham, L.; Heumann, K.; Koppelmann, F.: Zwangskommutierte Wechselrichter
       veränderlicher Frequenz und Spannung. ETZ-A 86, Nr.8 (1965) 268–274

8.8   Bystron, K.: Strom- und Spannungsverhältnisse beim Drehstrom-Drehstrom-Umrichter mit Gleichstromzwischenkreis. ETZ-A 87, Nr.8 (1966) 264−271

8.9   Abraham, L.: Der Gleichstrompulswandler (elektronischer Gleichstromsteller) und seine digitale Steuerung. Diss. TU Berlin 1967

8.10   Wagner, R.: Strom- und Spannungsverhältnisse beim Gleichstromsteller. Siemens-Z. 43, Nr.5 (1969) 458−464

8.11   Backhaus, G.; Möltgen, G.: Kommutierung beim sechspulsigen selbstgeführten Wechselrichter für Betrieb mit eingeprägtem Gleichstrom. ETZ-A 90, Nr.14 (1969) 327−331

8.12   Brenneisen, J.; Schönung, A.: Bestimmungsgrößen des selbstgeführten Stromrichters in sperrspannungsfreier Schaltung bei Steuerung nach dem Unterschwingungsverfahren. ETZ-A 90, Nr.14 (1969) 353−357

8.13   Heintze, K.; Tappeiner, H.; Weibelzahl, M.: Pulswechselrichter zur Drehzahlsteuerung von Asynchronmaschinen. Siemens-Z. (1971) 154

8.14   Keller, P.: Aufbau und Schaltungstechnik von statischen Wechselrichtern. Bull. SEV 63, Nr.21 (1972) 1234−1243

8.15   März, G.: Die ZDB-Schaltung, ihre Eigenschaften und Ihre Anwendung in der Leistungselektronik. ETZ-A 93, Nr. 10 (1972) 571−576

8.16   Forstbauer, W.: Unterbrechungsfreie Stromversorgung mit Wechselrichtern. Siemens-Z. 47, Nr.2 (1973) 123−126

8.17   Schmidt, J.: Der spannungsgesteuerte und selbstgeführte Wechselrichter. Diss. TH Aachen 1973

8.18   Depenbrock, M.: Einphasenstromrichter mit sinusförmigem Netzstrom und gut geglätteten Gleichgrößen. ETZ-A 94 (1973) 466−471

8.19   Kahlen, H. Thyristorschalter zum schnellen Abschalten von Gleichströmen. ETZ-A 94, Nr.9 (1973) 539−542

8.20   Förster, J.: Sektorsteuerung mit löschbaren Stromrichterbrücken. Techn. Rundschau Bern 65, Nr.3 (1973) 25−29

8.21   Meyer, M.: Über die Kommutierung mit kapazitivem Energiespeicher. ETZ-A 95, Nr.2 (1974) 79−85

8.22   Kahlen, H.: Gleichstromsteller für den motorischen und generatorischen Betrieb der Gleichstrom-Reihenschlußmaschine. ETZ-A 95, Nr.9 (1974) 441−445

8.23   Förster, J.: An- und Abschnittsteuerung mit Stromrichtern. Elektr. Bahnen 46, Nr.5 (1975) 124−126

8.24   Knuth, D.: Netzbelastungen von anschnitt- und abschnittgesteuerten Einphasen-Stromrichtern. ETZ-A 97, Nr.2 (1976) 78−83

8.25   Depenbrock, M.: Selbstgeführter Umkehrstromrichter zur Speisung von Drehstrommaschinen. Archiv f. Elektrotechnik 61 (1979) 215–220

8.26   Grant, D.A.: Technique for pulse dropping in pulse-width modulated inverters. IEE PROC., Vol. 128 No.1 (1981) 67−72

8.27   Sriraghavan, S.M.; Pradhan, B.D.; Revankar, G.N.: Three-phase pulse amplitude and width-modulated inverter system. IEE PROC., Vol. 128, No.3 (1981) 167−171

8.28   Becker, W.; Müller-Hellmann, A.: Analyse sektorgesteuerter Einphasenbrückenschaltungen. Archiv f. Elektrotechnik 63 (1981) 219–231

8.29   Alexa, D.: Umrichtersystem mit Pulswechselrichter und einem höheren Grundschwingungsgehalt der Ausgangsspannung. etz-Archiv Bd.3, H. 12 (1981) 433−436

8.30   Bhadra, S.N.; Nisit, K.; Chattopadhyay, A.K.: Regenerative Braking Performance Analysis of a Thyristor-Chopper Controlled DC Series Motor. IEEE Trans. on Industrial Electronics and Control Instrumentation, Vol. IECI-28, No.4 (1981) 342−347

8.31   Bieniek, K.: Neue Erkenntnisse zur Auslegung von Wechselrichtern mit Phasenfolgelöschung und eingeprägtem Zwischenkreisstrom. etz-Archiv, H. 2, Bd.4 (1982) 43−49

8.32   Holtz, J.; Wurm, H.-P.: A new type of voltage fed inverter for the megawatt range. EB-Elektr. Bahnen, 80. Jahrgang, H. 7 (1982) 214−221

8.33   Holtz, J.; Stadtfeld, S.; Wurm, H.-P.: A novel PWM technique minimizing the peak inverter current at steady-state and transient operation. EB-Elektr. Bahnen, 81. Jahrgang, H. 2 (1983) 55−61

8.34   Brychta,   P.:   Ein   neuartiger   selbstgeführter   Drehstromwechselrichter   mit
       Rückarbeitstransduktoren für Betrieb mit konstanter Eingangsgleichspannung. etz-Archiv
       Bd.5, H. 11 (1983) 359—363
8.35   Alexa, D.: Eine andere Variante des Umrichtersystems mit Gleichspannungs-
       Zwischenkreis und einem höheren Grundschwingungsgehalt der Ausgangsspannung. etz-
       Archiv Bd.5, H. 6 (1983) 203—205
8.36   Cho, G.H.; Jeon, S.J.; Park, S.B.: Optimum design of a new DC-SC circuit. IEE PROC.,
       Vol. 130, Pt. B. No.3 (1983) 171—180
8.37   Ziogas, P.D.: A Complementary Current Impulse Commutated Thyristor Inverter. IEEE
       Trans. on Industrial Electronics, Vol. IE-30, No.1 (1983) 29—34
8.38   Hildebrandt, N.: Dreistufige selbstgelöschte Brückenschaltung mit sehr geringen
       Netzrückwirkungen für Triebfahrzeugantriebe. Elektrie, H. 8 (1983) 430—433
8.39   Nestler, J.; Tzivelekas, I.: Kondensator-Löschschaltung mit Löschthyristor-Zweigpaar
       nach McMurray. Teil I: Beschreibung der Löschvorgänge etz-Archiv Bd.6, H. 2 (1984)
       45—50. Teil II: Analyse der Löschvorgänge. etz-Archiv Bd.6, H. 3 (1984) 83—90
8.40   Williams, B.W.: Current-impulse-displacement thyristor commutation with controlled
       trapped energy. IEE PROC., Vol. 131, Pt. B. No.2 (1984) 21—37

Chapter 9

9.1   Meissen, W.; Runge, H.; Schönung, A.: Anforderungen der Elektronik in der Energietech-
       nik an die Netzwechselspannung. ETZ-A 90, Nr.14 (1969) 343—346
9.2   I.E.E.: Sources and Effects of Power System Disturbances. International Conference.
       London, April 1974
9.3   Bretschneider, G.; Waldmann, E.: Zulässige Oberschwingungsspannungen in Strom-
       versorgungsnetzen. ETZ-A 97, Nr.2 (1976) 90—95
9.4   Heumann, K.; Schultz, W.; Schwarz, H.-G.: Bestehende und zukünftige Möglichkeiten,
       Netzrückwirkungen von Stromrichter-Anlagen zu beherrschen. ETZ-A 98, Nr.5 (1977)
       330—334
9.5   Becker, H.; Schultz, W.: Grundlagen zur Beurteilung von Oberschwingungsrückwirkungen
       in Versorgungsnetzen. ETZ-A 98, Nr.5 (1977) 335—337
9.6   Schmidt, H.: Netzrückwirkungen in einem Industrienetz mit einem hohen Anteil an
       Stromrichterleistung. ETZ-A 98, Nr.5 (1977) 341—345
9.7   Bonwick, W.J.: Voltage waveform distortion in synchronous generators with rectifier
       loading. IEE PROC., Vol. 127, No.1 (1980) 13—18
9.8   Hildebrandt, N.: Einphasige netz- und selbstgelöschte Gleichrichteranordnungen mit
       geringen Netzrückwirkungen. Elektrie 34, Nr.7 (1980) 367—370
9.9   Büchner, P: Über die Wirkungsweise von Saugkreisen in Netzen mit Stromrichter-
       Netzrückwirkungen. Elektrie 34, Nr.3 (1981) 115—118
9.10   Laber, H.: Phase Effects of Current-Source DC Link Converters on Power Systems.
       Siemens Forsch.- u. Entwickl.-Ber. Bd.10, Nr.6 (1981) 346—350
9.11   Möltgen, G.; Neupauer, H.: Ein netzfreundliches Verfahren zur Bahnstromversorgung
       über Direktumrichter. EB-Elektr. Bahnen, 79. Jahrgang, H. 7 (1981) 286—314
9.12   Krüger, K.H.; Kulicke, B.: Noncharacteristic Harmonics in a High Voltage Direct Current-
       Converter Station Caused by System and Firing Angle Asymmetry. Siemens Forsch.- u.
       Entwickl.-Ber. Bd.11, Nr.5 (1982) 241—244
9.13   Kloss, A.: Stromrichter-Oberschwingungen bei dynamischen Betriebszuständen. Elektro-
       niker Nr.14 (1982) 26—29
9.14   Sattler, Ph.K.; Strötgen, E.: Auswirkung der Versorgung einer stromrichtergespeisten
       Asynchronmaschine aus dem 16 2/3-Hz-Netz auf die Pendelmomententwicklung. etz-
       Archiv Bd.6, H. 1 (1984) 25—32

Chapter 10

10.1   Heumann, K.; Jordan, K.-G.: Das Verhalten des Käfigläufermotors bei veränderlicher
       Speisefrequenz und Stromregelung. AEG-Mitt. 54, Nr.1/2 (1964) 107—116
10.2   Leitgeb, W.: Zur Bemessung drehzahlveränderbarer Antriebe konstanter Leistung mit
       stromrichtergespeisten Drehfeldmaschinen. ETZ-A 94, Nr.10 (1973) 584—588

10.3   Weniger, R.: Einfluß der Maschinenparameter auf Zusatzverluste, Momentoberschwin-
         gungen und Kommutierung bei der Umrichterspeisung von Asynchronmaschinen.
         Archiv f. Elektrotechnik 63 (1981) 19—28
10.4.   Andresen, A.; Bieniek, K.: Der Asynchronmotor mit drei und sechs Wicklungssträngen
         am stromeinprägenden Wechselrichter. Archiv f. Elektrotechnik 63 (1981) 153—167
10.5   Andresen, E.Ch.; Bieniek, K.; Pfeiffer, R.: Pendelmomente und Wellenbeanspruchungen
         von Drehstrom-Käfigläufermotoren bei Frequenzumrichterspeisung. etz-Archiv Bd. 4,
         H. 1 (1982) 25—33

Chapter 11

11.1   Fryze, S.: Wirk-, Blind- und Scheinleistung in elektrischen Stromkreisen mit
         nichtsinusförmigem Verlauf von Strom und Spannung. ETZ Bd.53 (1932) 596—599,
         625—627, 700—702
11.2   Tröger, R.: Energetische Darstellung von Blindstromvorgängen. ETZ-A Nr.18 (1953)
         533—537
11.2a   Tröger, R.: Blindstromtarif auf energetischer Grundlage. ETZ-A 77, Nr.19 (1956)
         706—709
11.3   Mohr, O.; Hutschenreuther, G.: Die Leistungsdarstellung in Ein- und Mehrphasensyste-
         men durch Zeigerdiagramme. ETZ-A 83, Nr.8 (1962) 253—263
11.4   Oberdorfer, G.: Begriffserklärung und Erläuterung der Blindleistung. VDE Buchr. Bd.10:
         Blindleistung. Berlin 1963
11.5   Depenbrock, M.: Blind- und Scheinleistung in einphasig gespeisten Netzwerken. ETZ-A
         85, Nr.13 (1964) 385—390
11.6   Abraham, L.; Häusler, M.: Blindstromkompensation über Halbleiterschalter und Um-
         richter. VDE-Fachtag. Elektronik 1969, S. 100—114
11.7   Häusler, M.: Elektrotechnische Grundlagen des gleichspannungsseitig kommutierenden
         Stromrichters. ETZ-A 90, Nr.15 (1969) 363—367
11.8   Heumann, K.; Knuth, D.: Energieumformung mit Stromrichtern. ETZ-A 95, Nr.4 (1974)
         189—197
11.9   Förster, J.: Zur Stromrichter-Netzbelastung. ETZ-A 96, Nr.1 (1975) 52—57
11.10   Möltgen, G.: Der Leistungsfaktor bei Stromrichtern auf fahrdrahtgespeisten Schienen-
         fahrzeugen. Elektr. Bahnen 46, Nr.9 (1975) 207—213
11.11   Pfeiffer, E.: Netzrückwirkungsfreie Leistungssteuerung. ETZ-B 28, Nr.10 (1976)
         297—299
11.12   Schröder, D.: Betriebsergebnisse einer hochdynamischen Kompensationsanlage in einem
         Industrienetz. ETZ-A 98, Nr.5 (1977) 338—340
11.13   Krishnamurthy, K.A.; Mahajani, S.B.; Revankar, G.N.; Dubey, K.: Selective harmonic
         elimination and voltage control in thyristor pulse-width modulated inverters. Int. J.
         Electronics, Vol. 46, No.3 (1979) 321—330
11.14   Müller-Hellmann, A.: Pulsstromrichter am Einphasen-Wechselstromnetz. ETZ Archiv
         Nr.3 (1979) 73—78
11.15   Boehringer, A.; Brugger, F.: Transformatorlose Transistor-Pulsumrichter mit Aus-
         gangsleistungen bis 50 kVA. E u. M Nr.12 (1979) 538—545
11.16   Becker, W.: Pulsgesteuerter Einspeisestromrichter für Umrichter mit eingeprägtem
         Zwischenkreisstrom. ETZ 100, Nr.9 (1979) 434—436
11.17   Maier, R.: Auslegung von Filtern in der Starkstromtechnik. ETZ 100, Nr.9 (1979)
         438—439
11.18   Klinger, G.: Toleranzbandgeregelter Pulsstromrichter für eine Einspeiseschaltung der
         Lokomotive E 120. Elektr. Bahnen 78, Nr.4 (1980) 598—599
11.19   Palaniappan, R.G.; Vithayathil, J.: A Control Strategy for Reference Wave Adaptive
         Current Generation, IEEE Trans. on Ind. Electronis and Control Instrumentation, Vol.
         IECI-27, No.2 (1980) 92—96
11.20   It Bau Huang; Wei Song Lin: Harmonic Reduction in Inverters by Use of Sinusoidal
         Pulsewidth Modulation. IEEE Trans. on Ind. Electronics and Control Instrumentation,
         Vol. IECI-27, No.3 (1980) 201—207
11.21   Tenti, P.: A Quasi Analytical Procedure for Determining the Optimum Commutation
         Angles of PWM Converters. Archiv f. Elektrotechnik 62 (1980) 343—350

11.22   Kampschulte, B.: Der Einfluß der Energiespeicher im Zwischenkreisumrichter eines Asynchronmaschinenantriebs auf die Oberschwingungen. Archiv f. Elektrotechnik 62 (1980) 359–367

11.23   Edelmann, H.: Wirkleistung, Blindleistung, Scheinleistung bei periodischen Strömen und Spannungen in funktionsanalytischer Sicht. Siemens Forsch.- u. Entwickl.-Ber. 10, Nr.1 (1981) 16–14

11.24   Beck, H.P.: Fremdgeführter Zwischenkreisumrichter mit Spannungsrichter zur Speisung von Synchronmaschinen großer Leistung und hoher Drehzahl. Diss. TU Berlin, 1981

11.25   Beck, H.P.; Michel, M.: Spannungsrichter – ein neuer Umrichtertyp mit natürlicher Gleichspannungskommutierung. etz-Archiv Bd.3, H. 12 (1981) 427–432

11.26   Heckelmann, H.: Blindleistungskompensation bei nichtsinusförmiger Spannung. etz-Archiv Bd.4, H. 3 (1982) 85–89

11.27   Fischer, H.D.: Blindleistungskompensation bei nichtperiodischen Strömen und Spannungen. etz-Archiv Bd.4, H. 4 (1982) 127–131

11.28   Kühn, W.; Acharya, M.: Modulation der Gleichstromleistung bei paralleler Gleichstrom-Drehstrom-Übertragung. etz-Archiv Bd.4, H. 10 (1982) 315–319

11.29   Mazzucchelli, M.; Puglisi, L.; Sciutto, G.: Analysis and synthesis of ac static power controllers. etz-Archiv Bd.5, H. 10 (1983) 325–331

11.30   Appun, P.; Lienau, W.: Der Vierquadrantensteller bei induktivem und kapazitivem Betrieb etz-Archiv Bd.6, H. 1 (1984) 3–8

Chapter 12

12.1   Jötten, R.: Regelkreise mit Stromrichtern. AEG-Mitt. 48, Nr.11/12 (1958) 613–621

12.2   Jötten, R.: Die Berechnung einfach und mehrfach integrierender Regelkreise der Antriebstechnik. AEG-Mitt. 52, Nr.5/6 (1962) 219–231

12.3   Leonhard, W.: Regelkreis mit gesteuertem Stromrichter als nichtlineares Abtastproblem. ETZ (1965) 513

12.4   Schröder, F.: Untersuchung der dynamischen Eigenschaften von Stromrichterstellgliedern mit natürlicher Kommutierung. Diss. TH Darmstadt 1969

12.5   Schräder, A.: Eine neue Schaltung zur Kreisstromregelung in Stromrichteranlagen. ETZ-A 90, Nr.14 (1969) 331–336

12.6   Ben Uri, J.: Some aspects of the control of electric drives. Electric Power Applic., Vol. 1, No.3 (1978) 77–85

12.7   Chan, Y.T.; Chmiel, A.J.; Plant, J.B.: A Microprocessor-Based Current Controller for SCR-DC Motor Drives. IEEE Trans. on Ind. Electronics and Control Instrumentation, Vol. IECI-27, No.3 (1980) 169–176

12.8   Williams, B.W.: Microprocessor Control of DC 3-Phase Thyristor Inverter Circuits. IEEE Trans on Ind. Electronics and Control Instrumentation, Vol. IECI-27, No.3 (1980) 223–228

12.9   Athani, V.V.; Deshpande, S.M.: Microprocessor Control of a Three-Phase Inverter in Induction Motor Speed Control System. IEEE Trans. on Ind. Electronics and Control Inst., Vol. IECI-27, No.4 (1980) 241–298

12.10   Weihrich, G.; Wohld, D.: Adaptive Speed Control of DC Drives Using Adaptive Observers. Siemens Forsch.- u. Entwickl.-Ber. Bd.9, Nr.5 (1980) 283–287

12.11   Al-Nimma, D.A.; Williams, S.: Study of rapid speedchanging methods in ac motor drives. IEE Proc., Vol. 127, No.6 (1980) 382–385

12.12   Saupe, R.: Die drehzahlgeregelte Synchronmaschine – optimaler Leistungsfaktor durch Einsatz einer Schonzeitregelung. ETZ 102, Nr.1 (1981) 14–18

12.13   Sen, P.C.; Trezise, J.C.; Sack, M.: Microprocessor Control of an Induction Motor with Flux Regulation. IEEE Trans. on Ind. Electroncs and Control Inst., Vol. IECI-28, No.1 (1981) 17–21

12.14   Gupta, S.C.; Venkatesan, K.; Eapen, K.: A Generalized Firing Angle Controller Using Phase-Locked Loop for Thyristor Control. IEEE Trans. on Industrial Electronics and Control Instrumentation, Vol. IECI-28, No.1 (1981) 46–49

12.15   Tso, S.K.; Ho, P.T.: Decidated-microprocessor scheme for thyristor phase control of multiphase converters. IEE Proc., Vol. 128, No.2 (1981) 101–108

12.16 Heisterkamp, H.G.: Verfahren zur Steuerung der Schonzeit bei selbstgeführten Strom-
       richtern. etz-Archiv Bd.4, H. 1 (1982) 19 – 23
12.17 Tang, P.-C.; Lu, S.-S.; Wu, Y.-C.: Microprocessor-Based Design of a Firing Circuit for
       Three-Phase Full-Wave Thyristor Dual Converter. IEEE Trans. on Industrial
       Electronics, Vol. IE-29, No.1 (1982) 67 – 73
12.18 Grötzbach, M.: Dynamisches Verhalten leistungsstarker Stromrichter in vollgesteuerter
       zweipulsiger Brückenschaltung. etz-Archiv Bd.4, H. 2 (1982) 51 – 55
12.19 Owen, R.E.; McGranaghan, M.F.; Vivirito, J.R.: Distribution System Harmonics:
       Controls for Large Power Converters. IEEE Trans. on Power App. and Systems Vol.
       PAS-101, No.3 (1982) 644 – 652
12.20 Grötzbach, M.: Eigenzeitkonstante netzgeführter Stromrichter infolge natürlicher Kom-
       mutierung. etz-Archiv Bd.4, H. 11 (1982) 355 – 358

Chapter 13

13.1   Abraham, L.; Koppelmann, F.: Käfigläufermotoren mit hoher Drehzahldynamik. AEG-
       Mitt. 55, Nr.2 (1965) 118 – 123
13.2   Bystron, K.; Meissen, W.: Drehzahlsteuerung von Drehstrommotoren über Zwischen-
       kreisumrichter. Siemens-Z. 39, Nr. 4 (1965) 254 – 257
13.3   Hein, W.: Stufenschalter mit Thyristorlastumschalter für Wechselstrom-Triebfahrzeuge.
       Siemens-Z. 39, Nr.4 (1965) 269 – 271
13.4   Korb, F.: Einstellung der Drehzahl von Induktionsmotoren durch antiparallele Ventile
       auf der Netzseite. ETZ-A 86, Nr.8 (1965) 275 – 279
13.5   Skudelny, H.Ch.: Stromrichterschaltungen für Wechselstrom-Triebfahrzeuge. ETZ
       (1966) 249
13.6   Vogel, L.; Wiegand, A.: Thyristor-Stromrichter für Industrieantriebe. AEG-Mitt. 56, Nr.2
       (1966) 98 – 105
13.7   Germann, F.: Thyristorwechselrichter für gesicherte Stromversorgungsanlagen. AEG-
       Mitt. 56, Nr.7 (1966) 458 – 460
13.8   Elger, H.; Weiß, M.: Untersynchrone Stromrichterkaskade als drehzahlregelbarer
       Antrieb für Kesselspeisepumpen. Siemens-Z. (1968) 308
13.9   Pollard, E.M.; Flairty, C.W.; Hodges, M.E.; Laukaitis, J.A.: A 10 MW Thyristor AC
       Switch for Induction Heating Power Control and Protection. Power Thyristors and their
       Applications. IEE Conference Public. No.53, 177 – 184, London 1969
13.10 Kusko, A.: Solid-State DC Motor Drives. The M.I.T. Press 1969
13.11 Stiebler, M.; Zander, H.: Leistungselektronik zur Erregung großer Synchrongeneratoren.
       ETZ-A 90, Nr.14 (1969) 336 – 342
13.12 Keuter, W.: Kleinthyristoren und Triacs in der Haushalts- und Industrieanwendung.
       ETZ-B 21, Nr.19 (1969) 447 – 451
13.13 Bayer, K.H.; Waldmann, H.; Weibelzahl, M.: Die Transvektor-Regelung für den
       feldorientierten Betrieb einer Synchronmaschine. Siemens-Z. (1971) 765
13.14 Frank, H.; Landstrom, B.: Power-Factor Correction with Thyristor-Controlled Capaci-
       tors. ASEA Journal 44, Nr.6 (1971) 180 – 184
13.15 Neupauer, H.; Richter, E.: Parallelschwingkreisumrichter für die induktive Erwärmung.
       Siemens-Z. 45 (1971) 9
13.16 Förster, J.: Löschbare Fahrzeugstromrichter zur Netzentlastung und -stützung. Elektr.
       Bahnen 43, Nr.1 (1972) 13 – 19
13.17 Behmann, U.; Ingbert, St.: Elektrische Mehrsystem-Triebfahrzeuge in Europa. ETZ-B 24,
       Nr.3 (1972) 64 – 69
13.18 Rumpf, E.; Ronade, S.: Geräte und Verfahren für Steuerung und Regelung einer HGÜ
       und Gesichtspunkte für ihren Einsatz ETZ-A 93, Nr.3 (1972) 123 – 133
13.19 Lehmann, G.: Gestaltung von Typenreihen für Thyristor-Leistungsstromrichter. Techn.
       Mitt. AEG-TELEF. 62, Nr.6 (1972) 268 – 271
13.20 Foerster, J.; Schneider, G.; Stenzel, R.: Die größten Kesselspeisepumpen-Antriebe mit
       untersynchroner Stromrichterkaskade. Elektr. Wirtsch. 71, Nr.24 (1972) 695 – 699
13.21 Becker, H.: Beherrschung von Blindlastströmen in Verteilernetzen durch statische
       Kompensationseinrichtungen. VDE-Fachber. 27 (1972)

13.22 Boettger, K.; Schmidt, J.: Statische Wechselrichter für redundanten Parallelbetrieb. AEG-Mitt. 63, Nr.2 (1973) 71−72

13.23 Nitschke, H.-J.: Der bürstenlose Motor, ein neuer universeller, drehzahlregelbarer Drehstromantrieb. Techn. Mitt. AEG-TELEF. 63, Nr.2 (1973) 73−75

13.24 Knuth, D.; Müller, D.: Elektronische Motorschütze. Elektr. Ausrüstung 14, Nr.4 (1973) 17−20

13.25 Schneider, G.: Die untersynchrone Stromrichterkaskade. Techn. Mitt. AEG-TELEF. 63, Nr.5 (1973) 188−193

13.26 Ettner, N.; Käppner, A.: Stromrichtergespeiste drehzahlveränderbare elektrische Antriebe in der chemischen Industrie. Siemens-Z. 47, Nr.6 (1973)

13.27 Zielke, R.A.: A 50 MW Thyristor Controlled Power Converter. Michigan 1973

13.28 Betz, H.: Der Netzkupplungsumformer Neu-Ulm, eine Anlage zur Stromversorgung der Deutschen Bundesbahn. Techn. Mitt. AEG-TELEF. 63, Nr.7 (1973)

13.29 Lüns, F.; Scholtyssek, B.; Weber, J.: Regelbare Drehstromantriebe großer Leistung. BBC-Nachr. Nr.6/7 (1973) 155−161

13.30 Thomas, F.W.; Schmidt, W.: Einsatz von Direktumrichtern für das Elektro-Schlacke-Umschmelzverfahren. Siemens-Z. 47, Nr.9 (1973) 676−680

13.31 Glas, W.: Neuzeitliche Gleichstrom-Versorgungsanlage für Chlor-Elektrolysen. Chemie-Ing.-Techn. 45 (1973) 15

13.32 Murphy, J.M.D.: Thyristor Control of AC Motors. Oxford, London New York, Toronto, Sydney 1973

13.33 Matthes, H.G.: Über den Halbleitereinsatz in Umrichtern zur induktiven Erwärmung, Zürich 1973

13.34 Peneder, F.; Butz, H.: Erregersysteme für Drehstrom-Generatoren in Industrie- und mittleren Kraftwerken. BBC-Mitt. 61, Nr.1 (1974) 41−50

13.35 Lüns, F.: Gleichrichteranlage für eine Chlorelektrolyse mit direktem Anschluß an 110 kV. BBC-Nachr. 56, Nr.1/2 (1974) 36−42

13.36 Teich, W.: BBC-Asynchronmotor-Antrieb für Diesellokomotiven − Ein Baukastensystem für viele Leistungsklassen. ETR-Eisenbahntechn. Rdsch. Nr.5 (1974) 182−188

13.37 Kehrmann, H.; Lienau, W.; Nill, R.: Vierquadrantensteller − eine netzfreundliche Einspeisung für Triebfahrzeuge mit Drehstromantrieb. Elektr. Bahnen 45, Nr.6 (1974) 2−9

13.38 Becker, E.; Gammert, R.: Drehstromversuchsfahrzeuge − DE 2500 mit Steuerwagen − Systemerprobung eines Drehstromantriebes an 15 kV, 16 2/3 Hz. Elektr. Bahnen 47, Nr.1 (1976) 18−23

13.39 Nitschke, H.-J.; Putz, U.: Umrichter für Drehstromantriebe. Techn. Mitt. AEG-TELEF. 67, Nr.1 (1977) 2−5

13.40 Cießow, G.; Gölz, G.; Grumbrecht, P.: Drehstrom-Antriebssystem für Bahnfahrzeuge. Techn. Mitt. AEG-TELEF. 67, Nr.1 (1977) 35−42

13.41 Ellert, F.J.; Moran, R.J.: HVDC and Static VAR Control Applications of Thyristors. Invited Paper, IEEE/IAS Conf. Florida 1977

13.42 Gölz, G.: Converter-Fed Propulsion Systems with Asynchronous Traction Machines. World Electrotechical Congress. Moskau 1977

13.43 Heumann, K.: The Prospective Development of AC Thyristor Drives with Induction Motors. World Electrotechn. Congr. Moskau 1977

13.44 Gerlach, R.: Stromrichtererregung für schnellaufende Synchrongeneratoren. Techn. Mitt. AEG-TELEF. 68 (1978) 1/2

13.45 van Wyk, J.D.: Variable-speed ac drives with slip-ring induction machines and a resistively loaded force commutated rotor chopper. Electric Power Applic. No.5, Vol. 2 (1979) 149−160

13.46 Crowder, R.M.; Smith, G.A.: Induction motors for crane applications. Electric Power Applic. Vol. 2, No.6 (1979) 194−198

13.47 Kublick, Ch.: Unterbrechungsfreie Stromversorgungsanlage mit Pulswechselrichter. ETZ 100, Nr.11 (1979) 540−545

13.48 Pesch, H.: Die Hochspannungs-Gleichstrom-Übertragung Cabora Bassa-Apollo: Systemverhalten und Betriebserfahrungen. ETZ 100, Nr.26 (1979) 1492−1501

13.49 Coenrads, J.E.B.; Eriksson, S.: Frequenzumrichteranlauf von großen Synchronmaschinen für industrielle Antriebe. ASEA 25, Nr.1 (1980)

13.50 von Möllendorff, H.: Messung der Energieersparnis durch die Nutzbremse bei schienengebundenen Triebfahrzeugen. Elektr. Bahnen 78, Nr.1 (1980) 21−25

13.51 Hönig, J.: Umrichter zur Speisung des 16 2/3-Hz-Bahn-Netzes. Elektr. Bahnen 78, Nr.4 (1980) 92−97

13.52 Ziegler, W.: Drehstromantrieb mit Stromzwischenkreisumrichter für Bahnfahrzeuge. Elektr. Bahnen 78, Nr.5 (1980) 123−128

13.53 Gemmeke, K.; Müller, E.; Runge, W.; Schulze, H.; Steimel, A.: Drehstromantrieb für einen DT3-Triebwagen der Hamburger Hochbahn AG. BBC-Nachr. Nr.12 (1980)

13.54 Weber, H.H.: Stromrichter-Traktionstechnik bei den Schweizerischen Bundesbahnen und ihr prognostierter Nutzen. Elektr. Bahnen 78, Nr.12 (1980) 312−319 und 79, Nr.1 (1981) 23−31

13.55 Cossié, A.: Evolution de la locomotive à thyristors à la S.N.S.F. Elektr. Bahnen 79, Nr.1 (1981) 18−22

13.56 Dreimann, K.; Böhm, H.: Drehstrom-Kleinserie der Berliner Verkehrsbetriebe (BVG) − ein Meilenstein der Entwicklung der Drehstrom-Antriebstechnik bei AEG-TELEFUNKEN. Elektr. Bahnen 79, Nr.4 (1981) 110−116

13.57 Amler, J.: Energiesparwagen für die Nürnberger U-Bahn − die ersten serienmäßig hergestellten Drehstromtriebwagen. Elektr. Bahnen 79, Nr.5 (1981) 202−210

13.58 Bowles, J.P.: Multiterminal HVDC Transmission Systems Incorporating Diode Rectifier Stations. IEEE Trans. on Power Apparatus and Systems, Vol. PAS-100, No.4 (1981) 1674−1678

13.59 Gish, W.B.; Schurz, J.R.; Milano, B.; Schleif, F.R.: An Adjustable Speed Synchronous Machine For Hydroelectric Power Applications. IEEE Trans. on Power Apparatus and Systems, Vol. PAS-100, No.5 (1981) 2171−2176

13.60 Shibata, F.; Ohtsubo, A.; Tsuruta, K.; Kohrin, T.: Speed Control Of A Cascade Induction Motor With Three Sets Of Converters In Its Secondary Circuit. IEEE Trans. on Power Apparatus and Systems, Vol. PAS-100, No.6 (1981) 2946−2954

13.61 Fischer, J.; Leistikow, R.: Die Wechselrichter-Stromversorgung der Magnetbahn-Versuchsanlage Kassel. BBC-Nachr. Nr.2 (1981) 51−58

13.62 Körber, J.: Die Entwicklung der Drehstrom-Antriebstechnik für die Hochleistungslokomotive E 120. BBC-Nachr. Nr.5/6 (1981) 163−173

13.63 Gandert, H.J.: Schnelle Erregungssysteme und ihr Beitrag zur Netzstabilität bei großen Generatoren. ETZ 102, Nr.6 (1981) 299−302

13.64 Lidberg, K.: Frequenzumrichter zur Drehzahlsteuerung von Käfigläufermotoren. ASEA-Zeitschrift 26, H. 5−6 (1981) 107−111

13.65 Bröms, A.: Unterbrechungsfreie Stromversorgung. ASEA-Zeischrift 26, H. 5−6 (1981) 121−127

13.66 Kleinrath, H.: Drehstromantriebe mit Frequenzumrichtern. E u. M 98, H. 11 (1981) 452−458

13.67 Seefried, E.; Hofmann, W.: Wechselrichter zur Speisung von Asynchronmotoren auf der Basis von Leistungstransistoren. Elektrie 36, H. 5 (1982) 231−235

13.68 Appun, P.; Futterlieb, E.; Kommissari, K.; Marx, W.: Die elektrische Auslegung der Stromrichterausrüstung der Lokomotive 120 der Deutschen Bundesbahn. Elektr. Bahnen 80, H. 10 (1982) 290−294 und H. 11 (1982) 314−316

13.69 Törnerud, G.: Thyristor-Gleichstromsteller für die Stockholmer U-Bahn. Elektr. Bahnen 81, H. 9 (1983) 292−298

13.70 Kuhn, W.; Moll, K.: Umrichter nach dem Unterschwingungsverfahren für industrielle Antriebe. BBC-Nachr. 65, H. 11 (1983) 375−384

Chapter 14

14.1 Harrison, R.E.; Shemie, R.K.; Krishnayya, P.C.S.: A Proposed Test Specification for HVDC Thyristors Valves. IEEE Trans. on Power Apparatus and Systems, Vol. PAS-97, No.6 (1978) 2207−2214

14.2 Buri, H.; Leipold, Ph.: Anwendungsbezogene Prüfungen schneller Thyristoren. BBC-Nachr. Nr.12 (1979) 459−464

Subject Index

R. E. Hummel

Electronic Properties of Materials

An Introduction for Engineers

1985. 219 figures. XII, 319 pages. ISBN 3-540-15631-3

Contents: Fundamentals of Electron Theory: Introduction. –
Wave Properties of Electrons. The Schrödinger Equation.
Solution of the Schrödinger Equation for Four Specific Prob-
lems. Energy Bands in Crystals. Electrons in a Crystal. –
Electrical Properties of Materials: Electrical Conduction in
Metals and Alloys. Semiconductors. Electrical Conduction in
Polymers, Ceramics, and Amorphous Materials. – Optical
Properties of Materials: The Optical Constants. Atomistic
Theory of the Optical Properties. Quantum Mechanical
Treatment of the Optical Properties. Applications. – Magnet-
ic Properties of Materials: Foundations of Magnetism.
Magnetic Phenomena and Their Interpretation – Classical
Approach. Quantum Mechanical Considerations. Applica-
tions. – Thermal Properties of Materials: Introduction.
Fundamentals of Thermal Properties. Heat Capacity.
Thermal Conduction. Thermal Expansion. Appendices. –
Index.

W. Leonhard

Control of Electrical Drives

Translated from the German by the author in cooperation
with R. M. Davis, R. S. Bowes

(Completely revised and enlarged edition of "Regelung in der
elektrischen Antriebstechnik" published by Teubner,
Stuttgart 1974)

1985. 270 figures. XVI, 346 pages. (Electric Energy Systems
and Engineering Series). ISBN 3-540-13650-9

Contents: Introduction. – Some Elementary Principles of
Mechanics. – Dynamics of a Mechanical Drive. – Integration
of the Simplified Equation of Motion. – Thermal Effects in
Electrical Machines. – Separately Excited DC Machine. –
DC Motor with Series Field Winding. – Control of a Sepa-
rately Excited DC Machine. – The Static Converter Used as a
Power Actuator. – Control of Converter-supplied DC Drives.
– Symmetrical Three-Phase AC Induction Machine. – Power
Supplies for Variable Speed AC Drives. – Control of Induc-
tion Motor Drives. – Induction Motor Drive with Restricted
Speed Range. – Variable Frequency Synchronous Motor
Drives. – Some Applications of Controlled Electrical Drives.
– References. – Subject Index.

Springer-Verlag
Berlin Heidelberg
New York Tokyo

Time-Scale Modeling
of Dynamic Networks with
Applications to Power Systems

Editor: **J. H. Chow**

1982. X, 218 pages. (Lecture Notes in Control and Information Sciences, Volume 46). ISBN 3-540-12106-4

Contents: Time-Scales in Interconnected Systems. – Singular Perturbations and Time-Scales. – Modeling of Two-Time-Scale Systems. – Dynamic Networks and Area Aggregation. – Coherency and Area Identification. – Slow Coherency and Weak Connections. – Nonlinear Dynamic Networks. – Reduced Simulations of Nonlinear Power System Models. – Appendix A: Matrix M⁻¹K for 48 Machine System. – References.

This monograph develops a modeling methodology for a class of large scale systems with a network structure. The methodology provides the analytical tools for revealing time-scale properties and replacing hitherto heuristic means of defining decompositions into subsystems and control hierarchies. The time-scale methodology is based on the fact that responses of varying speeds are commonly observed in power systems and many other interconnected systems.

The book has been written for a broad audience of systems and control engineers. The background assumed does not exceed the basic linear system theory covered in most undergraduate programs. Not only will this monograph be useful to practicing engineers, but will also serve as a text for a graduate course in power system modeling.

Springer-Verlag
Berlin Heidelberg
New York Tokyo

Springer